W9-AQT-700

America's Secret Power

America's Secret Power

The CIA in a Democratic Society

LOCH K. JOHNSON

New York Oxford
OXFORD UNIVERSITY PRESS
1989

Oxford University Press

Oxford New York Toronto
Delhi Bombay Calcutta Madras Karachi
Petaling Jaya Singapore Hong Kong Tokyo
Nairobi Dar es Salaam Cape Town
Melbourne Auckland

and associated companies in
Berlin Ibadan

Published by Oxford University Press, Inc.,
200 Madison Avenue, New York, New York 10016

Oxford is a registered trademark of Oxford University Press

Library of Congress Cataloging-in-Publication Data
Johnson, Loch K., 1942–
America's secret power : the CIA in a democratic society /
Loch K. Johnson.
p. cm. Bibliography: p. Includes index.
ISBN 0-19-505490-3
1. United States. Central Intelligence Agency.
2. Intelligence service—United States. I. Title.
JK468.I6J63 1989
327.1'2'06073—dc19 88-23467 CIP

2 4 6 8 10 9 7 5 3 1

Printed in the United States of America
on acid-free paper

To the memory
of Kathleen W. Johnson

Preface

With a mixture of excitement and trepidation, I began my studies of the Central Intelligence Agency (CIA) in 1975 as a congressional investigator. I had been selected to serve on a U.S. Senate panel (the Church committee, chaired by Frank Church [D, Idaho]) formed to probe alleged abuses by the U.S. intelligence agencies.[1] As I made preparations to depart the quiet groves of academe in Ohio for the oblique streets and politics of Washington, D.C., a colleague dropped by my office to offer some friendly advice. Paraphrasing the nineteenth-century German philosopher Friedrich Nietzsche, he warned: "Stare not into the Abyss, lest the Abyss stare back at you!" Later that day another predicted with a broad grin: "You'll be run down in the streets of Georgetown at midnight by an unmarked milk truck." If the Abyss stared back, I was unaware of it, and, happily, no milk trucks came my way—though, come to think of it, I do remember a close encounter on the Washington Beltway with an eighteen-wheeler piloted by a grim-looking character who could have passed for a "knuckle-dragger" (CIA slang for a tough paramilitary officer).

While less fraught with danger than some anticipated, research on the CIA has certainly had its share of frustrations—yet undoubtedly less than when Professor Harry Howe Ransom of Vanderbilt University, a trailblazer among intelligence scholars, first turned his attention in this direction during the 1950s. "I used to go at this subject with a pick and shovel," he recalls. "Then [in 1975], congressional investigators used bulldozers to produce a mountain of new information."[2] The abundance of fresh documentation failed, however, to guarantee greater insight. Five years after the sweeping investigations conducted by the Senate, the House, and a presidential commission, national security expert Richard K. Betts accurately observed that "the increased importance of intelligence has not been matched by increased understanding of what it is or how it can and should be used to improve government decision."[3]

In 1987 a second crew of congressional investigators—this time with a more limited focus on operations involving Iran and Nicaragua (the celebrated Iran-contra affair)—dug into another rich vein of information about the secret workings of U.S. intelligence.[4] Yet Betts's observation remained valid; further efforts were still required to give meaning to the findings unearthed by the government sleuths, journalists, and scholars who have made the elusive workings of the CIA and its related agencies more accessible to outsiders in the past few years.

The purpose of this book is to carry on the search for a greater comprehension of how the CIA, with its preference for the shadows, might best exist in America's open society—especially how the United States might better manage the difficult task of balancing the genuine needs of national security, on the one hand, and the protection of individual liberties, on the other. Even in this new era of official inquiries into secret agencies, acquiring information about intelligence policy remains difficult for the outside researcher. "Intelligence is like witchcraft," writes one intelligence scholar, adding that "you have to be accepted by the community in order to practice it. No amount of wishful thinking will gain you entry to the trade. . . ."[5] In an attempt to overcome these investigative barriers, I have resorted to a range of methodologies. In addition to examining key documents on national intelligence policy, I spent thirteen years (1975–88) conducting interviews—some brief, others lasting several hours—with former and incumbent officials in the legislative and executive branches, including over five hundred intelligence officers, from CIA directors and their top lieutenants to those responsible for carrying out operations in the field. I scoured the scholarly literature and memoirs of former intelligence officers for insights and gathered and sifted the limited quantified data on intelligence policy available to the public.

Finally, as a congressional staff aide serving sequentially on the four Senate and House committees with official jurisdiction over intelligence and foreign affairs (1975–80), I had an opportunity to practice the valuable methodology of "participant observation." At close quarters I watched the interplay between the CIA and the Congress as a legislative overseer. This last technique holds a special danger for someone who aspires to detached, scholarly analysis; one can become too close to the subject and lose the most vital attribute of scholarship, namely, objectivity. I can only say that I have tried hard to avoid this pitfall; as always, the reader must be the final judge.

Most studies on intelligence examine only the domestic or, more often, the foreign-policy aspects. This book addresses both sides of the equation, for in reality they are often closely linked. During the Johnson and Nixon administrations, for example, the White House directed the CIA and other intelligence agencies to spy on domestic antiwar protesters—in clear violation of statutory limitations. The CIA has also intermittently established controversial ties with various domestic groups (among others, universities and the media) in order to fabricate a protective guise ("cover" in spy talk) for its foreign operations as well as for other purposes. To treat the CIA as an intelligence agency with a strictly foreign mission is to miss an important aspect of the debate over its proper role.

This book is divided into three parts. Part I, "The Intelligence Mission," begins with of an introduction to the dilemmas created by the presence of secret agencies in a democratic society (chap. 1). This introduction is followed by a discussion of why the United States has a central intelligence agency (chap. 2) and ends with a blueprint of the CIA's internal organization and how it relates to the rest of the U.S. "secret service," to use an old-fashioned description of America's intelligence establishment (chap. 3). Several first-rate histories and organizational studies of the CIA have already been published (see the Selected Bibliography). The intention in Part I is not to reinvent these wheels, but to provide a brief background or orientation for those readers without a deep knowledge of intelligence matters.

Part II, "Problems of Strategic Intelligence," begins with an overview of seven major dangers, or "sins," posed by the CIA's secret operations (chap. 4). Subsequent chapters—the heart of the book—consider in greater detail the most unsettling of these dangers. This critique opens with a look at certain pathologies in the CIA's collection and analysis of information—the Agency's most important mission (chap. 5)—and then explores its controversial attempts to disrupt or manipulate foreign governments through the use of "covert action" (chap. 6).

Part III, "The CIA and the Rights of Americans," examines the most chilling threat posed by secret intelligence agencies: their use against the very citizens they are meant to protect. This section begins with a study of how this danger took on ominous proportions in 1970 (chap. 7). In that year of domestic turbulence, Tom Charles Huston, a young, zealous White House aide in the Nixon administration, designed a master spy plan to turn the CIA and other agencies against Americans who opposed the war in Vietnam. Huston stands as a forerunner to Lt. Col. Oliver L. North, another eager White House aide who, in 1984–86, would, like Huston, unite with the CIA and other intelligence agencies in top-secret operations that soon crossed the boundaries of law and propriety. More subtle than these direct—though at the time carefully concealed—assaults against the law and, some would argue, more insidious still has been the CIA's careful nurturing of relationships with a variety of institutions and groups in American society. The ties between the CIA and the nation's universities are explored in chapter 8, followed by an examination, in chapter 9, of its connections with the media.

Part IV, "Intelligence in a Democratic Framework," examines recent efforts by Congress to supervise the CIA with a more serious intent than it had ever displayed in the past (chap. 10). As an illustration, chapter 10 presents a look at the U.S. House Committee on Intelligence in its first year of operation (1978). The appraisal of the "New Intelligence Oversight" continues in chapter 11 with a broad evaluation of the range of safeguards currently in place (in and outside Congress) to check future intelligence abuses. A concluding chapter offers a set of recommendations to improve the balance between the demands of national security, on the one hand, and the precious, fragile gift of civil liberties, on the other (chap. 12).

This book tries to portray the dilemmas faced by the citizens of the United States, and their government, as a result of living in a nation that values both secrecy for security and openness for democracy. My research into the hidden byways of America's most controversial secret agency has benefited substantially from the willingness of government officials (many of whom insisted upon anonymity) to discuss this subject with me and from my own service in the Congress, first as an American Political Science Association Congressional Fellow and subsequently as a staff aide. It has benefited as well from the many fine studies of intelligence policy included in the Selected Bibliography—particularly the scholarship of Professor Harry Howe Ransom, who has led the way for those concerned about the role of "dark" agencies in societies meant to flourish in the sunlight.

Athens, Ga.
December 1988

L.K.J.

Acknowledgments

I am pleased to thank several people for their help with this project. Richard K. Ashley, Stephen J. Flanagan, Glenn P. Hastadt, Arthur S. Hulnick, Michael Johnston, Thomas K. Latimer, and Harry Howe Ransom offered invaluable guidance on various portions of the manuscript. So have legislators and staff aides on the congressional intelligence committees and intelligence officials now or formerly with the executive branch; I have honored their preferences for anonymity when requested. My education on the CIA has been greatly assisted by discussions about intelligence issues with a wide circle of thoughtful people inside and outside the intelligence community. Though many preferred to remain anonymous, I would like to express my appreciation to David Aaron, Les Aspin, William B. Bader, Frederick D. Baron, Richard K. Betts, Scott D. Breckinridge, McGeorge Bundy, Ray S. Cline, William E. Colby, John T. Elliff, Harold P. Ford, Wyche Fowler, Michael J. Glennon, Allan E. Goodman, Morton H. Halperin, Michael I. Handel, David H. Hunter, Karl F. Inderfurth, George Kalaris, Anne Karalekas, Diane La Voy, William M. Leary, William G. Miller, Hayden B. Peake, Jay Peterzell, Walter Pforzheimer, Harry A. Rositzke, Dean Rusk, Frederick A. O. Schwarz, Jr., Frank Snepp, Britt Snider, Stafford Thomas, Stansfield Turner, Gregory F. Treverton, William Truehart, Richard R. Valcourt, H. Bradford Westerfield, and David Wise. Patient tutors, now past, included Jim Angleton, Seymour Bolten, J. Fred Buzhardt, Frank Church, and David A. Phillips. The editors of *Corruption and Reform, Foreign Policy, Intelligence and National Security, International Intelligence and Counterintelligence, International Studies Quarterly, Harvard Journal of Law and Public Policy, Legislative Studies Quarterly,* and *World Affairs,* as well as Brooks/Cole and Transnational Publishers, generously allowed me to draw upon work I have previously published in their pages.

I have carefully avoided the inclusion in this book of any classified information to which I was privy as a congressional aide. From my vantage point as a former "insider," I am convinced that a thorough examination of the CIA is possible through the use of interviews with intelligence officials and close study of documents in the public domain, without access to classified materials. Inside congressional experience does offer some advantages, though, especially a practical education in a field of policy-making that is both remote and esoteric and access to a network of knowledgeable contacts. Staff on the House Permanent Select Committee on Intelligence and on the Senate Select Committee to Study Governmental Operations with Respect to Intelligence Activities (the Church Committee) kindly reviewed key sections of the manuscript to ensure that I had not inadvertently disclosed material of a classified nature. Various intelligence officers, active and retired, also graciously read the chapters with this problem (and textual accuracy) in mind.

The Department of Political Science and the Office of the Vice President for Research at the University of Georgia kindly provided financial support at critical moments. Margaret O'Connor and Barbara Holder lent their fine typing skills. My research assistants—Andrew Hally, David Price, and Muffie Wiebe—helped check sources. Leena and Kristin Johnson once again created the ideal setting for reflection, writing, and happiness. Finally, at Oxford University Press I would like to thank Valerie Aubry for encouragement, and Henry Krawitz and Rosalie West for skillful editing. Naturally, errors of fact or judgment in this book are my responsibility alone.

Contents

Abbreviations

ABM	anti-ballistic missile
AFIO	Association of Former Intelligence Officers
AG	Attorney General
ASC	American Security Council
ASNE	American Society of Newspaper Editors
BNE	Board of National Estimates
CA	covert action
CAS	Covert Action Staff
CBJB	Congressional Budget Justification Book
CE	counterespionage
CI	counterintelligence
CIA	Central Intelligence Agency
CIG	Central Intelligence Group
CNSS	Center for National Security Studies
COINTELPRO	FBI Counterintelligence Program
COMINT	communications intelligence
COMIREX	Committee on Imagery Requirements and Exploitation
COS	chief of station
CRS	Congressional Research Service
DCD	Domestic Contact Division
DCIA	Director of the Central Intelligence Agency
DCI	Director of Central Intelligence
DDA	Deputy Director for Administration
DDCIA	Deputy Director of the Central Intelligence Agency
DDI	Deputy Director for Intelligence
DDO	Deputy Director for Operations
DDS & T	Deputy Director for Science and Technology
DEA	Drug Enforcement Agency

DIA	Defense Intelligence Agency
DINSUM	Defense Intelligence Summary
DoD	Department of Defense
DOD	Domestic Operations Division
DS & T	Directorate for Science and Technology
ELINT	electronic intelligence
ERDA	Energy Research and Development Administration
FBI	Federal Bureau of Investigation
FBIS	Foreign Broadcast Information Service
FOIA	Freedom of Information Act
FRD	Foreign Resources Division
GAO	General Accounting Office
GRU	Soviet military intelligence
HPSCI	House Permanent Select Committee on Intelligence
HUMINT	human intelligence (espionage)
IAD	Internal Affairs Division
ICBM	intercontinental ballistic missile
ICS	Intelligence Community Staff
IG	Inspector General
IIM	Interagency Intelligence Memoranda
INR	Bureau of Intelligence and Research (Department of State)
IOB	Intelligence Oversight Board
IRS	Internal Revenue Service
ITT	International Telephone and Telegraph
I & W	Indicators and Warning
JCS	Joint Chiefs of Staff
JPRS	Joint Publication Research Service
KGB	Soviet secret police
KIQ	Key Intelligence Question
MRBM	medium-range ballistic missiles
NATO	North Atlantic Treaty Organization
NCD	National Collection Division
NCSI	National Center for Strategic Intelligence
NDI	no derogatory information
NFAC	National Foreign Assessment Center
NFIB	National Foreign Intelligence Board
NFIC	National Foreign Intelligence Council
NFIP	National Foreign Intelligence Program
NIC	National Intelligence Council
NIC/AG	National Intelligence Council/Analytic Group
NID	National Intelligence Daily
NIE	National Intelligence Estimate
NIO	National Intelligence Officer
NIS	National Intelligence Survey
NISC	National Intelligence Study Center
NIT	National Intelligence Topic
NOC	nonofficial cover

NPIC National Photographic Interpretation Center
NSA National Security Agency
NSA National Student Association
NSC National Security Council
NSAM National Security Action Memorandum
NSCID National Security Council Intelligence Directive
NSDM National Security Decision Memorandum
NSIC National Strategy Information Center
NSPG National Security Planning Group
NSSM National Security Study Memorandum
NTM National Technical Means
OAG Operations Advisory Group
OB Order of Battle
ODE Office of Development and Engineering
OMB Office of Management and Budget
OPA Office of Public Affairs
OPC Office of Policy Coordination
ORD Office of Research and Development
OSS Office of Strategic Services
OTA Office of Technology Assessment
OTR Office of Training
PAC Political Action Committee
PDB President's Daily Brief
PFIAB President's Foreign Intelligence Advisory Board
PHOINT photographic intelligence
PLO Palestine Liberation Organization
PM paramilitary
PRC People's Republic of China
PRC Policy Review Committee
PRG Policy Review Group
RADINT radar intelligence
SALT Strategic Arms Limitation Talks
SAM surface-to-air missile
SCC Special Coordination Committee
SIG Senior Interagency Group
SIGINT signals intelligence
SIRC Security Intelligence Review Committee
SNIE Special National Intelligence Estimate
SO Special Operations (the CIA's paramilitary staff)
SRG Senior Review Group
SSCI Senate Select Committee on Intelligence
TECHINT technical intelligence
TELINT telemetery intelligence
USC United States Code (a statutory identification system)
USIA United States Information Agency
USIB United States Intelligence Board
VCI Viet Cong Infrastructure

Figures

Tables

Chronology

Era of Trust, 1947–74

1947	National Security Act passes
1947	CIA covert actions throughout Europe
1949	Central Intelligence Agency Act passes
1953	CIA-sponsored coup in Iran
1954	CIA-sponsored uprising in Guatemala
1954–55	U-2 spyplane developed
1961	Bay of Pigs
1961	CIA begins covert actions in Vietnam; new Headquarters in Langley, Va.
1962	Operation Mongoose against Fidel Castro
1962	First spy satellite
1964	CIA begins covert actions in Chile
1964–65	CIA pessimistic estimates on Vietnam ignored
1966–67	*Ramparts* exposes CIA
1967	Katzenbach Report
1970	Huston Plan
1972	Watergate; détente
1974	*New York Times* exposes CIA; Hughes-Ryan Act

Era of Skepticism, 1975–76

1975	"Year of Intelligence": Church, Pike, and Rockefeller inquiries; Angola covert action denied by Congress
1976	Ford Executive Order on intelligence (No. 11905); Senate Intelligence Committee established

Era of Uneasy Partnership, 1977–86

1977	House Intelligence Committee established; Harvard University issues faculty-CIA guidelines
1978	Electronic Surveillance Act passes; Omnibus Charter fails; Carter Executive Order on intelligence (No. 12036)
1979	Shah of Iran falls from power; Soviet invasion of Afghanistan
1980	Intelligence Accountability (Oversight) Act passes
1981	Reagan Executive Order on intelligence (No. 12333)
1982	Intelligence Identities Protection Act passes; first of the Boland amendments passes
1984	Nicaraguan mining flap; assassination manual
1986	Gates's guidelines on CIA-academic relations; Iran-contra revelations

Era of Distrust, 1987–

1987	Iran-contra investigations
1988	Intelligence Oversight Act passes the Senate
1989	House withdraws Intelligence Oversight Bill

I

THE INTELLIGENCE MISSION

ONE

Democracy and the CIA

Firestorm

With the National Security Act of 1947, and amendments in 1949, the U.S. government established the contemporary intelligence community.[1] In the years to follow, the public and even most government officials were privy to little information about its workings. This state of ignorance changed dramatically in the wake of a series of articles published by the *New York Times* throughout December 1974. These reports claimed that the Central Intelligence Agency (CIA)—the most prominent of the U.S. intelligence agencies—had engaged in unsavory covert operations during the 1960s against the democratically elected government of Chile and, more startling, had resorted to espionage operations at home (on a "massive" scale, concluded *Times* correspondent Seymour M. Hersh).[2] Among other charges, the newspaper accused the CIA of compiling files on over ten thousand American citizens, despite the language of the National Security Act barring "the Agency" (as the CIA is called by insiders) from any security or police functions within the United States.

These allegations appeared on the heels of the Watergate trauma, the name given those White House crimes of the Nixon administration that included a burglary against the Democratic National Committee housed in the Watergate Hotel offices in Washington (and whose attempted cover-up drove the president from office in August 1974). As a result, Hersh's exposes had a particularly explosive effect. "The time has come," concluded a leading member of Congress, Sen. Hubert H. Humphrey (D, Minnesota), "for Congress to face up to a responsibility it has shirked for too many years."[3] He announced that he would offer legislation to establish a new congressional committee with a mandate to ensure closer supervision of all intelligence operations. The government soon launched major executive and legislative investigations, beginning with a presidential commission chaired by Vice President Nelson A. Rockefeller in January 1975 and quickly followed by two congressional committees, led in the Senate by Frank Church (D, Idaho) and in the House by Otis Pike (D, New York). What became known as "The Year of Intelligence" (1975) was underway. Among intelligence officers, these months are still remembered painfully as the "Year of the Firestorm" or the time of the "Intelligence Wars."[4]

Controversy was hardly a new experience for the CIA. Over the years critics had excoriated the Agency for, among other things, the secret distribution of money to private American groups, including the National Students Association; for attempts, largely inept, to overthrow unfriendly regimes abroad (the most conspicuous being an invasion of Cuba via the Bay of Pigs in 1961 using Cuban expatriates as surrogate soldiers); for participation in a "pacification" operation in Vietnam, known as the "Phoenix Program," which involved a number of "illegal killings" (that is, assassinations outside the realm of military combat; see Chap. 2); for murder plots against foreign leaders, reported as early as 1967 by Washington muckraker Drew Pearson; for the distribution of disguises and other paraphernalia to former CIA personnel, who later used them in Watergate-related capers (now known as the "red-wig problem" at the CIA, after the coppery toupee provided to Watergate conspirator E. Howard Hunt, the spy novelist and onetime CIA officer); and for sundry other intrigues around the globe reported (some truthfully, some falsely) in the print and electronic media on a regular basis.[5]

A cartoon published in the *New Yorker* magazine in 1965 depicted Asian villagers fleeing an erupting volcano. "The CIA did it. Pass it along," said an onlooking Communist guerrilla leader. The drawing humorously portrayed a widespread belief (being exploited in this instance by an enemy soldier): Behind a great many turbulent events abroad lurked the shadowy figure of the CIA.

Yet the public and its leaders tolerated all the earlier allegations—even the proven errors of judgment and performance—with, at worst, a mild slap on the wrist. In 1967, for instance, government officials placed the National Students Association as well as other educational and private voluntary organizations off limits to the CIA. In 1974, though, the eyes of Washington officialdom were less easily averted, for, claimed the *Times,* the American public itself had become the object of the CIA's dark trades. The Orwellian vision of Big Brother had moved from the pages of *Nineteen Eighty-Four* to the headlines of America's major newspapers. Toppling Marxist regimes was one thing, even if the plans ran amok; widespread surveillance of American citizens—read voters—went too far.

A succession of events had prepared the nation for a strong reaction to charges of CIA duplicity. The misleading public statements and the failures of the Vietnam War raised serious doubts about the wisdom of trusting officials in the executive branch. The involvement of the CIA with Watergate conspirators—regardless of how peripheral the contact—added further questions. Watergate also produced a reform-minded Congress, along with a wolf pack of investigative journalists ready to attack at the slightest hint of government impropriety. Further, the Nixon administration had ushered in a new and unexpected era of detente with the

Soviet Union. If the cold war were over, did the United States need so aggressive a CIA? Public attitudes toward the Agency seemed to be in a state of flux. As William E. Colby, director of the CIA from 1973 to 1976, has written: "All the tensions and suspicions and hostilities that had been building about the CIA since the Bay of Pigs and had risen to a combustible level during the Vietnam and Watergate years, now exploded."[6]

One result of the investigations conducted by the Rockefeller commission and the legislative committees was the release of a torrent of information on the U.S. intelligence agencies in place of the previous trickle. The three panels held extensive hearings and issued reports that, together, stood several feet high.[7] For the first time, the citizens of the United States—not to mention enemies abroad—were given an opportunity to examine the hidden side of their government. What they found in these reports should have made it clear that intelligence agencies, if misused, pose a chilling threat to freedom. Among other findings, the investigations revealed that:

- a CIA program to open mail to or from selected American citizens generated 1.5 million names stored in the Agency's computer bank
- intelligence units within the Federal Bureau of Investigation (FBI) created files on over one million Americans
- the FBI carried out five hundred thousand investigations of so-called subversives from 1960 to 1974, without a single court conviction
- computers in the National Security Agency (NSA) monitored every cable sent overseas, or received, by Americans from 1947 to 1975
- Army intelligence units conducted investigations against one hundred thousand American citizens during the Vietnam War era
- the CIA engaged in drug experiments (the MK/ULTRA Project) against unsuspecting subjects (two of whom died from side effects)
- at least two foreign leaders were the direct targets of CIA assassination plots (none successful)
- letters written anonymously by FBI agents were designed to incite violence among blacks
- the FBI COINTELPRO (Counterintelligence Program) targeted civil rights activists and Vietnam War dissidents, disrupting family and friendship ties
- the FBI attempted to blackmail civil rights leader Martin Luther King, Jr., and encouraged him to commit suicide
- the CIA manipulated elections in democratic regimes (Chile was but one of several)
- the Internal Revenue Service (IRS) allowed tax information to be misused by intelligence agencies for political purposes
- intelligence agencies carried out burglaries in the homes and offices of suspected "subversives"
- the CIA infiltrated religious, media, and academic organizations

If anyone had forgotten the dangers of hidden and unfettered power—despite the recent revelations of Watergate—here were some unpleasant reminders.

Secret Power in an Open Society

The central theme presented in this book is that democracy and secret intelligence organizations—despite their seemingly inherent antithesis—can exist safely and effectively within the same society, but only with the most careful precautions. The fundamental objective of democracy is to allow the people an opportunity to participate in decisions that determine their fate. Democracy, as Charles Douglas Lummis observes, "is a word that joins 'the people' and 'power' . . . It describes an ideal, not a method of achieving it. It is not a *kind* of government but an *end* of government." Lummis continues, "If the word means what it says, there is democracy where the people have the power."[8] This seemingly uncomplicated idea has proven to be fraught with difficulties. In contemporary mass society, no nation has been purely democratic by this definition, for in no country have all the people been able to participate in the decisions that determine their fate. Rather, in E. H. Carr's words, democracy has been "a matter of degree."[9] In some countries citizens have had a greater opportunity to participate than in others.

The extent of this gap between the people and their leaders provides an index of democracy in each society. A widening gap, as Carr noted over thirty years ago, is the "major threat to mass democracy."[10] In light of the hardships associated with governing large and complex societies, comprising millions of citizens, a gap of some width is apt to remain, even in those few countries with a cultural ethic favoring participation by the masses. Its width, though, can be narrowed to a point where, as with a spark plug in an automobile engine, the people can "fire" their representatives into responsiveness. This narrowing of the gap between political elites and masses is neither an easy nor a one-time task; instead, it is laborious and ongoing. As the historical record shows, usually the people fail: democracies have been both rare and fragile. Few have existed in the past; few exist today—only nineteen in a world of some 184 nations, according to a recent count.[11]

In most cases elites have simply refused to share power with the people. Autocracies or oligarchies have been the most prevalent forms of rule, and often these governments have been brutal dictatorial regimes. Even in nations like the United States that have exhibited the accepted hallmarks of a democratic regime—above all, a free press and open elections in which parties or groups may compete for office—the job of monitoring the fealty of elected leaders and their subordinates has been an enduring problem. America's founders were well aware that control over government abuse would remain a demanding challenge to the new nation. "A dependence on the people is, no doubt, the primary control of government," stated *The Federalist*, No. 51, bowing to the central role of elections; but, the paper continued, "experience had taught mankind the necessity of auxiliary precautions"—that is, the internal checks and balances established within the government as a safeguard against what the founders feared most: a dangerous concentration of powers in one branch of government. "Ambition must be made to counteract ambition," *The Federalist* warned.[12]

Americans continue to enjoy a free press and the opportunity to participate in open elections (even though most citizens fail to exercise the precious right to vote), and the transfer of power and authority from one administration to another has been remarkably smooth when one considers that revolution and violence have characterized the changing of the guard in most societies throughout world history. But how well has the second safeguard of democracy worked—the checks and balances advocated by the founding fathers and raised high on the altar of veneration in the high school civics books?

Less visible than the press and elections, this safeguard is also less easy to examine and evaluate. Yet the day-in, day-out checks within the government on the misuse of high office may mean even more to the preservation of democracy than a free press or open elections—especially with an electorate that chooses to stay home in large numbers on election day. Abuses of power warranting punishment at the polls may never come to light unless the government itself—aided, to be sure, by citizens in the media and elsewhere—helps maintain a steady vigilance over its own parts, alerting citizens to encroachments and wrongdoing. The people outside the government need help from those within to pinpoint responsibility for improper decisions and poor administration; officials must be held answerable—or "accountable" (a favorite word of political scientists)—for their actions. This accountability, in turn, rests upon a continuous review ("oversight") of governmental decisions and how well they are carried out.

The responsibility for oversight is sufficiently large to require many hands, for the organs of government have multiplied and swollen dramatically in contemporary society. A former assistant secretary of state suggests how difficult the assurance of accountability can be when he writes that: "in action after action, responsibility for decision is as fluid and restless as quicksilver, and there seems to be neither a person nor an organization on whom it can be fixed. At times the point of decision seems to have escaped into the laybrinth of governmental machinery, beyond layers and layers of bureaucracy."[13] Though this observation was made in the 1960s, it applies well to the events of the Iran-contra affair in 1987, in which many key foreign-policy decisions seem to have been made by sleight of hand with little or no authority from elected officials. President Ronald Reagan, among other top officials, claimed to have had no knowledge of the diversion of funds from arms sales in Iran to the CIA-backed contras, a counterrevolutionary group (deriving its name from the Spanish *contrarevolucionarios*) at war against the Marxist Sandinista regime in Nicaragua.

The standing committees of Congress, responsible whistle-blowers inside the executive agencies, special investigative panels, interest groups, the media—all are necessary in order to shine lights into the multiple crevices of modern government. Only in this way may the people drive from office those who have violated their vows of trust and used the state for their own benefit, or in the zealous pursuit of some "higher cause" over and above the wishes of the people as expressed in laws made by their elected representatives.

The question of accountability and the proper conduct of oversight is vast and complicated, and it remains in a state of infancy as a scholarly subject.[14] This book concentrates on a small but important portion of the whole. The focus here is on the record of accountability and oversight in what may be its supreme test in a democracy: the task of monitoring and supervising the behavior of officials who work within the least accessible recesses of government, the secret intelligence agencies.

A few democracies (Canada, for one) have followed the American lead to establish rigorous intelligence oversight committees. In most, however, accountability for intelligence agencies has been left to a limited number of people within the executive domain. Britain, for example, entrusts this obligation to its cabinet ministers, placing a premium on the enhanced secrecy fostered by this doctrine of "ministerial responsibility." Traditionally, in Britain, to call civil servants—certainly intelligence officials—to account has been beyond the pale of legislative prerogative. Moreover, Britain has several Official Secrets Acts that further enclose intelligence policy tightly within the ambit of those ministries dealing with national security. The United States also has laws to protect intelligence secrets (including the Intelligence Identities Protection Act of 1982, which makes disclosing the names of intelligence employees a crime[15]); but something as sweeping as the Official Secrets Acts, which virtually seal off intelligence policy completely from the British citizenry, has no counterpart in the United States.

In America responsibility for intelligence accountability has been considerably more diffuse. Inspectors general and various oversight boards within the executive branch, as well as the Congress, the courts, and the media, all attempt to monitor intelligence activities—with varying degrees of enthusiasm and success. Preeminent among these overseers, in theory at least, has been the Congress with its subcommittees and, since 1976–77, full committees assigned to supervise intelligence policy and to review each year the budgets of the intelligence agencies.

How energetic has the American experiment in intelligence oversight been? How closely has the CIA been monitored, by what means, and to what effect? Has supervision of the CIA curbed abuses, or merely shackled a once imaginative and aggressive protector of these United States against foreign threats? How well has accountability—and, therefore, democracy—worked within this sensitive policy domain? Here are the key questions that guide this inquiry.

The answers in these pages suggest three broad conclusions. First, intelligence oversight has varied in intensity over the years, from benign neglect in its earliest stages (1947–74) to a marked assertiveness in the post-Watergate, post-Vietnam War period when Congress began to demand a restoration of its authority across the board. Second, in the decades from 1976 to 1986 intelligence policy became more accountable on the whole and therefore more democratic—without losing its effectiveness. On this point several readers will disagree—especially within the CIA—but I am convinced of its correctness and, more importantly, so are many experienced intelligence officials, including former CIA directors. And, third, the Iran-contra scandal revealed that serious flaws continue to exist in the established safeguards against the abuse of power by the intelligence agencies and the National Security Council (NSC). The checks put in place from 1975 to 1980 failed to stop—or even alert Congress to—this unfortunate operation, planned and executed by the NSC staff with support from the CIA. The American experiment in balancing the intelligence mission with accountability had been dealt a serious blow.

Before the Iran-contra affair, modern intelligence in the United States had passed through a three-stage evolution toward greater democratic control over the CIA. The first phase, the Era of Trust (1947–74), was a time when the intelligence agencies were permitted almost complete discretion to chart their own courses, free of meaningful scrutiny by overseers in the Congress. The second phase, the Era of Skepticism (1974–76), saw the agencies reel under the impact of investigations led by public officials suddenly skeptical about the "honorable men" (a description of Agency personnel advanced by CIA directors Richard Helms [1966–73] and William E. Colby [1973–76], among other insiders) who had been allowed to run the invisible side of government.[16] The third phase, the Era of Uneasy Partnership (1976–86), witnessed experimental forms of closer legislative supervision over the intelligence community and heightened public awareness of the CIA and its missions—a "democratization" of intelligence, which, critics complained, undermined the effectiveness of America's secret power. By making the intelligence agencies more open to the Congress and the public, argued the critics of reform, the country had made itself more vulnerable to its enemies.[17]

The fourth and current phase in the evolution of modern American intelligence, the Era of Distrust (1986–), began with press disclosures in November 1986 revealing the secret sale of U.S. arms to Iran. Since the House and Senate intelligence committees had no knowledge of this operation before the media accounts emanating from the Middle East (with their allegations of CIA involvement), the revelations raised serious misgivings among legislators about the Agency's willingness to honor the new oversight arrangements. These arrangements required, by law (the Hughes-Ryan Act of 1974 and the Intelligence Accountability Act of 1980), formal reports to the intelligence committees on secret arms sales and other "covert actions" involving the intelligence agencies. Further charges that the profits from the arms sales had been channeled through Swiss bank accounts to finance the contras in Nicaragua, despite an act of Congress (the Boland amendment) limiting government involvement in the supply of weapons to the counterrevolutionaries, added fuel to the fires of criticism against the CIA that were gathering strength on Capitol Hill. The Iran-contra affair thus renewed public concern over the danger that secret intelligence agencies might run out of control. With the echoes of the 1975 intelligence investigations still faintly lingering, the nation took up the debate once more between those who advocated increased democratic controls over the CIA and its sister agencies and those who favored a grant of special discretionary authority for intelligence—indeed, ideally, a turning of the clock back to the Era of Trust.

This tension between the desire to supervise the intelligence agencies and a willingness to let them operate in full secrecy lies at the heart of the dilemma addressed in this book, for democracy and intelligence represent values that remain in conflict, pulling one against the other. "While there is a strong public interest in the public disclosure of the functions of governmental agencies," a high-ranking CIA official once put it, "there is also a strong public interest in the effective functioning of an intelligence service."[18]

Democracy rests on the assumption that government should be conducted openly: that decisions should be preceded by wide public debate, that the rule of law is more trustworthy than the rule of man, that sneakiness is incompatible with the ideals of a virtuous society. Ours is "a government of laws and not of men," wrote John Adams into the Massachusetts state constitution in 1780, expressing a belief widely held in the new nation. In contrast, intelligence depends upon secrecy and sharply limited debate, and it usually entails the violation of ethics and laws in those countries overseas where U.S. agents operate, not to mention the use of tactics or "dirty tricks" that are far removed from the accepted philosophical tenets of democratic theory—lying, sabotage, even clandestine warfare and assassination in times of peace.

One can trace the lines of tension between these oppositive premises at the highest levels of government. "Gentlemen do not read other's mail," declared Secretary of State Henry L. Stimson in 1929 with innocent exuberance as he closed down the code–breaking section of the Department of State.[19] In dramatic counterpoise, a more common perspective during the early years of the cold war was one advanced in 1954 within a then top-secret report to President Dwight Eisenhower (part of the Hoover commission review of government organization):

> It is now clear that we are facing an implacable enemy whose avowed objective is world domination by whatever means and at whatever cost. There are no rules in such a game. Hitherto acceptable norms of human conduct do not apply. If the U.S. is to survive, long-standing American concepts of "fair play" must be reconsidered. We must develop effective espionage and counterespionage services. We must learn to subvert, sabotage and destroy our enemies by more clear, more sophisticated and more effective methods than those used against us. It may become necessary that the American people will be made acquainted with, understand and support this fundamentally repugnant philosophy.[20]

One possible response to the democracy-vs.-intelligence dilemma is, in the manner of Stimson, to eliminate or at least sharply curtail all intelligence operations. Yet today's world—bristling with nuclear warheads and scarred with acts of terrorism—is a place of unprecedented danger. The United States is no longer "remote from the wrangling world," in Thomas Paine's happy phrase of 1776.[21] Trucks filled with explosives and driven by suicidal zealots have carried a fiery death to Marines and U.S. embassy personnel in Lebanon; Soviet submarines lurk silently off American shores, carrying weapons of quick and total annihilation; lethal drugs pour across this nation's borders and into the veins of young Americans; Middle East oil ministers convene in secret cartel meetings, drafting schemes to manipulate the economies of the industrial nations. Only with accurate information from every point on the compass can the United States hope to cope with these and many other threats. Perhaps in Stimson's day the nation could afford the luxury of Cinderella ethics, but today Americans seem to have no choice other than to engage in espionage operations: to collect and analyze to the best of their ability data on the capabilities and intentions of present or potential adversaries. "The winds and waves are always on the side of the ablest navigators," wrote Gibbon.[22] Skillful navigation is dependent upon the gathering of precise information. Without a competent intelligence service, a nation is doomed to steerage beneath cloudy skies.

To abolish or emasculate the CIA would be an act of folly, for while the Agency can pose a threat to democracy from within, it provides a vital protection for democracy against serious threats from abroad. Here is the paradox. And from this paradox comes the central challenge: to guard against CIA excesses that are anathema to democracy while at the same time holding high the Agency's shield against dangers from beyond these shores. In a rekindled public debate, the citizens of the United States and their representatives in government have to decide which secret intelligence operations the nation must tolerate in order to protect itself in this world of enmity and doomsday weapons, and which operations lie beyond the pale of acceptability.

Assassination plots in times of peace would seem to qualify as an example of the unacceptable, while the reading of other people's mail might have to be tolerated—if the proper authorities obtain a legal warrant for this purpose by demonstrating probable cause that the security of the nation truly requires intrusive measures. And when intelligence operations are deemed acceptable (ideally, according to criteria spelled out in statutory language drafted in close consultation between the executive and legislative branches), responsible officials—elected and appointed—must guarantee through vigorous oversight that its spymasters and agents are held accountable and operate firmly within the established boundaries. The core purpose of this book is to examine the question of proper boundaries for intelligence activities, and the chances that overseers can maintain them.

TWO

The Purpose of American Intelligence

The Shock of Pearl Harbor

In *Alice's Adventures in Wonderland* the White Rabbit sought counsel from the King of Hearts on the art of presenting evidence. "Begin at the beginning," the King sensibly advised, "and go on till you come to the end: then stop."[1] To begin at the beginning, the origins of American intelligence coincide with the birth of the nation.[2] Even during these early stirrings, the leaders of the Revolution were well aware of the vital role intelligence operations would play in a successful war of independence. In 1776 the Continental Congress established America's first intelligence service, the Committee of Secret Correspondence, and Gen. George Washington (whose own secret code number was "711") made use of an effective network of spies led by Paul Revere. Perhaps the most famous officer in this network was young Nathan Hale, whose memorable declaration ("I only regret that I have but one life to lose for my country") was uttered before being hanged by the British in 1776 for espionage. Today a statue of Hale near the entrance to CIA headquarters honors his memory.

The interest of the founders in matters of intelligence went well beyond espionage (that is, the secret collection of information) and the interpretation of information gained either in this manner or through open sources—what modern practitioners call "analysis."[3] Benjamin Franklin and Thomas Jefferson, among others, encouraged the use of intelligence operations for another objective: to influence other nations in favor of American foreign-policy objectives—an approach known as "covert action" or "special activities" in modern parlance. Franklin urged the government of France to join with the colonists in the creation of a secret conduit for the supply of military aid in support of the revolutionary war. To conceal this relationship from the British, a front (or "proprietary") was formed, ostensibly a private commercial enterprise called the HORTALEZ Company.[4] As president, Jefferson approved a covert action to supply arms for a coup designed to place on the throne of Tripoli a man friendlier toward the United States than the ruling Bashaw had proven to be.[5]

The American revolutionaries were concerned, too, about protecting their army and the new nation from hostile foreign spies, an important and difficult responsibility known today as "counterintelligence." Through counterintelligence operations, many British espionage agents were apprehended during the War of Independence, including, reported the CIA snidely in 1975 as Church committee investigators poured over its files, one by the name of Dr. Benjamin Church.[6]

These three types of activities—collection (espionage, when the information is secretly acquired) and analysis, covert action, and counterintelligence—have comprised the nation's intelligence mission since 1776. This book does not attempt to trace their evolution; several good histories have performed this service.[7] Instead, it skips forward to the more immediate historical wellsprings of America's modern intelligence establishment. Here the beginnings may be found in that unforgettable day, December 7, 1941, when the Japanese Air Force struck the United States with a surprise attack against the military base at Pearl Harbor, Hawaii. In establishing a modern intelligence service after World War II, U.S. government officials hoped above all to avoid another military shock like this one.[8]

Almost one hundred Navy ships and some three hundred airplanes were based in Pearl Harbor at that fateful hour. All eight battleships were hit and five sank. So did two destroyers and several other ships. Over two hundred aircraft were damaged and many were destroyed. Luckily, the two aircraft carriers in the Pacific Fleet were at sea and escaped. Less fortunate were the 2,330 servicemen killed and the 1,145 wounded, along with 100 civilian casualties.

The blow stunned the nation. It represented the most disastrous intelligence failure in the history of the United States. American leaders had failed to appreciate both the capabilities and the intentions of the enemy. They had no idea the Japanese had developed aerial torpedos that, when dropped into the sea, could navigate in the relatively shallow waters of Pearl Harbor. The greatest damage to American ships came from these weapons.[9] The leaders of the United States were in error, too, in their belief that a Japanese attack against the Philippines was infinitely more likely than one against Hawaii.[10] Washington officials were confused by the buzz of ambiguous information about Japanese military plans, information that was often inconsistent and irrelevant—the "noise" that frequently engulfs and sometimes drowns out the important facts the nation needs to know.[11] Moreover, the U.S. government failed to analyze with any degree of thoroughness the fragments of data that were available regarding the possibility of a Japanese attack on Hawaii. Perhaps most inexcusably, officials failed to distribute in a timely way to key Naval officers what was known, apparently for fear that the sensitive source of this information (Operation MAGIC, the U.S. intelligence program that cracked the Japanese military codes) might be revealed—"compromised" in spy argot—if shared too widely.[12]

The Coordination of Intelligence

This last contribution to the failure at Pearl Harbor—poor coordination of intelligence—was a problem Harry S. Truman vowed to address soon after he became president. The exigencies of war, though, delayed a major overhaul of U.S. intelligence—not that much existed to overhaul. "On the eve of Pearl Harbor the United States had no strategic intelligence system worthy of the name," conclude two foreign policy specialists.[13] At the time only about $3 million was spent annually on intelligence across the board.[14] For the time being, the United States made do during the war (under the helpful tutelage of British intelligence officers) with a loosely defined, but at least somewhat more centralized, intelligence apparatus than had existed before the war. This apparatus was called the Office of Strategic Services (OSS)—an acronym with the second meaning of "Oh So Social" in the view of some wags who looked upon the organization as a haven for adventurous blue bloods from the northeastern region of the United States.[15]

Despite various notable accomplishments by the fledgling OSS, American intelligence continued to suffer immediately after the war from uneven coordination as well as from poorly defined lines of authority. During this period, President Truman found himself perturbed chiefly by the tropical downpour of intelligence reports that descended on his desk from different parts of Washington, sometimes directly contradicting one another. He preferred a single report from one central organization.

"So I got a couple of admirals together," Truman once recalled, "and they formed the Central Intelligence Agency for the benefit and convenience of the President of the United States. . . ." As a result, Truman continued, "instead of the President having to look through a bunch of papers two feet high, the information was coordinated so that the President could arrive at the facts."[16]

In 1946 Truman issued an Executive Order establishing a Central Intelligence Group patterned after the OSS.[17] Then, on July 26, 1947, the National Security Act gave statutory authority to the idea, replacing the Central Intelligence Group with the CIA. This law placed the CIA under control of a new White House structure that would henceforth coordinate American foreign policy: the National Security Council. By law, the NSC consisted of only four principal members: the president, vice president, secretary of state, and secretary of defense—though other officials would be invited from time to time to take part in its deliberations. Unlike other ad hoc cabinet committees that it resembled on the surface, the NSC had a permanent staff and a director, eventually called the president's assistant for national security affairs. This council would soon become the most important forum in the government for discussing major intelligence proposals and formulating advice to the president on national security issues. Whether or not the president chose to accept the advice and order the implementation of the recommended policy was a matter for him to decide; the NSC was (and remains) strictly an advisory committee, albeit one whose exclusive membership makes it a panel with unparalleled leverage over decisions within the government.[18]

Although the concept of centralized coordination lay at the heart of these efforts to create a new intelligence system, strong forces resisted endeavors to subsume all existing intelligence entities under the plenary control of the CIA. Two authorities provide this description of the bureaucratic struggles that soon beset the infant agency:

> The CIA immediately ran into difficulties of the sort which seems to confront all new agencies in Washington. It inherited feuds going back to the days of the OSS. Over the next three years, the CIA and G-2 [the Army intelligence staff] clashed over which was to control secret agents abroad. The former finally won. A running battle raged between the State Department and the CIA over the question of whether CIA personnel attached to diplomatic establishments abroad were to be under the jurisdiction of the heads of missions. Bitterness broke out between the FBI and the CIA when the latter took charge of foreign field operations; the CIA claimed that the FBI agents did not cooperate in the changeover and the FBI claimed that CIA personnel were rank amateurs, careless about security. The Atomic Energy Commission refused for a long while to share its scientific information with CIA personnel, who were deemed unable to interpret it properly.[19]

The CIA was eventually able to consolidate its position among the rivaling agencies, a result attributable in large part to the extraordinary administrative skills of Gen. Walter Bedell Smith, former ambassador to Moscow, who became the CIA director in 1950, and to the enviable ties the CIA enjoyed with the NSC under the leadership of Allen Dulles, who succeeded General Smith as CIA director in 1953 and was the brother of the secretary of state in the Eisenhower administration, John Foster Dulles. The improved reputation of the CIA acquired during the Smith-Dulles years buoyed its recruitment efforts immeasurably. To many bright young people interested in public service, the CIA seemed the place to be—an exciting command post from which to wage the cold war. Aided by former OSS analysts who had returned to academic careers, the CIA developed an effective campus recruitment network to spot and enroll the most promising youth for intelligence careers (see chap. 8).

During the 1940s and 1950s, though, the CIA failed to achieve undisputed dominance over intelligence policy in the American government as some hoped it would. As Ransom states, "the principle of federation prevailed over the concept of tight centralization in shaping the structure of the intelligence community."[20] This principle of federation endured. Even when the CIA had reached a stage of maturity, one of its deputy directors, Adm. Rufus Taylor (1966–69), referred to the various intelligence agencies as little more than a "tribal federation."[21]

The centrifugal forces in the intelligence system are, however, less intense today than in the days before World War II. The numerous other intelligence agencies have accepted the CIA director (DCIA) as at least the titular head of the intelligence establishment (or "community," in the accepted mythology). In this role, he is known as the director of central intelligence (DCI, the acronym by which he is usually referred to in Washington) and is charged with the mission of trying to coordinate information from the different agencies (the "producers") for presentation to the principal decision makers in the government (the "consumers").

Just how difficult the job of DCI is can be easily appreciated by considering just one of his—so far all DCIs have been males—responsibilities: military intelligence. Beneath him, in theory, are the military intelligence agencies, yet they have another boss, too: the secretary of defense, who clearly outranks the DCI in the language of the 1947 National Security Act and in Washington protocol. The DCI was not even a member of the president's cabinet until 1981, when President Ronald Reagan elevated his appointee, William J. Casey, to this level. (Casey's successor, William H. Webster, formerly the director of the FBI, asked President Reagan to remove him from the cabinet in order to avoid the hint of political involvement that comes with membership in this high forum.) Moreover, imagine any DCI attempting to tell the long-time head of the FBI—the imperious J. Edgar Hoover, with his host of close ties in the Washington political establishment—how to run Bureau intelligence operations. The few forays ventured by DCIs into this perilous territory so angered Hoover that his relations with CIA plummeted, especially during the late 1960s when he and DCI Richard Helms clashed over jurisdictional questions.[22] The two men soon rarely spoke to one another. (Relations among lower-level officials, though, were often cordial and indeed, between CIA and FBI counterintelligence officers, even close.) So, the centralism sought by reformers in 1947 was a relative thing, a balancing of their ideal with what was bureaucratically possible.

Missions

Beyond structural changes designed to tilt American intelligence more toward centralism stood the still more important matter of defining its mission. What tasks were these new institutional relationships expected to accomplish? An examination of the 1947 National Security Act and the accompanying legislative history indicates that, as President Truman had insisted, the CIA was established—above all else—to collect, evaluate, and coordinate intelligence, and to provide for its proper dissemination within the government. The CIA would be a clearinghouse for all the American intelligence agencies. The Agency would also be expected, in the words of Allen Dulles, "to weigh facts, and to draw conclusions from those facts, without having either the facts or the conclusions warped by the inevitable and even proper prejudices of the men whose duty it is to determine policy. . . ."[23] In other words, the collection of information (either through espionage or from the public record), its analysis, and its coordination—what we may call, for short, the collection mission—would be the CIA's primary responsibility. This was hardly a new venture, but one that reformers hoped would now be carried out more effectively. The nation could ill afford another Pearl Harbor.

The National Security Act of 1947 gave the CIA (under the direction of the NSC) five specific authorities, of which the last gestured—in slippery language—toward duties beyond the collection mission. The statute charged the CIA with a duty to advise the NSC on intelligence activities related to national security; make recommendations to the NSC for the coordination of such activities; correlate, evaluate, and disseminate intelligence within the government; carry out services for existing agencies that the NSC decides might be best done centrally; and, in the ambiguous catchall phrase of the act, "perform such other functions and duties related to intelligence affecting the national security as the National Security Council may from time to time direct."[24] So while the founding statute overwhelmingly emphasized the collection mission (to which we return in chaps. 4 and 5), the door was left ajar to use the CIA for "other functions and duties"—an invitation quickly accepted by the NSC as an opportunity for launching the new Agency on a wide range of covert actions around the world.

Covert Action

As the cold war against the Soviet Union intensified in 1947, the executive branch drew upon this fifth authority to direct the CIA and its contacts around the world toward the advancement of President Truman's famous doctrine. In the president's words to Congress that year, the United States would "help free peoples to maintain their free institutions and their national integrity against aggressive movements seeking to impose on them totalitarian regimes."[25] The United States would, in short, stop communism. This would be done with every instrument at the government's disposal, not only the diplomatic corps and the Marine Corps, but CIA covert action as well—the hidden hand of secret intervention.

Nomenclature

Covert action (CA) is sometimes referred to as the "quiet option" within the CIA, which is usually the agency expected to design and execute this policy option. The phrase comes from a supposition that covert action is apt to be less noisy and obtrusive than some other instruments of American foreign policy, such as the landing of a Marine brigade. This supposition has frequently been proven false, most notoriously with the failed Bay of Pigs operation against Cuba in 1961.[26]

Sometimes professional practitioners also refer to covert action as the "third option," one between diplomacy and open warfare. As former secretary of state and national security adviser Henry Kissinger once put it: "We need an intelligence community that, in certain complicated situations, can defend the American national interest in the gray areas where military operations are not suitable and diplomacy cannot operate."[27] Still others prefer the additional euphemisms "special activities" or "special operations" (though, technically, "special operations" is a term more properly applied to Department of Defense unconventional warfare, using such units as the Army Green Berets or the Navy Seals, or narrowly to CIA paramilitary operations; see later discussion).

Formal definitions of covert action may be found in several public documents. An executive order on intelligence signed by President Jimmy Carter in 1978, for instance, refers to covert actions ("special activities" in this document) as operations "conducted abroad in support of national foreign policy objectives which are designed to further official United States programs and policies abroad and which are planned and executed so that the role of the United States government is not apparent or acknowledged publicly, and functions in support of such activities, but not including diplomatic activity or the collection and production of intelligence or related support functions."[28] An earlier formal statement, contained in the language of the 1974 Hughes-Ryan Act and reiterated in the 1980 Intelligence Accountability Act, provides this simple definition of covert action, and the one most widely used in government circles: "operations in foreign countries, other than activities intended solely for obtaining necessary intelligence."[29] Or, in short, operations other than espionage.

Though the examples of covert action offered later in this book further refine these definitions, "covert action" remains a complex—and sometimes slippery—phrase. The executive regulations and laws that address covert action, for instance, are keyed to secret intervention abroad by U.S. intelligence agencies; yet the NSC staff—a White House entity not normally considered an intelligence agency (though it routinely offers advice on, and coordination for, intelligence policy)—has attempted on occasion to influence events overseas in a clandestine fashion. When Kissinger directed the NSC during the Nixon years, he carried out major foreign-policy initiatives in secret, among them negotiations with the North Vietnamese and the crafting of a Sinai disengagement between the Israelis and the Egyptians. Did this constitute "quiet diplomacy"—or secret intervention and, therefore, covert action? Most experts would say diplomacy in these cases, but the examples illustrate how fine the line between the two can sometimes be when it comes to exercising secret influence over events abroad.

In another more recent—and notorious—example, the Reagan administration turned to the NSC in 1984 as a means for carrying out a secret arms sale to Iran in exchange for that nation's influence over the release of American hostages held in the Middle East. The NSC staff then diverted the profits from this sale to U.S.-backed contras in Nicaragua—an action now known as the Iran-contra affair. Both of these operations fit the standard definitions of covert action, for they involved secrecy, the use of the CIA, and the goal of influencing another country; yet subsequently the Reagan administration attempted to argue that since the NSC was the prime mover in this operation, not the CIA (though it participated) or some other intelligence agency, the normal decision procedures for covert action became unnecessary, even irrelevant. This represented a significant blurring of the normal ways of authorizing and reporting CA operations (see chap. 6).

So the seemingly straightforward official definitions of covert action belie sophisticated nuances that have made efforts to isolate and examine this policy all the more difficult for both overseers and scholars. In the wake of the Iran-contra affair, some former senior CIA officials went so far as to reject the description of this controversial operation as a covert action, arguing that since Iran knew about the arms it was secretly buying the operation fell into the category of "secret diplomacy" instead. A true covert action, they maintained, requires that the target country be completely unaware of the CIA's operation against it. Legislative overseers present at the Symposium on the Management Intelligence (Georgetown University, Washington, D.C., March 3, 1988), including a member of the Senate Intelligence Committee, Arlen Specter (R, Pennsylvania), roundly dismissed this interpretation, noting that its secrecy, reliance on the CIA, and goal of influencing Iran made the operation a covert action in traditional parlance.

A further blurring of distinctions has occurred with attempts by the Reagan administration to "privatize" covert action, that is, to raise funds for this option outside the government's established appropriations process—from foreign nations, including Brunei and South Africa, among others, and from wealthy private citizens, like the owner of the Coors brewery in Colorado (see chap. 6). Moreover, some private groups (among them the Alabama-based Civilian Military Assistance organization, a spinoff of a network nurtured by the Reagan administration) have taken it upon themselves to raise funds for anticommunist guerrillas like the contras, outside U.S. government channels but with the knowledge and encouragement of the president and the NSC staff. With the admitted clouding effect that these ambiguities present, most practitioners and outside experts on the subject nonetheless treat covert action as essentially those secret operations carried out overseas by U.S. intelligence agencies; and, no doubt, the overwhelming majority of the covert actions implemented by the United States are of this sort—in excess of 95 percent of the total number, according to those interviewed for this book, with almost all of these operations conceived of and carried out by the CIA.

Whatever the variation in terminology, the goal of covert action remains constant: to influence events overseas, secretly and in support of American foreign policy. The theory behind the use of covert action has been straightforward, though "theory" is no doubt too strong a word; its practitioners have been less theorists than practical men with an interest in advancing American interests around the world. Their main concern has been, in the words of a former chief of the CIA Covert Action Staff, "the global challenge of communism . . . to be confronted whenever and wherever it seemed to threaten our interests."[30]

Driven by a zero-sum view of U.S.-Soviet rivalry, as well as by threats posed more recently by a proliferation of international terrorist organizations, the government of the United States turned increasingly toward covert action as one of many methods designed to win this "game." Though the catchall phrase in the 1947 National Security Act had been conceived of originally by its drafters as but a tail to the section on intelligence, the tail soon began to wag the dog. Quite probably some CIA officers from the beginning viewed this open-ended CA authority as the Agency's most important mission, even if the avowed primary purpose of the CIA was to prevent another Pearl Harbor through improved collection, analysis, and coordination.[31] Personnel, prestige, and, most important, money flowed increasingly from the collection mission to the CA mission, as the possibility of altering world affairs through a secret quick fix soon captured the attention of presidents and their advisers—especially following early successes in Europe (1947–49), Iran (1953), and Guatemala (1954).[32]

According to former DCI William E. Colby, the early Soviet military threat to Europe was successfully met by the North Atlantic Treaty Organization (NATO), the economic threat by the Marshall Plan, and the political and subversive threat in part by the CIA. "As you look ahead to the next ten or twenty years, we don't know when another kind of political crisis might arise in the world," he warns, "and I think it is better that we have the ability to help people in these countries where that will happen, quietly and secretly, and not wait until we are faced with a military threat that has to be met by armed force."[33]

Elsewhere Colby has elaborated further on the rationale for covert action. He has argued that

> in a number of instances, some quiet assistance to democratic and friendly elements enabled them to resist authoritarian groups in an internal competition over the future direction of their countries. Post-war Western Europe resisted communist political subversion and Latin America rejected Cuban-stimulated insurgency. They thereby thwarted at the local level challenges that could have escalated to the international level. . . . That there can be debate as to the wisdom of any individual activity of this nature is agreed. That such a potential must be available for use in situations truly important to our country and the cause of peace is equally obvious.[34]

Whether or not the West African nation of Angola qualifies as "truly important" to the interests of the United States is a matter of debate, but the plans of the Ford administration, in 1975, to assist the pro-Western side in the Angolan civil war illustrates a common argument in favor of covert action. According to a senior CIA official responsible at the time for CA planning, the motivation for entry into this war (eventually blocked by Congress through a law prohibiting covert action in Angola, the Clark amendment) was, first, to prevent the Soviet Union from expanding its presence and influence in Africa. Especially worrisome was the perceived strategic or geopolitical threat; the Soviets had already gained a toehold in Guinea and Somali, and now Angola offered them valuable port facilities on the South Atlantic seaboard. Second, the Ford administration hoped to achieve a negotiated settlement of internal differences in Angola, one that would place moderate groups in power. "Ultimately, the purpose was to throw the Soviets out," the CIA official concluded, "at which point we would leave, too."[35]

Appearing before the Senate committee investigating the intelligence community in 1975 (the Church committee), legislators again asked Colby to justify the use of covert action. In response, he paraphrased the conclusion reached by the Murphy commission that "there are many risks and dangers associated with covert action. . . . But 'we must live in the world we find, not the world we might wish.' Our adversaries deny themselves no forms of action which might advance their interests or undercut ours. . . . In many parts of the world a prohibition on our use of covert action would put the U.S. and those who rely on it at a dangerous disadvantage. . . . Therefore . . . covert action cannot be abandoned. . . ."[36]

Other witnesses before the Church committee also accepted the necessity of a CA capability, but for use only in carefully restricted circumstances. "The guiding criterion," advised Clark Clifford, former secretary of defense and an author of the National Security Act, "should be the test as to whether or not a certain covert project truly affects our national security." Cyrus Vance, who would soon become secretary of state in the Carter administration, told the committee that he believed "it should be the policy of the United States to engage in covert actions only when they are absolutely essential to the national security."[37] Just one witness before the Church committee opposed covert action altogether. "Such operations," testified Morton H. Halperin, a former NSC staffer, "are incompatible with our democratic institutions, with Congressional and public control over foreign policy decisions, with our constitutional rights, and with the principles and ideals that this Republic stands for in the world."[38] Senator Church's own views were supportive of covert action, but only if the operations were "consistent with either the imperative of national survival or with our traditional belief in free government"[39]—a rather nebulous prescription.

Normative views aside, the fact remained that the covert action shop had become a place for rapid promotion within the Agency. Directors Dulles, Helms, and Colby each headed the CA arm of the CIA on their way to the directorship. In contrast, no Agency intelligence analyst has ever made his way to the top, despite the emphasis on this task in the 1947 National Security Act (though after the death of DCI Casey in 1987, his deputy, Robert M. Gates, a leading Agency analyst, was nominated by President Reagan as his replacement; however, during the confirmation hearings, Gates's ambiguous answers regarding his knowledge of the Iran-contra scandal eventually forced the administration to withdraw his nomination[40]).

By the late 1960s the CIA had hundreds of CA operations underway, and over half of the Agency's annual budget was dedicated to this mission, with a large percentage of these resources targeted against U.S. opponents in the Vietnam War. Only as this war began to taper off were these aggressive operations cut back. In 1969 the number of positions and dollars assigned to covert action went into a precipitious decline as DCI Helms shifted resources slowly back toward the collection mission. By 1978 the share of the intelligence budget reserved for covert action had plummeted to less than 5 percent—a slide that suddenly halted and went into reverse during the last two years of the Carter administration (following the Soviet invasion of Afghanistan) and then continued an upward acceleration during the "Cold War II" days of the Reagan administration (see chap. 6).[41]

Methods of Covert Action

While most observers seem to agree that some form of CA capability is necessary for the protection and advancement of the nation's interests, opinions differ on the question of exactly what kinds of operations ought to be permitted. Some examples, past and present, serve to illustrate the scope of this option since the establishment of the CIA. Covert-action operations may be grouped according to four broad categories: propaganda, and political, economic, and paramilitary (PM) covert action. They represent, respectively, about 40, 30, 10, and 20 percent of the total number of covert actions over the years—though PM operations have been by far the most expensive and controversial.[42]

PROPAGANDA No form of covert action is used more extensively than propaganda (sometimes called psychological warfare, or simply "psy war").[43] The government has an overt propaganda agency within the Department of State, the United States Information Agency (USIA), which has offices in American embassies around the world. The USIA conducts a wide range of public relations activities, from courting the local press to hosting lectures, luncheons, and the like, as well as maintaining libraries where foreigners can peruse the latest American periodicals. To supplement this flow of information about the United States, much of which is often viewed cynically by locals as too "official," the CIA secretly provides a flood of supportive propaganda distributed through its vast, hidden network of "media assets": reporters, newspaper and magazine editors, anchormen, television producers, cameramen, broadcast technicians—the whole range of media personnel. Whatever foreign policies or slogans the White House may be pushing at the time—the virtues of neutron bombs or Pershing missiles for Europe, the pernicious role of Moscow in the European peace movement, (there have been hundreds of such propaganda "themes" over the years)—the CIA will likely be advancing the same ideas through its covert channels.

In Country X, for example, the CIA chief of station (COS) will receive from Headquarters one or more themes that he is expected to circulate. He will contact his media assets, say, a newspaper editor, three correspondents, a television reporter, and two magazine writers—all on the CIA payroll—and they will plant the theme into their news outlets. (In most cases the asset will write the actual copy; only about one third of the Agency's propaganda is written at CIA Headquarters or in its offices overseas.) The end result in each case is that a native citizen, ideally one with respected media credentials, endorses the U.S. position in his or her native tongue and through the country's own news outlets. This approach, argue those who advocate this form of covert action, is apt to be far more persuasive than a perfunctory USIA news release on official government parchment.

In addition to this support for Department of State propaganda themes, the CIA will use its media assets to help or harm foreign political leaders or aspirants, depending upon how these individuals are positioned to affect the interests of the United States. The classic example in the public domain is the CIA's effort to discredit the Chilean Marxist-socialist leader, Salvador Allende.[44] The Nixon administration feared that the Soviet Union might use Chile as a base to spread communism throughout the Western Hemisphere, that Allende would become another Fidel Castro. In an attempt to stop his rise to power, the White House directed the CIA to use its network of assets (paid agents) in Chile against Allende. To help protect their own interests in Chile, the International Telephone and Telegraph (ITT) Corporation and other American businesses poured $1.5 million into the CIA pot—an example of a privatization of covert action in the 1960s. Indeed, the corporations strongly encouraged President Nixon to intervene covertly in the first place, out of a concern that Allende might nationalize their holdings.

In the 1964 election alone, the CIA secretly spent $3 million in Chile to blacken the name of Allende and his party, and over $3 million in fiscal year

1972. An expenditure of $3 million in Chilean presidential elections was roughly equivalent on a per capita basis to an expenditure of $60 million in an American presidential election during this period, according to estimates by the Church committee—a staggering amount of money at the time. Between 1963 and 1973, the CIA spent over $12 million in Chile on propaganda alone, not to mention additional monies spent on other forms of covert action. The Church committee described the forms of propaganda employed in the 1964 election: "Extensive use was made of the press, radio, films, pamphlets, posters, leaflets, direct mailings, paper streamers, and wall paintings. It was a 'scare campaign,' which relied heavily on images of Soviet tanks and Cuban firing squads and was directed especially to women. Hundreds of thousands of copies of the anti-communist pastoral letter of Pope Pius XI were distributed by Christian Democratic organizations. . . ."[45]

In other uses of propaganda, the CIA routinely attempts to infiltrate Western literature (books, magazines, newspapers, and the like) into totalitarian regimes, to be distributed covertly by Agency assets. "This has maintained several independent thinkers in the Soviet bloc," argues one high-ranking CIA official, "has encouraged the distribution of ideas, and has increased the pressures on totalitarian regimes."[46] Occasionally, the CIA has resorted to leafleting from airplanes, trying transistor radios to balloons lofted toward hostile nations, and broadcasting from makeshift radio stations in remote jungles.

One of the most successful CIA propaganda operations took place in Central America in 1954. The Agency set up a radio station in the mountains of Guatemala, where local assets began broadcasting the fiction that a revolution was taking place and that the masses had risen up against the procommunist dictator Jacobo Arbenz. The skillful broadcasts became something of a self-fulfilling prophecy and Arbenz, believing the reports of a mythical people's army of five thousand men advancing toward the capital, resigned without firing a shot.[47]

Until the affiliation leaked to the press in the early 1970s, Radio Free Europe and Radio Liberty were CIA operations. Now they are run by the USIA. As part of its propaganda program, the CIA also funded the National Student Association until 1967 when the connection was leaked to the counterculture magazine *Ramparts* and caused a scandal. Critics were angered by the existence of covert CIA funding for a domestic group and feared the corrosive influence such relationships between secret government agencies and private organizations might have on the fabric of American society. The purpose behind the funding of the National Students Association, according to the CIA, had been to encourage trips abroad by American students so they could help rebuff Soviet efforts to manipulate international student conferences.

Another program that became controversial was the CIA's sponsorship of anticommunist books written by Soviet and East bloc defectors as well as by American authors.[48] Praeger Publishers, among others, knowingly ("wittingly," in CIA lingo) assisted this operation, again raising questions in the minds of critics about the possibility that the CIA might be involved in propaganda directed toward an American audience, contrary to the prohibitions against domestic operations spelled out in the National Security Act. The CIA has also encouraged its contacts within the American media to write negative reviews and comments about books critical of the Agency and give positive evaluations to books favorable to the Agency or the philosophy of its leaders.[49]

While the CIA has never acknowledged these practices, it has an ally outside the government who is not so shy: the Association of Former Intelligence Officers (AFIO), an interest group supportive of the intelligence community. Its domestic propaganda operations can be entirely overt. Publications that strike the fancy of its board, for instance, are purchased in bulk and distributed to opinion leaders throughout the United States and abroad. While the objective remains the same for both the CIA and AFIO—to advance the viewpoint of intelligence professionals—the latter's approach has the virtue of being an activity widely viewed as legitimate and, indeed, protected by the First Amendment.

Propaganda operations conducted by the United States overseas come in various shades, depending upon whether the source is accurately revealed ("white" propaganda), unattributed ("gray"), or deceptively attributed to an enemy ("black"). William Colby defines the differences as follows:[50]

1. White propaganda is acknowledged openly by its source, the U.S. government, in which case it would not be an operation conducted by the CIA but rather by the USIA or some other overt American agency.
2. Gray propaganda is either unattributed or attributed to some ostensible third source, that is, neither the United States nor anyone within the target group.
3. Black propaganda is planted by the United States but in such a way that it seems to be the product or even an internal document of the target group, say, false Soviet "documents" in support of *apartheid* in South Africa.

"White," "gray," and "black," then, refer to the reputed source of the propaganda. Viewed in terms of the validity or "truthfulness" of the content, propaganda in addition may be true or false, with various intermediary gradations. Propaganda that is untrue or largely false is often referred to as "deception" or "disinformation."[51]

These combinations produce the possibilities outlined in table 2.1. According to the CIA, roughly 2 percent of recent CIA propaganda projects fall into the lower right cell (that is black-false propaganda). The great majority reportedly are of the gray-true and gray-mixed variety.[52]

TABLE 2.1. The Form and Validity of U.S. Propaganda Operations

Source	*True*	*Mixed*[a]	*False*
White (overt)			
Gray (covert) } CIA	←≅98%	→	
Black (covert) } CIA			≅2%

[a]exaggerated or selective "truth"

The CIA's extensive propaganda capability produces a great tide of information flowing secretly into hundreds of hidden channels around the world. Some seventy to eighty covert media insertions are made into different parts of the system each day. Once released, this information cannot be bottled up or directed to only one spot on the globe, as one might apply an antiseptic to a sore. Rather, the propaganda is free to drift here and there, even back to the United States. This can lead to "blow back" or "replay," whereby information directed toward America's adversaries abroad finds its way back home to deceive citizens in this country.

POLITICAL COVERT ACTION The "quiet assistance" advocated by William Colby sometimes takes the form of financial aid to friendly politicians and bureaucrats abroad—bribes, if one wishes to put a harsh light on the technique, or stipends to advance the cause of democracy, if one prefers a rosier interpretation. Whatever one likes to call this assistance ("King George's cavalry," is the expression used by British intelligence), the record is clear that through the CIA the United States has provided, off and on, substantial sums of money for political purposes to a number of groups and individuals overseas, including members of the Christian Democratic Party in Italy, King Hussein of Jordan, and pro-Western factions in Greece, West Germany, Egypt, Sudan, Suriname, Mauritius, the Philippines, Iran, Ecuador, and Chile, to mention only some examples in the public record.[53] An important part of political covert action has been funding for anticommunist labor unions in Europe, an objective of high priority soon after the end of World War II.

The CIA maintains an extensive stable of "agents of influence" around the world, that is, individuals—from valets and mistresses to personal secretaries and key ministerial aides—who have sufficient access to high-ranking political figures to influence their decisions. Propaganda and political covert action are meant to work hand in glove, and both are sometimes combined under the "political" heading. Both have the common purpose of persuading important foreign officials to become more favorably disposed toward the United States and less so toward the Soviet Union or other American adversaries.

At times the Covert Action Staff at CIA Headquarters has resembled nothing so much as a group of campaign consultants, producing slick materials for foreign candidates partial toward the United States: brochures, speeches, placards, even campaign buttons and bumper stickers—at times for remote regions of the world with no such traditions and precious few automobiles. The CIA recently sent to one anticommunist faction struggling for advantage in an African civil war fifty thousand political lapel buttons proclaiming the partisan affiliation of the wearer: "I am a member of the ———Party." In this instance, however, battlefield results proved more important than campaign buttons, and the CIA's side was forced to retreat back to the hinterland.

So while the aspiring American politician will have the acronym PAC (Political Action Committee) close in mind, for the foreign politician (if he is anticommunist or at least neutral) CIA may be an acronym to remember. The Soviets and others play this game (and all the other CA ploys), too, of course, and elections and other less peaceful means of changing the guard in foreign countries are often little more than hidden struggles between the CIA and the KGB, which on behalf of their respective governments employ money, propaganda, and at times violence in an attempt to place their favorites into positions of national authority.

ECONOMIC COVERT ACTION Another approach to covert action (often included under the broad heading of "political") is the attempt through secret means to disrupt the economies of U.S. adversaries. In one instance during the Kennedy years—though evidently without the knowledge of the president—the CIA tried to spoil Cuban-Soviet relations by lacing sugar bound from Havana to Moscow with an unpalatable, though harmless, chemical substance. A White House aide caught wind of the operation and informed the president. Kennedy rejected the plan and quickly had the 14,125 bags of sugar confiscated before they were shipped to the Soviet Union.[54]

During the efforts to undermine Salvador Allende, the CIA used various overt and covert measures to disrupt the Chilean economy. By heightening the level of social unrest in Chile, the U.S. government hoped that local military forces would finally decide to strip President Allende of his power. Inciting labor strikes was one proposal considered by the CIA. This option was eventually rejected by the Agency's leaders, but the CIA continued to provide monies to groups within Chile directly involved in strike tactics (especially against the trucking industry) in an effort to impede the flow of commerce.[55]

Additional economic operations used by the CIA (and, no doubt—as with all of these operations—the KGB) against other countries reportedly have included: counterfeiting foreign currencies; depressing the world price of agricultural products vital to adversaries in the developing world (especially those with one-crop economies); trying to control the rainfall over enemy territory by cloud seeding; preparing parasites for the destruction of foreign crops; contaminating oil supplies (as in North Vietnam during the 1960s); diluting pesticides bound for hostile nations; dynamiting electrical power lines and oil-storage tanks; and mining harbors to discourage commercial shipping (as occurred in Nicaragua during the Reagan years).[56]

PARAMILITARY COVERT ACTION No covert actions have held higher risk or been more controversial than paramilitary (PM), or warlike, operations. They often involve rather large-scale "secret" wars—as if anything large-scale could remain secret for long. From 1963 to 1973 the CIA backed the Meo hill tribes of North Laos in a war against the North Vietnamese puppets, the Communist Pathet Lao, in what was essentially a draw until the United States withdrew from the struggle.[57] The CIA's Special Operations unit has sponsored other guerrilla wars, providing support at one time or another for insurgents in the Ukraine, Poland, Albania, Hungary, Indonesia, China, Oman, Malaysia, Iraq, the Dominican Republic, Venezuela, North Korea, Bolivia, Thailand, Haiti, Guatemala, Cuba, Greece, Turkey, Vietnam, Afghanistan, Angola, and Nicaragua, to recall examples in the public record.[58]

In addition to support for groups engaged in insurgency fighting, the CIA has: funded various PM training activities, including counterterrorist training; provided military advisers (usually "sheepdipped," that is, borrowed from the Pentagon and outfitted in nonofficial battlefield gear); and transported abroad—directly or indirectly—arms, ammunition, and other military equipment to pro-U.S. factions, including, recently, sophisticated Stinger and Blowpipe missiles for Afghan rebels. Further, the CIA paramilitary program has included assistance to the Department of Defense in the development of its own unconventional warfare capability ("Special Ops") and has provided weapons to the Pentagon for covert sales abroad. Some of the weapons sold to Iran by the Department of Defense in the notorious arms-sale scandal of 1986–87 had their origin in the CIA's PM arsenal. The PM branch has also provided training for Third World military and police units responsible for the protection of their leaders ("executive driving" is a part of the curriculum, designed to teach the pupil how to maneuver an automobile through terrorist roadblocks and other perils in the style of movie stuntmen.)[59] During the first two years of the Carter administration, when PM operations were sharply curtailed, specialists in this trade spent most of their time—partially in an attempt to justify their existence—in training, equipment maintenance, and (the overwhelming work load) support to intelligence collection programs—particularly those involving specialized equipment and delivery techniques.

A special case within the realm of PM operations is the murder of individual enemies: the assassination option, sometimes referred to euphemistically as "executive action" or "termination with extreme prejudice" and, at one time within the CIA, screened by a "Health Alteration Committee." Insofar as the Church committee was able to ascertain, this option has been resorted to infrequently and, at least with heads of state, never successfully—despite numerous attempts against Fidel Castro.

The Cuban president received the full attention of the Covert Action Staff and the Special Operations unit during the Kennedy administration: propaganda and political, economic, and PM operations. The CIA emptied its medicine cabinet of drugs and poisons in various attempts to kill or debilitate Castro: depilatory powder in his shoes (which presumably would enter his bloodstream through his feet and cause his famous beard to fall out); the hallucinogenic drug LSD and deadly botulinum toxin in his cigars; Madura foot fungus in his underwater diving suit; and highly poisonous Blackleaf-40 prepared for injection into his skin through the needle-tip of a special ballpoint pen. When these and similar efforts failed (for Castro is elusive and well guarded by an elite counterintelligence corps trained by the KGB), the CIA upped the ante. The Agency hired the Mafia, which still had contacts in Cuba from pre-Castro days when Havana was a world gambling center, to kill the Cuban leader. Several assassins were dispatched, but none succeeded or even came close.[60]

A second major target for assassination by the CIA during the Kennedy years was Patrice Lumumba, the Congolese leader. Through diplomatic pouch, CIA Headquarters forwarded to the COS in the Congo (now called Zaire) an unusual assortment of items: rubber gloves, gauze masks, a hypodermic syringe, and lethal biological toxins. The enclosed instructions explained how to inject the poison into Lumumba's food or toothpaste to bring about his quick death. Before the

COS was able to carry out the operation, however, Lumumba was murdered at the hands of a rival Congolese faction.

Rafael Trujillo of the Dominican Republic, Ngo Dinh Diem of South Vietnam, and Gen. Rene Schneider of Chile were all national leaders killed by assassins who once had connections with the CIA, but the Church committee concluded after an exhaustive inquiry that, at the time each was murdered, the CIA no longer had control over the assassins. The CIA also gave weapons to dissidents who then may have plotted murder against President Sukarno of Indonesia and François "Papa Doc" Duvalier of Haiti, but again the plots seem to have gone forward without the imprimatur of the Agency.

The CIA's Special Operations unit has distributed weapons throughout the developing world. The Church committee came across these shipments to pro-Western dissidents in several small nations: high-powered rifles with telescopes and silencers, suitcase bombs, fragmentation grenades, rapid-fire weapons, 64-mm antitank rockets, .38-caliber pistols, .30-caliber M-1 carbines, .45-caliber submachine guns, tear-gas grenades, and enough ammunition to equip an army. More recently the Iran-contra investigations of 1987 documented the shocking and ironic sale of CIA and Defense Department weapons by the Reagan administration to a nation hardly considered a friend to the United States and a well-known patron of terrorist groups in the Middle East. Evidently, President Reagan hoped this move would encourage the Iranians to assist in the release of U.S. hostages held in Lebanon.[61] These examples uncovered by congressional investigators represented relatively small amounts of weapons distributed to groups here and there who the incumbent administration, the CIA, or both thought might be helpful to American interests abroad. In contrast, for the CIA's major PM operations, like those involving the hapless Meo tribesmen of Laos and the Kurds of Iraq, the contras in Nicaragua, and the mujahideen in Afghanistan, the amount of ordnance provided by the CIA has been enormous—dwarfing that affordable to most countries in the world.[62]

Less well understood than high-level assassination plots is CIA involvement in the incapacitation or murder of lower-level officials. Here the public record is largely blank, though some indications exist that such operations have occurred. The Church committee discovered a cable sent to CIA Headquarters from a Middle East division chief regarding an obstreperous Iraqi colonel with Soviet connections. The division chief recommended that the colonel be disabled for several months by exposing him to an incapacitating chemical. "We do not consciously seek subject's permanent removal from the scene," read the cable. "We also do not object should this complication develop."[63]

The most well-known operation designed to remove large numbers of lower-level officials from the scene was the Phoenix Program, a counterintelligence program carried out in South Vietnam as part of the U.S. war effort to subdue the influence of the Communists in the South Vietnamese countryside (the so-called Viet Cong Infrastructure, or VCI). According to William Colby, who for a time was in charge of this project, some twenty thousand VCI leaders and sympathizers were killed as a result of Phoenix; but, Colby hastened to emphasize, about 85 percent of those killed were engaged in military or paramilitary combat with South Vietnamese or American soldiers. Another 12 percent died at the hands of South Vietnamese security forces, and none died through an authorized program of "assassination." Critics, though, consider Colby's distinctions a thin line, and even he has conceded that assassinations might have taken place by overzealous Vietnamese (or even American) participants in the Phoenix Program.[64]

As repugnant as the Phoenix Program was, at least it took place in a war zone where the United States had openly and massively committed itself to an overt (if undeclared) war. More recently, in 1984, CIA personnel distributed a manual in the northern provinces of Nicaragua that instructed its guerrilla fighters (the contras) in the arts of "neutralizing" local civil officers.[65] In its investigation of CIA assassination plots a decade earlier, the Church committee had discovered that the verb "neutralize" in Agency cables often seemed to translate into "murder."[66] Despite a strong public reaction against assassination plots when they were revealed by the Church committee; despite executive orders signed by presidents Ford, Carter, and Reagan to prohibit assassination; despite the lack of a commitment by the Congress or the American people to an open warfare against Nicaragua (though there was some modest support in the polls and a limited amount of money from the Congress for covert action there), the CIA seemed to advocate with this manual a Phoenix-like plan of systematic assassinations against village representatives of the ruling Sandinista regime.

Once the manual was revealed in the press, the leadership of the CIA denied that the word "neutralize" necessarily meant murder or, for that matter, that they had authorized the use of the manual in the first place. As in the case of Phoenix excesses, they laid the blame at the feet of overzealous officers in the field. The director of the CIA at the time, William J. Casey, eventually apologized publicly for the manual and reprimanded five officials near the bottom of the Agency hierarchy for permitting its distribution. In President Reagan's opinion, the entire matter was "much ado about nothing."[67]

Counterintelligence

Like covert action, counterintelligence (CI) went without specific mention in the 1947 National Security Act but, by the early 1950s, it had similarly achieved a status of considerable importance as a mission within the CIA. Counterintelligence specialists soon waged nothing less than a secret war against antagonistic intelligence services. Explaining why this warfare evolved, a CI expert points out that "in the absence of an effective U.S. counterintelligence program, [adversaries of democracy] function in what is largely a benign environment."[68]

Led from 1954 to 1974 by the mysterious and now legendary James Jesus Angleton, the Idaho-born and Yale- and Harvard-educated chief of counterintelligence, the CIA's Counterintelligence Staff developed its own global network of assets. Rather than merely settle for catching foreign spies as they plied their "tradecraft" (a CIA term for "methods") against the United States, the CI Staff conducted aggressive operations to confuse and thwart hostile intelligence services.

Certainly the adversaries of American democracy have been numerous and widespread, providing counterintelligence with a formidable challenge. According to FBI figures, over a thousand Soviet officials are on permanent assignment in the United States alone. Among these, more than 50 percent have been positively identified as members of the KGB or GRU, the Soviet civilian and military intelligence units. Conservative estimates for the number of unidentified intelligence officers raise the figures to over 60 percent of the Soviet representation; some defector sources have estimated that as many as 70 to 80 percent of Soviet officials in the United States have an intelligence connection. Without even counting the Soviet mission to the United Nations in New York City, the USSR in 1984 had over one hundred more officials in this country than the United States had in the Soviet Union.

Furthermore, the number of Soviet visitors in the United States has tripled since 1960 and continues to increase. The opening of American deep-water ports to Russian ships in 1972 gave Soviet intelligence services "virtually complete geographic access to the United States," observes a CI specialist. In 1974, for example, over two hundred Soviet ships with a total crew complement of thirteen thousand officers and men called at forty deep-water ports in this country.

Various exchange groups provide additional opportunities for Soviet intelligence gathering within the United States. In 1974, for example, some four thousand Soviets entered the United States as commercial or exchange visitors. During one ten-year period (1964–74), the FBI identified over a hundred intelligence officers among the approximately four hundred Soviet students who attended American universities as part of an East-West student exchange program. Also, in the fourteen-year history of this program, more than one hundred American students were the targets of Soviet recruitment efforts in the USSR.

Counterintelligence officers worry, too, about the sharp increase in the number of Soviet immigrants to the United States (less than 400 in 1972 compared to 4,000 in 1974); the rise in East-West commercial exchange visitors (from 641 in 1972 to 1,500 in 1974); and the growing number of Soviet bloc officials in this country (from 416 in 1960 to 798 in 1975). The most serious threat is from "illegal" agents who have no easily detectable contacts with their intelligence services:

> The illegal is a highly trained specialist in espionage tradecraft. He may be a [foreign] national and/or a professional intelligence officer dispatched to the United States under a false identity. Some illegals are trained in the scientific and technical field to permit easy access to sensitive areas of employment. . . .
>
> Once they enter the United States with either fraudulent or true documentation, their presence is obscured among the thousands of legitimate emigres entering the United States annually. Relatively undetected, they are able to maintain contact with [their foreign control] by means of secret writing, microdots, and open signals in conventional communications which are not susceptible to discovery through conventional investigative measures.[69]

In a word, the espionage activities of the Soviet Union and other Communist nations directed against the United States are extensive and relentless. The official FBI figures presented above have all increased in recent years. To combat this threat, American CI officers have developed a two-pronged response: obtaining information about hostile intelligence services and, based upon this information, counteracting their plans.

To guard against hostile intelligence operations aimed at this nation CI officers require a vast amount of information. They must know the organizational structure of the enemy service, the key personnel, the methods of recruitment and training, and the specific operations. This information must be gathered within the United States and in all the foreign locations to which U.S. vital interests extend. The efforts of intelligence services throughout the world to conceal such information from one another, through various security devices and elaborate deceptions, creates what Angleton has referred to poetically as a "wilderness of mirrors."[70]

The practice of counterintelligence consists of two matching halves: security and counterespionage. Security is the passive or defensive side of counterintelligence, involving the establishment of static defenses against all hostile and concealed operations directed toward the United States, regardless of who may try to carry them out. These defenses include the screening and clearance of personnel and the development of programs to safeguard sensitive intelligence information—what the professionals refer to as the administration of security controls. The American intelligence services, like their counterparts throughout the world, try to defend three goods: their personnel, their installations, and their operations. Among the defensive devices employed to this end are security clearances, polygraphs (lie-detection machines), locking containers, security education, document accountability, censorship, camouflage, and codes. Additional methods employed for physical security include fences, sentries, lighting, alarms, badges and passes, and watchdogs. The protection of specific locales is augmented by the use of curfews, checkpoints, restricted areas, and border-controls.

Counterespionage (CE) is the offensive or aggressive side of counterintelligence. It involves the identification of a specific adversary and a knowledge of the operation he is conducting. Counterespionage personnel must then attempt to undermine these hostile operations by infiltration of the enemy intelligence service (a ploy called "penetration") and through various forms of disruption. Ideally the thrust of the hostile operation is turned back against the enemy, say by blackmailing a KGB agent and forcing him to work against the USSR, or at least by using him as a conduit for misleading "intelligence" (a disinformation operation). The Soviet operation in 1986–87 to recruit Marine guards stationed at the U.S. embassy in Moscow, in order to gain access to sensitive intelligence locked in embassy safes, provides an unhappy illustration of a recent and successful KGB counterespionage operation (and the failure of the U.S. security methods outlined earlier). The sexual companionship of attractive KGB vamps—"swallows" in the language of spies—provided the lure in this case.

Thus the security side of counterintelligence is, according to one of America's top CI specialists, "all that concerns perimeter defense, badges, knowing everything you have to know about your own people," while the CE side "involves knowing all about intelligence services—foreign intelligence services—their people, their installations, their methods, and their operations."[71] Several types of operations exist under the rubric of counterespionage. One, though, transcends all others in importance: the penetration. Since the primary goal of counterintelligence is to contain the intelligence service of the enemy, it is eminently desirable to know his plans in advance and in as much detail as possible. This important but difficult task may be accomplished through a high-level infiltration of the opposition service. As a DCI has written, "Experience has shown penetration to be the most effective response to Soviet and Bloc [intelligence] services."[72]

Moreover, a well-placed infiltrator in a hostile intelligence service may be better able than anyone else to determine whether one's own service has been penetrated. A former director of the Defense Intelligence Agency (DIA) has observed that the three principal programs used by the United States to meet, neutralize, and defeat hostile intelligence penetrations are successful U.S. penetrations abroad, the careful screening and clearance of personnel, and efforts at home to safeguard sensitive intelligence information with good physical defenses.[73] The importance of the penetration is further emphasized by another seasoned CIA counterespionage officer: "Conducting counterespionage with penetration can be like shooting fish in a barrel"; in contrast, "conducting counterespionage without the act of penetration is like fighting in the dark."[74]

Methods used to infiltrate enemy services take several forms. Usually the most effective and desirable penetration is the recruitment of an "agent-in-place,"[75] a citizen of an adversary nation who is already in the employ of its intelligence service. Ideally, he (perhaps she) will be both highly placed and venal. The individual, say a KGB officer in Bonn, is approached and asked to work for the intelligence service of the United States. Various inducements—ideology, money, sex, and whatever else human beings may find valuable—may be used to recruit him against his own service. If the recruitment is successful, the operation may prove to be of enormous worth since the agent is presumably already trusted within his organization and his access to documents may be unquestioned. Jack E. Dunlap, who during the 1960s was employed by and eventually spied against the largest of the American intelligence agencies, the National Security Agency (NSA), is a well-known example of an agent-in-place successfully recruited and "run" (handled or supervised) by the Soviets. His handler was an Air Force attaché at the Soviet embassy in Washington. A single penetration can result in an intelligence gold mine, as were agents-in-place Kim Philby (a high-ranking British intelligence officer) for the Soviet Union, and Col. Oleg Penkovsky and Lieut. Col. Pyotr Popov (strategically situated GRU officers) for the United States.[76]

Another method of infiltration is the "double agent," someone engaged in intelligence operations for two or more secret services simultaneously. He collects information for one service about the other, or, if his panache quotient is high, about both services for one another, and is knowingly or unknowingly ("wittingly" or "unwittingly") manipulated by one or both services against the other. Double-agent operations are costly and time-consuming, and like many aggressive intelligence and counterintelligence operations, they are risky. Human lives are at stake, and sometimes the leaders of intelligence services can be callous about the fortunes of officers and agents far removed from the air-conditioned sanctuary of Agency headquarters. ("The gentlemen run the business, and gentlemen have short memories," says a retired British intelligence officer with bitterness.[77]) The running of double agents also involves a large amount of pure drudgery, with few dramatic results, as new information is checked endlessly against existing files. On top of this tedium comes the continuous, laborious task of double-checking against a double cross.

Passing credible documents is no simple task either. The bait must be made interesting to the opposition. To assure that fake papers are plausible, agents must provide the genuine article now and then. These real documents are, of course, classified and must be cleared—often a painstakingly slow process. Also, complains a CI officer with considerable experience, "this means letting a lot of good stuff go to the enemy without much in return."[78]

Each of these tasks requires hard work, careful planning, and considerable time. Because of the extraordinary personnel requirements of double-agent operations the British were restricted in their ability to run them during the Second World War—they handled approximately one hundred and fifty double agents during the entire period of the war, and no more than about twenty-five at any one time.[79] Moreover, their mission was eased greatly by the ability of the British to read the German cipher throughout most of the conflict; otherwise even fewer agents could have been handled.

Almost as good as the agent-in-place, and less troublesome than the double agent, is the "defector with knowledge." In this case the first step is interrogating the defecting agent and validating his *bona fides* (that is, his avowed credentials). With the defector, the CIA is freed from the worrisome, ongoing task of providing a careful mix of false and genuine documents and other logistical support that are necessary when an agent is kept in place and turned against his own intelligence service. The task is usually the simpler one of debriefing the agent at length in a secure place of the CIA's or FBI's choosing. Though an agent-in-place is preferable because he can continue to provide useful information, often a person does not want to risk his life by staying "in-place," especially in an organization like the KGB where the security is sophisticated. Defection is therefore more common from the Soviet bloc countries, whereas agents-in-place are more apt to be recruited in the developing countries, which lack the resources to finance the complex security arrangements of a modern CI defense.[80]

The interrogation of intelligence defectors who have fled their former service while within the United States is primarily the responsibility of the FBI, though the CIA may request follow-up sessions with the individual. CIA-recruited defectors abroad are occasionally brought to the United States and resettled. The FBI is notified, and after the CIA completes its interrogation, the FBI starts its own. The CIA does not bring all defectors to the United States, only those who can be expected to make a significant contribution and who are willing to come. The CIA generally handles the resettlement not only of defectors from abroad, but (at the request of the FBI) also those who defect inside the United States.

Sometimes the *bona fides* of a defector remain a matter of dispute among U.S. government officials for years, sometimes forever, with the CI professionals within the CIA and within the FBI—or, more often, between the two agencies—unable to agree on whether the person is the genuine article he claims to be, now ready to commit treason against his own country, or is actually a KGB plant (a deception agent). One of the most celebrated cases involved Yuri Nosenko, a KGB defector. Though FBI counterintelligence officers accepted his reliability, James Angleton never believed in him. The CIA chief of counterintelligence felt sure that Nosenko was a plant whose purpose was to divert mounting suspicion in the United States that the USSR may have been behind the assassination of President John F. Kennedy.[81] Angleton had Nosenko confined to a small building at a CIA facility in southern Virginia for 1,277 days in spartan conditions, where, according to CI insiders, he was interrogated relentlessly and treated in a shabby manner. As a newsletter for intelligence professionals noted recently, Nosenko "never broke down under the stress and was finally released and nominally cleared. Yet the questions linger. . . ." This newsletter, staffed by retired CIA hands, finds Angleton's hypothesis "more likely" to be true.[82]

Beyond their *bona fides* lies another problem with defectors: about half of them return to their native land. Some re-defect because they become homesick and find it hard to adapt to a new life in the United States, while others have been "deception agents" all along. The most spectacular recent re-defector was Vitaly Sergeyevich Yurchenko, a KGB officer, who entered the U.S. embassy in Rome in June 1985, provided valuable information to the CIA and the FBI when flown to the United States and debriefed, only to walk into the Soviet embassy in Washington in November, claiming he had been abducted by the CIA in Rome and tortured in the United States. Whether Yurchenko was a clever "fake defector," acting out his part to probe and embarrass the CIA and the FBI, or a lonely, confused man who fled back to the KGB on an impulse remains, like the Nosenko case, a matter of dispute among CIA professionals.

The deception agent (as the false defector or "defector with knowledge" is sometimes called by CI professionals) is a close cousin to the penetration or double agent in the pantheon of CE personae. Simply stated, deception is an attempt to give the enemy a false impression about something, causing him to take action contrary to his own interests. Fooling the Germans into the belief that D-Day landings were to be in the Pas de Calais rather than in Normandy is a classic example of a successful deception operation in World War II.

Deception is related to penetration because U.S. assets operating within foreign intelligence agencies can serve as excellent channels through which misleading information can flow to the enemy. So double agents serve both as collectors of intelligence and channels for deception (and in this latter sense the double agent and the defection agent are one and the same). Additional opportunities are available for deception other than through recruitment of assets from hostile intelligence services; in fact "an infinite variety" exists, according to an experienced practitioner.[83] For example, the United States can allow a penetration of its own intelligence service and then feed false information to the enemy through him.

Other CE operations include provocation, surreptitious surveillance of various kinds (for instance, audio, mail, physical, and "optical"—that is, photographic), and interrogation (sometimes incommunicado for long periods of time as in the case of Nosenko). Provocation involves the harassment of opposition intelligence services in sundry ways, such as publishing the names of their personnel or sending a defector into their midst who is in reality a double agent. Part and parcel of CE tradecraft, too, are the surveillance techniques of clandestine radio transmission and letters with messages written between the lines in invisible ink, as well as trailing suspected agents, observing "dead drops" (the exchange of material, like documents or instructions, between a spy and his handler), conducting wiretaps, and photographing individuals entering opposition embassies or at other locations. During the Arlington Cemetery burial of CIA officer Richard Welch (cowardly gunned down in 1975 by terrorists in front of his home in Athens, Greece), Agency security discovered two Eastern European "diplomats" among the press corps, blithely snapping photographs of CIA intelligence officers who were attending the funeral.

Counterintelligence is, as one seasoned observer has written, a "murky world"—one "full of risks, dangers, personal jealousies and never-ceasing suspicions that the man in the office next to yours may be a Soviet agent. It is a situation that creates paranoia, corroding men's characters."[84] Little wonder that the brilliance of James Angleton—attorney, poet, scholar, master orchid grower and fisherman, burned by his one-time friend Kim Philby and twice shy, in his later years convinced that every defector was a deception agent, that even the Sino-Soviet confrontations were but a sham designed to lull the West into complacacy—was matched only by his profound paranoia.

Fears of a Gestapo

To return to where this chapter began, President Truman's demand for better intelligence collection, evaluation, and coordination in the wake of the surprise attack at Pearl Harbor stimulated the creation of a new, more centralized intelligence service, and this objective drew the overwhelming attention of legislators and intelligence planners during the drafting of the CIA portion of the 1947 National Security Act. The unstated CA and CI missions did not remain dormant for long, however. The cold war seemed to demand better ways of dealing with Communist opponents than the slow and often stagnant use of diplomacy, on the one hand, and the risky deployment of overt military force, on the other. Covert action and aggressive forms of counterintelligence proved irresistible.

The executive branch included the catchall phrase in its draft of the 1947 act to allow for the possibility of imaginative options that might arise in the future, and Congress gave its approval, along with the necessary funds in the coming years, with few questions asked. The cold war nurtured counterintelligence, too, as the U.S. government—alarmed by the theft of atom-bomb plans by the Soviets and other instances of espionage—sought means to protect the nation's secrets. As James Angleton became the confidant of Dulles, Helms, and other stanchions at CIA Headquarters, counterintelligence began to flourish along side covert action. But all of this was to come. In July 1947, attention focused on the basic question of Mission No. 1: the coordination of more reliable facts and better analysis for presentation to the chief executive and other top-level policymakers.

While the momentum in Washington behind intelligence reform seemed inexorable in 1947, some legislators remained wary about the possible dangers to democracy raised by the proposed centralization of intelligence authority. Implicit in the calls for better coordination was a new centralism that might contain the seeds of an internal police force, led by an intelligence czar—conceivably a Beria or Göring within the government of the United States. Marine Corps Gen. Merritt A. Edson explicitly warned of this danger in testimony before a committee in the House of Representatives, cautioning that the establishment of a central intelligence agency "opens the door to a potential Gestapo."[85] With the specter of the Third Reich fresh in mind, these legislators—though few in number—were quick to voice their concerns.

And just as quickly, the Truman administration stepped forward to mollify them. "The prohibition against police power or internal security functions will assure," said Gen. Hoyt Vanderberg, soothingly, in Senate hearings, "that the Central Intelligence Group [of which he was the director] can never become a Gestapo or security police."[86] The reference was to President Truman's Executive Order that created the Central Intelligence Group (CIG) in 1946. "No police, law enforcement or internal security functions shall be exercised under this directive" read one of its provisions. The Congress set into the concrete of law a similar stipulation when drafting the 1947 act; CIA domestic espionage was now strictly prohibited in explicit, statutory language.[87] Fears on Capitol Hill about a budding Gestapo seemed laid to rest.

Within a year, however, the alarm bells went off again within the Congress as legislators reacted to a request from the CIA for additional—sweeping, thought some critics—authority. Through the Central Intelligence Act of 1949 (essentially a series of amendments to the National Security Act of 1947), the CIA hoped to strengthen dramatically the powers of its director. Legislators friendly toward the new agency drafted a proposed law worded in such a way as to give the director freedom from Civil Service regulations in hiring and firing, exemption from the usual public reporting requirements, and wide discretion over the covert spending of CIA funds without the check of normal auditing procedures. The Agency wanted complete invisibility, not only from other nations but, some feared, from overseers within the American government as well. In Ransom's words, the amendments "were introduced to permit [the CIA] a secrecy so absolute that accountability might be impossible."[88]

The reaction from some members of Congress was sharp.[89] Let us take no chances, advised Sen. William Langer (R, North Dakota), "against the establishment of a Gestapo in the United States." The proposal represented "very radical legislation," cautioned Sen. Edwin C. Johnson (D, Colorado): the CIA "might send its men inside the United States."

The CIA rallied its defenders within the Congress. "The pending bill," retorted Millard Tydings (D, Maryland), the bill's floor manager in the Senate, "has nothing to do with the internal affairs of the United States. All these men work outside of the United States. . . . They cannot work in the United States." On the House side, Tydings's counterpart urged members to "bear in mind that the CIA is prohibited by law from any internal security function." Moreover, he continued, the "CIA functions exclusively under the powers granted it by the National Security Act of 1947 and not under any Executive order whatsoever." Thus reassured, the Cassandras backed away, the bill passed (on June 20, 1949), and a thick curtain of secrecy descended over the Central Intelligence Agency.

As the legislators debated the expansion of CIA powers, apparently few, if any, were aware that the Agency had already gone beyond the primary mission assigned to it in 1947: to collect, interpret, and coordinate intelligence. By the time of the second debates in 1949, the CIA had already been secretly charged by the president and his chief national security advisers to engage in propaganda (psychological warfare) and conduct political, economic, and paramilitary covert operations around the globe.[90] Covert action had taken flight, beyond the view of the small coterie of uneasy democratic theorists on Capitol Hill. From the very beginning of the cold war, Congress had only a fuzzy understanding of—and little desire to supervise—the operations of the modern American intelligence service.

THREE

The Design of American Intelligence

The Intelligence Colossus

Today over forty federal agencies have responsibilities for intelligence operations, including a dozen like the CIA with primarily a foreign mission. Together, according to various published reports, they employ around 150,000 individuals and spend upwards of $13 billion each year.[1] They have become "the largest consumers and producers of information in the world—and thus in history."[2] The primary job of these agencies is to provide intelligence for the National Command Authority—that is, the president, the cabinet, and the Joint Chiefs of Staff—and for the Congress. The various programs and operations dedicated to this objective (see later discussion) are known collectively as the National Foreign Intelligence Program (NFIP), a description used by intelligence professionals rather than a statutory phrase. Among the many intelligence agencies, a few are preeminent in terms of budgets, personnel, technology, and influence; they comprise what is widely referred to as the "intelligence community"—a misnomer belying the tensions and competition for influence among a diverse collection of government entities.[3] In figure 3.1 the main components of the "community" are presented as they existed in the 1980s; because organizational detail is highly perishable in secret agencies, this "wiring diagram" is offered only as an approximation of how the boxes may be named or joined together in the future.

Each intelligence agency is ultimately responsible to the president through his assistant for national security affairs, the coordinating chief of staff for the National Security Council (NSC). The NSC has several subcommittees for intelligence decision making, some of which are shown in figure 3.1: the Net Assessment Group for comparing American military capabilities against those of U.S. adversaries; the Verification Panel to monitor the fidelity of other nations to

FIGURE 3.1 The United States Intelligence Community

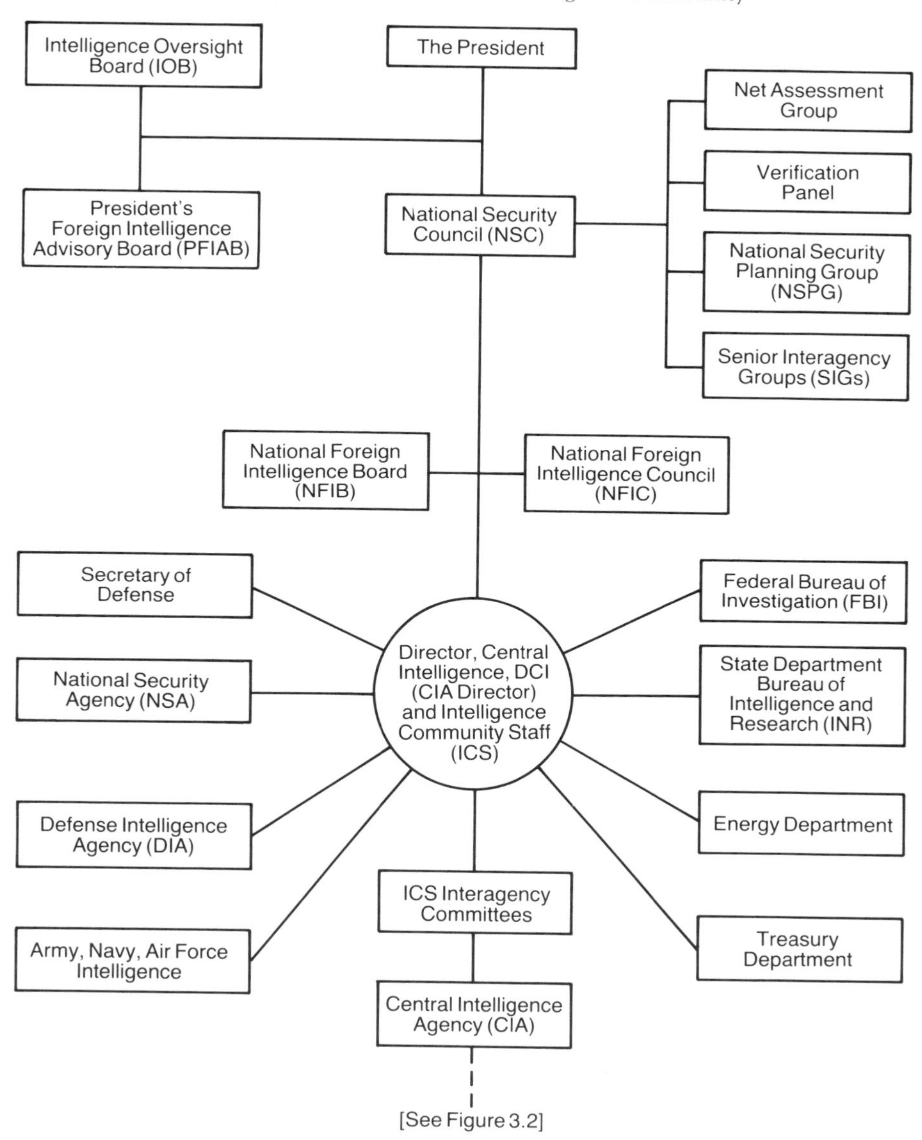

military and intelligence agreements reached through international negotiations (from treaties to executive agreements); during the Reagan administration the National Security Planning Group (NSPG), whose duties over the years—the review of CA proposals and other significant intelligence activities—have been carried out by the 10/12 and 10/13 committees (Truman), the 5412 Committee (Eisenhower), the Special Group (Kennedy), the 303 Committee (Johnson), the 40 Committee (Nixon), the Operations Advisory Group (Ford), and the Special

Coordination Committee and Policy Review Committee (SCC and PRC, Carter); and some twenty-two Senior Interagency Groups (SIGs), including ones for Foreign Policy, Defense Policy, Economic Policy, Energy Policy, Space Policy, and Intelligence Policy. The SIG for Intelligence Policy is where collection requirements are debated at the highest levels and where the key resource allocations are established.

On the NSPG (and its earlier incarnations) sit the four statutory members of the NSC: the president, the vice president, the secretary of state, and the secretary of defense. In addition, attendance at most meetings of the NSPG includes the president's assistant for national security affairs, the CIA director, and the chief of staff to the president, with others invited from time to time. Represented at the SIG level are senior representatives (deputy undersecretaries, assistant secretaries, and equivalent ranks) from the Departments of State and Defense, the CIA, the NSC, and the Intelligence Community Staff (ICS)—the latter an organization designed to assist in the coordination of the broad intelligence community. On June 9, 1987, President Reagan created two additional entities to coordinate national security policy: a Senior Review Group (SRG) to work as a cabinet-level interagency panel for the consideration of intelligence and other national security issues for recommendation to the NSPG, and a Policy Review Group (PRG) to serve as a senior sub-cabinet-level interagency group in support of the SRG.

As the president's primary advisory board for national security policy, the NSC produces documents that are among the most highly classified in the government. Its key papers—reports and recommendations to the president—are known as National Security Council Intelligence Directives or NSCIDs (pronounced "n-skids"), National Security Decision Memoranda or NSDMs ("niz-dems"), National Security Study Memoranda or NSSMs (niz-sims"), and National Security Action Memoranda or NSAMs ("niz-ams"). Emblazoned with the stamp "top-secret," these documents—the intelligence community's "secret charter"—are the foundation of America's covert activities around the world, the basic operational authority for decisions made by the president and the other statutory members of the NSC. The vast majority of all intelligence policy proposals recommended to the NSC originate within the CIA. Those proposals that are approved by the NSC then take the form of NSCIDs—broad marching orders from on high for important intelligence operations. Operational specifics are left for the most part to CIA officials to work out.

Beneath the NSC in the chain of command comes the director of the CIA (DCIA), who, in his capacity as chief of the entire intelligence community, is also more commonly referred to as the Director of Central Intelligence (DCI). The deputy director of the Central Intelligence Agency (DDCIA) is second in command in the community and serves for all practical purposes as the top day-to-day administrator for the CIA at its headquarters in Langley, Virginia (see fig. 3.2). The DCI is in charge (at least in the organizational diagrams) of the National Foreign Intelligence Program, which includes: CIA and its operations; satellite and airplane reconnaissance programs and other remote surveillance operations; intelligence codemaking and codebreaking; strategic intelligence programs carried out by the Departments of State, Defense, Treasury, Energy, and Justice; and the overseas CI responsibilities of the Department of Defense.

Helping both the DCI and his deputy within the Office of the Director at CIA Headquarters are a whole series of subsidiary units: an Office of General Counsel, home of the Agency's litigants as well as its Publications Review Board (a censorship panel established to conduct prepublication reviews of writings by former or active intelligence officials); an Office of Inspector General, expected to investigate everything from affirmative action complaints to major allegations of unlawfulness or impropriety by Agency officials; a Public Affairs Office, the public relations branch of the CIA; an Office of Legislature Liaison to handle relations with Capitol Hill; a Comptroller's Office for budgetary control; and, perhaps most important of all, a National Intelligence Council (NIC), where reside the sixteen National Intelligence Officers (NIOs). Working hand in hand with senior analysts throughout the community under the umbrella of the NIC Analytic Group (NIC/AG, essentially a staff that drafts research reports), the NIOs prepare the major analytic papers for government decision makers.

Several ICS interagency committees have been established to aid the DCI in his uphill coordination efforts, including panels for human intelligence (HUMINT), which is classic espionage in contrast to the use of machines for spying (technical intelligence or TECHINT); signals intelligence (SIGINT), which is a generic term for describing the interception and analysis of communications intelligence (COMINT), other electronic intelligence (ELINT), and telemetry intelligence (TELINT—chiefly missile emissions); counterintelligence (CI); photographic intelligence (the Committee on Imagery Requirements and Exploitation, or COMIREX); long-range community planning; and budget coordination.

The DCI and these ICS panels are guided by two overarching inter-agency groups: the National Foreign Intelligence Board (NFIB—formerly the U.S. Intelligence Board, or USIB) and the National Foreign Intelligence Council (NFIC). The distinction between the two is twofold: first, the NFIB is chaired by the DCI, and its participants are high-level agency chiefs from throughout the community, whereas agency deputy directors serve on the NFIC; and second, the NFIB examines the substance of policy, while the NFIC supposedly concentrates more on budget and personnel matters—though often this distinction is lost as deputies find themselves waging the policy *and* budget battles of their principals in the NFIB forum. Indeed, for all the boxes in figures 3.1 and 3.2, the reality of the informal procedures frequently replaces the official wiring diagrams, manning tables, and written guidelines. As a former secretary of state remembers, "When we members of the NSC *really* wanted to talk, we met informally without all the staff around. The statutory members of the NSC used to meet with President [Lyndon] Johnson on Tuesdays over lunch, and I usually met weekly with [Secretary of State Robert S.] McNamara on Saturday mornings."[4]

How misleading formal diagrams can be is illustrated further by two seemingly important committees appended directly to the president in organizational charts: the President's Foreign Intelligence Advisory Board (PFIAB, pronounced "piff-e-ab"), established in 1956, and the Intelligence Oversight Board (IOB), established in 1976. Both are comprised of prominent private citizens whose past ex-

FIGURE 3.2 The Central Intelligence Agency

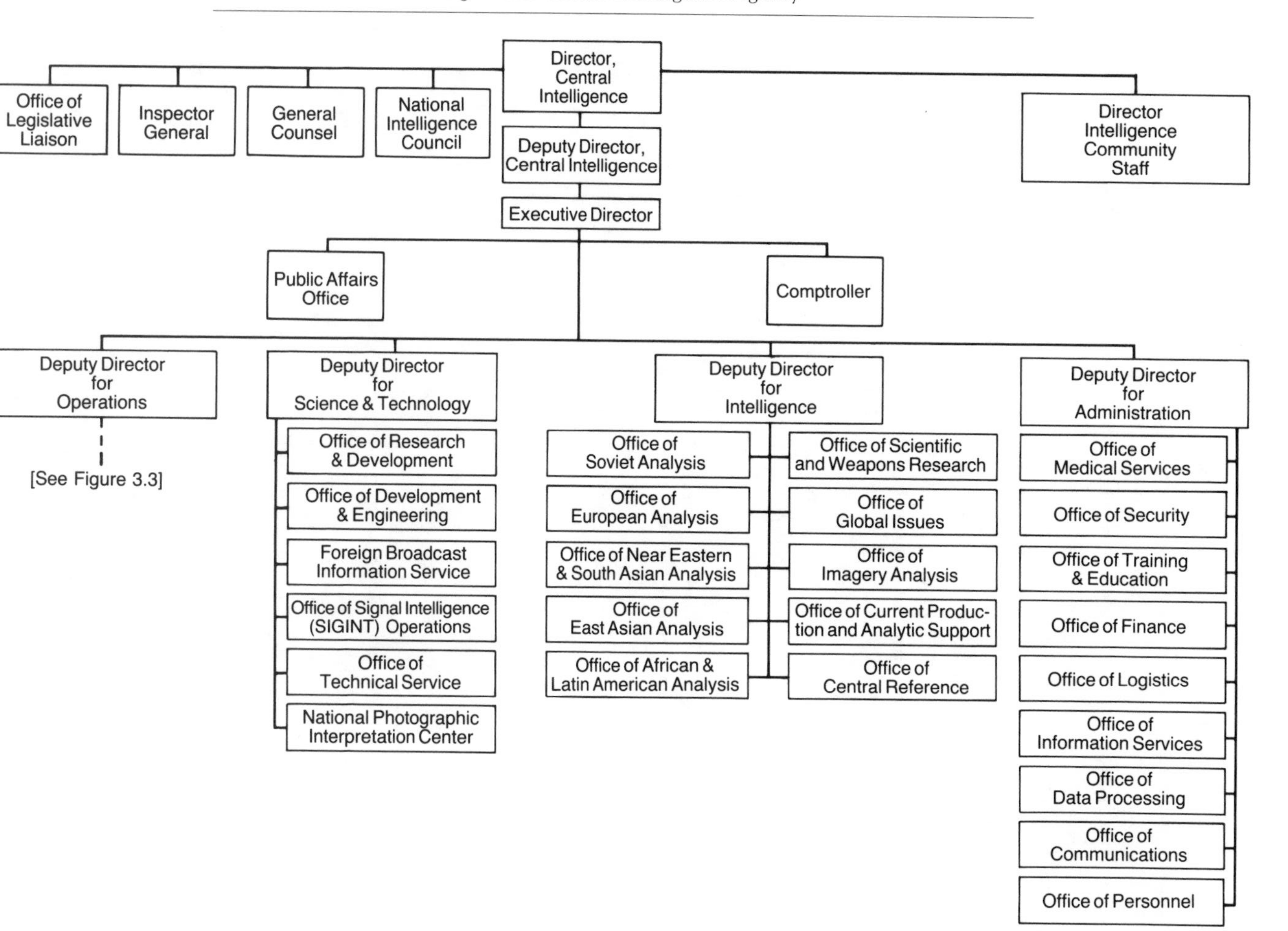

From *Fact Book on Intelligence*, Office of Public Affairs, Central Intelligence Agency, April 1983, p. 9.

periences equip them theoretically to oversee and guide intelligence policy. In fact, both panels for all intents and purposes have been largely ignored—though in PFIAB's early history the scientific expertise of some of its members provided excellent advice on the development of the U-2 reconnaisance airplane and other technical matters.

In recent years the part-time members of PFIAB and the IOB have come to Washington for infrequent meetings (two days every two months during the Reagan years), bent the ear of the president occasionally on some intelligence scheme they may have heard about somewhere or other, and generally enjoyed the status of a presidential appointment free of much effort. These individuals are often distinguished, but not for their work on these panels (Texas billionaire H. Ross Perot, for example, appointed to PFIAB by President Reagan in 1982). The staffs on both panels are small, their jurisdictions limited (covert action is one of many subjects normally outside their purview), and their findings are never made public. Their members are typically well-to-do elderly men in retirement who serve without pay. As one former U.S. senator said as a member of the IOB, "I'm no longer on the Hill; I'm over the hill."[5] The IOB and the PFIAB have indeed had little more than honorific significance. The important decisions of the intelligence community take place where NSC members gather (formally or informally), on the interagency panels, and within the distant recesses of the agencies themselves—the most well known of which is the Central Intelligence Agency.

The CIA

In a strongly fenced and guarded region of Langley, Virginia, in the midst of some two hundred acres of beautifully forested land along the Potomac River twelve miles from downtown Washington, stands the CIA, hidden by trees from the busy George Washington Memorial Parkway a few hundred meters away. The "Pickle Factory," the "Company," or simply the "Agency"—as CIA Headquarters is variously called by insiders—rises seven stories high and is topped by elaborate radio antennae for worldwide communications.

On the seventh floor are the adjoining offices of the DCI and his deputy, with windows opening onto a panoramic view of woods that march down to the Potomac. Now and then deer can be seen grazing beneath the trees. The offices are long and spacious, decorated with memorabilia from past successful operations (including in 1987 a Soviet pistol on the wall, captured by CIA-financed Afghan rebels north of Kabul). On the ground floor, the glass doors at the front of the building stand before a wide lobby. On one wall of the lobby, chiseled into marble, is the biblical injunction "And ye shall know the truth and the truth shall make you free" (John 8:32). Perhaps Allen Dulles, under whose leadership the construction of the elaborate headquarters was completed in 1963, mused occasionally as he walked by: "And ye shall know the truth—if ye are me, or the president." Against the opposite wall rests a "Book of Honor," which records year by year the number of intelligence officers who have died in the line of duty since 1950 (over forty, with seven paying nature's debt in 1965 alone, no doubt in Vietnam), sometimes listing their names next to a star ("1975 ☆ Richard S.

Welch"), sometimes leaving a blank in those cases—a majority—when the officer's name must be kept secret.

On the marble floor of the lobby a large emblem of the Agency is inlaid with the words "Central Intelligence Agency" over the head of a golden eagle. Armed guards stand just beyond the symbol, ready to inspect the identification cards of anyone entering the building. Between the eagle on the ground floor and the eagle's nest on the seventh are sandwiched seemingly endless hallways and doors of different colors. Behind the doors sit the thousands of country analysts, scientists, clerical assistants, computer technicians, cartographers, and others who make up the work force of the Agency.

Organizationally, the CIA is divided into five sections: the Office of the Director and, below him, four functional directorates. Wearing his two hats—DCIA and DCI—the director is expected not only to run his own shop, but also to coordinate and generally supervise every other major intelligence agency. No CIA director has been able to wear the DCI hat with much confidence. The CIA reportedly accounts for less than 15 percent of the total community's expenditures and personnel; the other agencies—particularly military intelligence—receive the lion's share of resources. These other agencies have behaved like independent feudal barons, fiercely resistant to strong management from the head of the CIA. One study concludes that the community "has developed into an interlocking, overlapping maze of organizations, each with its own goals."[6]

The CIA director is somewhat more assured of control over his own four directorates—though the Agency's chief day-to-day manager, Frank Carlucci, while serving as DDCIA in the Carter administration, expressed despair over never feeling truly in control of the Agency. He reportedly felt as though he were "operating a power plant from a control room with a wall containing many impressive levers that, on the other side of the wall, had been disconnected."[7] These main divisions are known as the Directorate of Operations headed by a deputy director for operations (DDO), which is the home for CIA clandestine operations (and is often still referred to as the Clandestine Services by insiders, recalling its earlier name); the Directorate of Science and Technology, headed by a deputy director for science and technology (DDS&T); a Directorate of Administration, with its chief, the deputy director for administration (DDA), responsible for housekeeping and security chores; and a Directorate of Intelligence (labeled the National Foreign Assessment Center, or NFAC, during the Carter administration), headed by a deputy director for intelligence (DDI), responsible for the analysis of foreign intelligence.

DIRECTORATE OF OPERATIONS The largest and most controversial division is the Directorate of Operations, home of what insiders call "the spooks"—the Agency's spyhandlers and CA operatives. Roughly two-thirds of the personnel in the Operations Directorate are involved in espionage, counterintelligence, and liaison work with intelligence services in allied nations. The rest are engaged in some form of covert action, such as mounting PM operations or secretly financing friendly politicians overseas. As part of his efforts to obtain foreign intelligence, the DDO has a subsidiary division that operates within the United States, supposedly to gather information from foreign visitors and to "debrief" selected American travelers upon their return to the United States. This unit, now known as the National Collection Division (NCD) and earlier as the Domestic Contact Division (DCD), has made CIA critics especially wary because of its at-home operations. Of concern, too, is a second unit, the Foreign Resources Division (FRD), which attempts to recruit as agents foreigners living or traveling in the United States. (See chapter 8 for further discussion of the DCD and FRD.) For reasons that will become obvious in chapter 6, the Operations Directorate is sometimes known as the "dirty tricks" division of the CIA.

As figure 3.3 shows, the Operations Directorate is subdivided into geographic and other specialized staffs. The geographic staffs include, according to one authoritative study, the Soviet bloc, Near East, Europe, East Asia, Africa, and the Western Hemisphere.[8] Separate staffs within the directorate are in charge of covert action, counterintelligence, and a few other tasks such as counterterrorism and counternarcotics. Covert-action decisions are reached within the Covert Action Staff (CAS) and, for paramilitary endeavors, a Special Operation (SO) unit. Here, according to one recent DDO, resides the CIA's "tribal knowledge" of what covert actions might work, though most detailed proposals rise through the Agency bureaucracy from the stations abroad.[9] Even within the Operations Directorate, bureaucratic wars have been fierce at times, especially so when James Angleton headed the Counterintelligence Staff (CIS) and was notorious for concocting and carrying out his own operations abroad—sometimes more covert action than counterintelligence—without the knowledge of CAS or SO personnel.

Each of the area divisions at Headquarters is organizationally tied to apposite CIA personnel in the field. The European Division, for instance, is responsible for the CIA's officers and agents in each European nation, from Norway to Spain and Greece. The top CIA man or woman in each country is called, the reader will remember, the chief-of-station, or COS—the DCI representative abroad and the DDO's country spymaster. Beneath the COS serve the CIA "case officers," who are Americans, and their native agents ("assets").

The CIA emphasis overseas is on the recruitment of spies from the "hard targets" like the USSR and other leading Communist states. Over half of the time case officers spend abroad is devoted to observing (technically referred to as "spotting and assessing"), gaining access to, recruiting, and finally handling an agent from one of these hard targets. Simply meeting and becoming better acquainted with a potential recruitment can be a painstaking and labor-intensive task. As an official in the Operations Directorate has explained: "Often, when we target a person and say, 'All right, there is an individual who has access to information we want,' often our case officer doesn't have a direct approach to him, so he may have to recruit several 'access agents' in order to finally develop [the relationship] to the point where he can engage in a dialogue with that individual and hopefully bring him around to recruitment."[10]

Several of the CIA assets abroad have more than one capability, or at least a potential for carrying out more than one kind of assignment. A single CIA agent may collect intelligence, provide counterintelligence information, and conduct covert action of various kinds; therefore, it is not always clear how to count a

FIGURE 3.3 The Directorate of Operations

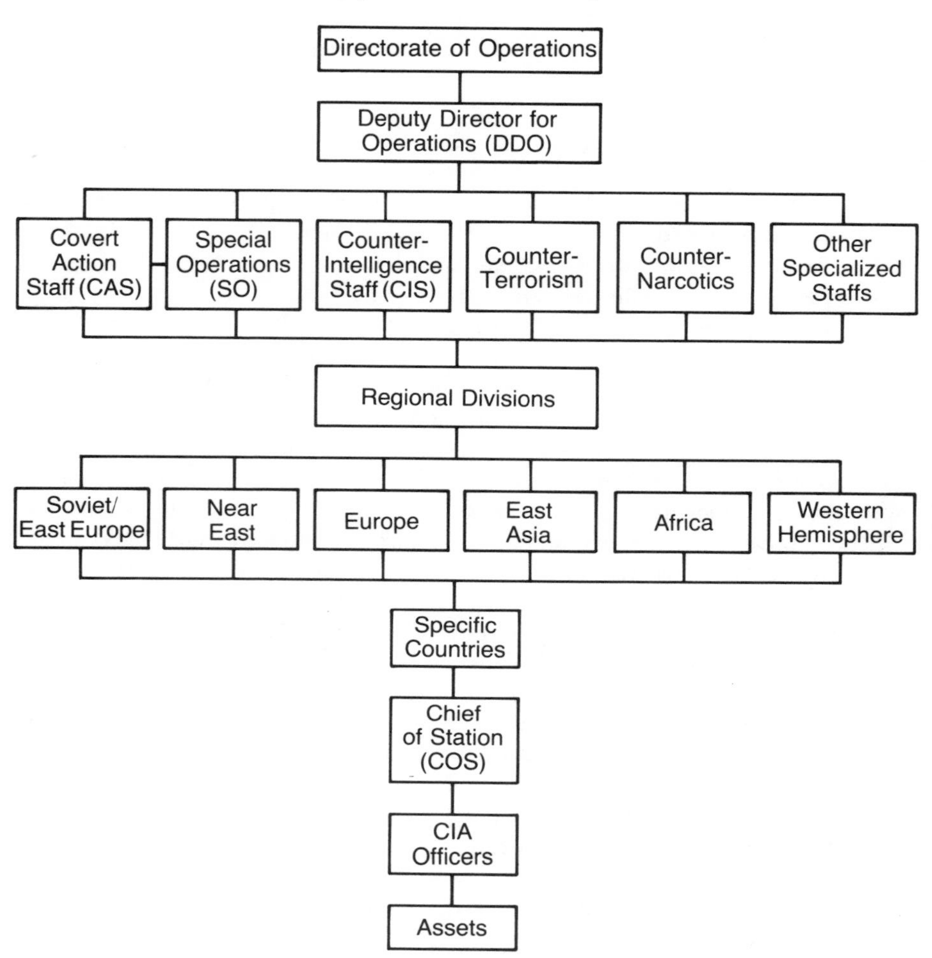

person in any sort of tidy audit. Several media assets may also serve, for example, as so-called agents of influence—that is, they make media placements, but they might also possess an easy access to government officials or party leaders that might be used to exert political influence (say, through a staff aide). As one former CIA officer has noted with reference to the Agency's network of assets: "These confidential collaborators around the globe are the main day-in-day-out instruments of covert political action."[11]

The Operations Directorate provides the CIA stations abroad with specific operating instructions that serve as an "activities blueprint" for the year. The case officers and a few technical officers abroad are expected to carry out the three primary missions of collection, counterintelligence, and covert action under the guidance of the COS, who serves in essence as the CIA equivalent of an ambassador. The COS reports to his country desk housed in one of the geographic divisions of the Operations Directorate at Headquarters, supposedly without bypassing the local U.S. ambassador (the chief of mission) to whom he is, in theory at least, subordinate. The link between the COS and the ambassador in the field has often been delicate and has led to considerable tension at times between the CIA and the Department of State. This relationship is important and is examined at greater length in chapter 11 to illustrate the bureaucratic strains that exist within the government between entities with shared responsibilities for the conduct and supervision of American foreign policy. The other three directorates have been less subjected to the roils of public criticism.

DIRECTORATE OF SCIENCE AND TECHNOLOGY This directorate, the newest and smallest of the CIA's major divisions, is devoted to the improved application of technology to espionage—notably in spying with satellites and high-altitude airplanes. The DS&T does the research and development on James Bond gadgetry used by agents in the field; experiments with drugs and other chemicals (the deadly poisonous pill, encased in a hollow silver dollar, which U-2 pilot Francis Gary Powers was supposed to swallow if shot down over the Soviet Union, but chose not to, is an example of S & T handicraft); constructs such handy objects as fake rocks and trees with hollowed out spaces for hiding messages or listening and sensing devices (say, for registering the presence of radioactive materials—nuclear weapons—on passing trucks); and handles many of the Agency's computer-processing tasks.

Not everything works out as planned. For example, a DS&T mechanical condor was supposed to flap around Latin America gathering intelligence but crash-landed in test flights and was abandoned. And a chemical substance mixed with the mud of the Ho Chi Minh trail in Vietnam was supposed to make it extra-slippery and impassable.

The DS&T is also the home of the Foreign Broadcast Information Service (FBIS), charged with the responsibility for monitoring foreign radio and television broadcasts, and the National Photographic Interpretation Center (NPIC), where analysts pore over photographs of Soviet missile bases and other valuable data gathered by satellite and reconnaissance aircraft. Housed in Washington, D.C., NPIC is run jointly by the CIA and the Defense Intelligence Agency (DIA).

Through its Office of Technical Assistance (OTA), one of the most well-concealed units in the CIA, the DS&T provides a clandestine eavesdropping capability for the U.S. government overseas (in coordination with the National Security Agency, discussed later). The OTA is notably skilled in the development of agent-communications technology—secret writings, "dead-drop" paraphernalia (for hiding messages to and from agents), minature cameras, "audio-bugs" for electronic surveillance, disguises, and the like—all quite helpful for such work as counterterrorism and counternarcotics operations where the objective is to infiltrate dangerous organizations and report to law enforcement officials on their criminal plans. Occasionally, the capabilities of OTA have been misused; here, for example, is where the Watergate burglars obtained disguises. More recently, however, the director of OTA has stressed, "We're no longer afraid to say no. We've learned a lot from the period of history when some people failed to say no."[12]

DIRECTORATE OF ADMINISTRATION Agency housekeeping, hiring, training, computer processing, worldwide communications and logistics, and various other administrative duties are carried out by this directorate. While on the surface it may seem the most innocuous shop of all, the fact that it provides training, communications, and logistical support for clandestine collection and CA operations abroad has brought this directorate close to the Operations Directorate and imposed upon it some share of the "dirty tricks" stigma. Together, these directorates have formed, in the words of two authorities, "an agency within an agency . . . [which] like the largest and most dangerous part of an iceberg, float along virtually unseen."[13]

Within the Directorate of Administration is the Office of Security, a key element in the Agency's CI defenses. This office is responsible for the physical protection of Agency facilities at home and abroad from infiltration by foreign spies or American traitors. It also administers the polygraph tests given to all new CIA recruits and, at least every five years, to seasoned Agency employees. Personnel in the Office of Security are highly trained in surveillance techniques. These skills were turned against American antiwar protestors during the Vietnam War with Operation CHAOS, in violation of the Agency's statutory authority. So what appeared to be, on the organizational charts, a harmless directorate of broom carriers and recruiters harbored a unit that went beyond the bounds of law and propriety in the 1960s and 1970s—though not without persistent encouragement from senior officials in the CIA and in the White House (see chap. 7).

DIRECTORATE OF INTELLIGENCE Here is the vital directorate where the CIA's research work is conducted—the scrubbing, sorting, and interpretation of data that comprises the essence of the intelligence mission. In light of the attention devoted to this mission in the next two chapters, suffice it to say here simply that the Intelligence Directorate also performs various tasks for the wider intelligence community, including sharing data from the enormous CIA computerized files on foreign personalities. Published reports suggest that only about 20 percent of Agency employees are engaged in information processing and analysis, spending less than 10 percent of the total budget. In contrast, two-thirds of the CIA personnel and budget are devoted to covert—or clandestine—operations (terms often used interchangeably to encompass secret intelligence collection, counterintelligence, and covert action). Nonetheless the CIA still has the largest analytic staff of all the agencies within the intelligence community.

Far beyond the campuslike grounds of CIA Headquarters at Langley are scattered the other headquarters of agencies within the intelligence community, from the FBI in the center of the District of Columbia to the mysterious National Security Agency some forty miles away at Fort Meade, Maryland. Together they present a supervisory challenge that neither the DCI nor the Congress has met with more than limited success.

The Department of Justice

The Federal Bureau of Investigation (FBI) and the Drug Enforcement Administration (DEA) are subdivisions within the Justice Department. The job of the DEA, as its name implies, is to halt the flow of dangerous drugs into the streets of the United States. This is a difficult and hazardous job; drug dealers play for high stakes and are inclined to shoot first and ask questions later when anyone attempts to disrupt their lucrative business. Tracking the flow of narcotics is hard enough, let alone successfully curbing it. The CIA has a counternarcotics division and assists the DEA in gathering intelligence on narcotics smugglers, but despite the expenditure of considerable resources and the establishment of a worldwide network of CIA and DEA counternarcotics agents, illegal drugs continue to flood this country. Only about 10 percent of the harmful substances brought into the United States are intercepted by the government.

The antinarcotics effort is not a responsibility the CIA relishes. For a number of reasons its professionals would just as soon leave this sticky task to the DEA. Chief among them are the hazards involved, the discouraging record of successful busts, and the occasional involvement of Agency personnel in drug profiting in this sordid but lucrative underworld.

The difficulty of remaining completely free of the drug tar baby while conducting foreign intelligence operations has been demonstrated time and again. During the 1960s selected CIA contacts in Indochina—important as assets for collection, CI, and CA—reportedly also had separate contacts and business alliances with drug lords (or were drug traffickers themselves) in the heroin-rich "Golden Triangle" of northeast Burma, northern Thailand, and northern Laos. During the 1970s and 1980s General Manuel Antonio Noriega of Panama proved useful for various U.S. intelligence operations; at the same time he was reportedly making millions through drug trafficking. In Afghanistan during the 1980s the CIA achieved one of its greatest CA successes, helping to drive the Soviet military out of this nation with PM assistance to the anti-Communist mujahideen. Several of the rebel factions inhabit regions of Afghanistan dotted by the colorful white and purple poppies from which opium and heroin are extracted, and some have been involved in one way or another in the trafficking of narcotics. The CIA, in a word, may have one agenda, while its assets—and, now and then, an officer gone bad—may have side agendas of their own.[14]

Rising an imposing eleven stories, filling an entire city block, and resembling a fortress constructed of monstrous concrete waffles, the J. Edgar Hoover FBI Building stands on Pennsylvania Avenue as an overpowering monument to the man who led the "Bureau" for forty-eight years (1924–72). The most expensive office building ever constructed by the federal government ($125 million), "Fort Hoover" houses some 7,500 employees. While some of these individuals chase bank robbers, hunt for the "Top Ten Wanted" criminals, and (mostly) investigate white-collar crime, others are engaged in more obscure intelligence operations that are little known or understood by most Americans and were not anticipated when the FBI was created.

Established in 1906, the Bureau was the brainchild of Charles J. Bonaparte, attorney general to President Theodore Roosevelt and descendant of the French emperor's family. During its early history, the FBI was concerned chiefly with enforcing national banking and antitrust laws and various fraud statutes and preventing smuggling. Later, under Hoover's leadership, civilian authority for domestic intelligence became centralized in the FBI—primarily at the urging of President Franklin Roosevelt and his four attorneys general. As war clouds gathered on the European horizon, President Roosevelt and his aides grew increasingly alarmed about the possible effects of communism and fascism on American society. A presidential directive of September 1939 ordered the Bureau to be alert for any information relating to "subversive activities" and "espionage." These twin concerns led ultimately to the establishment within the Bureau of two key intelligence branches: Internal Security and Counterintelligence. Under Hoover's successor, Clarence M. Kelly, the internal security functions of the Bureau were transferred to the Criminal Investigative Division, which left counterintelligence as the sole responsibility of the FBI Intelligence Division.

The Internal Security Branch concentrated chiefly on "extremists" of all kinds: Black Panthers, American Indian extremists, white "hate" organizations (such as the Klu Klux Klan), and various revolutionary groups (especially the American Communist Party). The problem confronted by the Bureau's internal security mission was the proper identification of those who were genuinely "subversive." The FBI used a wide net, taking the role of—in the opinion of one thorough study—"a kind of ideological security police, an arbiter of what was inside the boundaries of legitimate political discourse and what outside."[15] The FBI has an estimated two thousand agents on political investigative assignments.[16] They are, in turn, responsible for thousands of undercover informers (sometimes referred to as "informants"), who serve as the primary means of domestic intelligence gathering for internal security investigations.

The purpose of the Bureau's Counterintelligence Branch was to discover terrorist activities and hostile foreign intelligence operations in the United States and destroy their effectiveness. Counterintelligence (CI) is, as discussed in chapter 2, the business of catching terrorists, spies, and "agitators" in the service of foreign powers, groups, or movements. A misconception of the genuine, indigenous dissent against the Vietnam War as foreign-inspired agitation provoked both the Johnson and Nixon administrations to misuse the government's CI capabilities against American citizens, especially students (see chap. 7).

The Bureau works with other intelligence agencies to monitor the activities of suspect individuals; generally the CIA and the military intelligence units conduct the monitoring overseas, and the FBI assumes these duties within the United States. Liaison among the agencies is maintained to assist the smooth coordination of the CI mission, though the liaison has occasionally broken down in the past as bureaucratic disagreements and personality clashes interfered. As mentioned earlier, defector cases can be particularly troublesome since sometimes the CIA and the FBI may disagree over the *bona fides* (authenticity) of the defector, one perhaps viewing him as a clever double agent still in the employment of a hostile intelligence service and the other satisfied that he is an important and reliable new source of intelligence recently within the enemy camp.[17]

Foreign espionage obviously continues to be a threat to the United States. Estimates by the intelligence community indicate, for example, that over five hundred agents of the KGB or GRU are presently within the United States under diplomatic cover alone,[18] and, as discussed in chapter 2, many other Soviet "illegals" using false identities are also assumed by the FBI to be scattered across the country. Ferreting out the spies—some of whom may be potential terrorists intent on the destruction of American targets—without violating the rights of innocent American citizens is an often delicate challenge faced by FBI counterintelligence officials. As congressional investigations discovered in 1975, the Bureau has sometimes failed to achieve this balance—at the expense of civil liberties.[19]

The Department of State

One of the smallest and most open of the American intelligence agencies is the Department of State's Bureau of Intelligence and Research (INR), located in the rambling State Department complex known as Foggy Bottom. When President Truman disbanded the OSS following the Second World War, some of its intelligence officers went to Army Intelligence and others to the State Department where they formed the nucleus of INR. This intelligence agency is responsible for briefing the secretary of state on current intelligence developments and produces analytic reports and studies based largely on information collected by other agencies and by U.S. embassies abroad.

Former secretary of state, Dean Rusk, recalls his relationship with INR during the years 1961–69: "The INR staff would underline key sentences and longer passages in intelligence reports prepared for me by Department analysts and by agencies throughout the intelligence community, drawing my attention to important findings. I would read this material and often ask INR to research certain aspects further. The INR staff would also provide me with periodic oral briefings, and during each day would send to my office snippets of information about significant developments abroad."[20]

A primary objective of INR is to introduce into national policy-making councils a diplomatic sensitivity to intelligence reports. Its own intelligence reports are among the most highly regarded in the government—some say the best. One reason INR has been able to produce high-quality, reliable analysis may stem in part from its relatively slow turnover of analysts. "The Polish account [that is, desk] in INR is headed by a man of twenty years experience," comments an envious CIA official. "We have no such corporate memory at CIA."[21]

The Department of Defense

The Pentagon, a massive five-sided brown structure housing the Department of Defense (DoD), sprawls on the Virginia side of the Potomac River close to the Fourteenth Street Bridge leading to Washington. The Defense Department is the primary producer and consumer of intelligence information and controls from 75 to 90 percent of the nation's intelligence budget (the figure varies from year to year within this range). Under the Defense Department chain of command are the National Security Agency (NSA), the Defense Intelligence Agency (DIA), and the three military service intelligence units (Army, Navy, and Air Force).

The NSA, the largest and most secretive of the American intelligence agencies ("NSA means Never Say Anything," says one wry insider), is physically located halfway between Washington, D.C., and Baltimore, Maryland, on the grounds of a major military base, Fort Meade. The NSA has the largest floor space, the longest corridors, the most electrical wiring, and the biggest computers of any agency within the intelligence community. Compared to the technological paraphernalia and extraordinary physical security visible at NSA headquarters, the CIA could pass for an ordinary office building. Atop the NSA structure are huge antennae and odd-looking shapes that provide rapid communications throughout the world. Surrounding its headquarters are concentric rectangles of barbed and electrified fences ten feet high; guards with attack dogs patrol these fences night and day. Closed-circuit television cameras scan the parking lots, and soldiers armed with machine guns stand ready at the four gatehouses. The CIA has two armed and uniformed guards in its lobby and none visible in the hallways; in contrast, the NSA lobby and corridors overflow with Marines and military policemen—some of whom stare down like drill sergeants with bad digestive conditions from elevated control points along the corridors.

The security procedures at the NSA make it seem like a storage site for atomic bombs. Instead, here is the location of the nation's cryptological, or code-breaking, center.[22] Nations communicate with their diplomats (and their spies) overseas through elaborate codes. Within this enormous beehive of technology, NSA cryptologists, computer programmers, mathematicians, engineers, linguists, electronic and radar experts, and communications specialists practice the science and art of breaking codes and intercepting international communications around the globe.[23] They also work to make America's own secret communications as inviolate as possible.

From its creation in 1952 by presidential order, the NSA has grown into a vast mechanical octopus, reaching sensitive tentacles into every continent in search of information on the intentions and capabilities of other nations. The NSA staffs some two thousand fixed listening posts to intercept and decipher coded messages sent by foreign governments and military units. In Europe and the Pacific, several circular antennae—each larger than a Super Bowl stadium—assist in the scanning of the airwaves for coded messages, as do flying and seaborne intercept platforms used by the NSA in cooperation with the Air Force and the Navy. The interception of various signals and communications by the NSA falls under the rubric of signals intelligence or SIGINT, which, the reader will recall, professionals further subdivide chiefly into COMINT, ELINT, TELINT, and radar intelligence (RADINT). As part of the SIGINT process, NSA personnel conduct "signals analysis" (an examination of the types of signals emitted by the adversary), "cryptanalysis" (the content of the signals), and "traffic analysis" (the structure of communications—who is communicating with whom, how frequently, and at what time of day or night).

The end result of having so many global listening posts is a virtual torrent of information pouring from the skies into the Fort Meade facility. One former director of the NSA, Adm. Noel Gayler (pronounced "guy-ler"), said he often felt lilke a man with "an open firehose nozzle held to his mouth."[24] In a 1970 review of military intelligence collection, a commission concluded that NSA "collection efforts are driven by advances in sensor technology, not requirements filtering down from consumers of the community's products."[25]

Unfortunately, this awesome technology has also been turned against the American people, as when for over three decades the NSA used its powers to read cable communications sent abroad or received by American citizens.[26] During the Iran-contra operation of 1986, the NSA also assisted the CIA and Lieutenant Colonel North of the NSC in the controversial arms sale to Iran. The NSA refused at the time to respond to an inquiry by the secretary of defense about its possible involvement, on grounds that the operation was not within the realm of his "need-to-know"—a startling commentary on the independence, not to say arrogance, of the NSA in light of its organizational setting within the framework of the Department of Defense.[27]

While the NSA is the chief technological intelligence arm of the Department of Defense, the Defense Intelligence Agency was designed to be its arm for the improved coordination and analysis of intelligence collected by Army, Navy, and Air Force intelligence units.[28] Located in a new facility at Bolling Air Force Base in Washington, D.C., the DIA was established in 1961—making it the newest of the major intelligence agencies—by an internal directive from Secretary of Defense Robert S. McNamara. (The executive branch has found it easier to create intelligence agencies with a stroke of the pen rather than through legislation; executive guidelines are infinitely easier to alter than laws, too, sometimes in secret—not to say ignored.)

The secretary wanted the DIA to coordinate military intelligence training and operations, which were badly fragmented at the time, as well as to arbitrate and resolve conflicting intelligence reports produced by the separate military intelligence units. Within three years the DIA had succeeded, temporarily, in shifting the emphasis of the service units away from analysis toward the collection of raw intelligence; it sought to centralize and control the analytic and interpretive tasks essential to the production of polished ("finished") intelligence regarding the capabilities and intentions of hostile military forces. Tension and competition between the DIA and the three service intelligence units continued, however, as the latter remained reluctant to give up traditional duties and privileges. The service units expressed particular concern about a dilution of their capabilities for providing solid tactical intelligence for field-level combat responsiveness.

The DIA soon found itself pulled between two poles: high-level Washington policymakers wanted broad, strategic military information from the DIA that would fit into their global planning, while military commanders demanded more narrow, tactically relevant data applicable to the battlefield. (The appropriate balance between strategic—or national—intelligence and tactical intelligence is a vexing problem of priorities for many of the intelligence agencies.) Trying to serve widely divergent consumers, the DIA often failed to satisfy either. As a "blue ribbon" Defense Department investigative panel reported in 1970, "The principal problems of the DIA can be summarized as too many jobs and too many masters."[29]

One result of this bureaucratic tug-of-war has been for the individual services to continue budgeting and staffing ever larger intelligence units of their own. Army intelligence even grew so expansive during the Vietnam War era that it crossed the line of the law to spy on American antiwar dissenters.[30] These efforts to maintain service autonomy and preserve wartime capabilities have led to the duplication of analytic work meant to be performed by the DIA. Moreover, the fact that most of DIA analysts come from, and return to, the military services after a short tour with the agency encourages the analysts to advocate the perspective of their home service, rather than to assume a broader identity with the DIA as the Defense Department's advocate on intelligence matters. "Powerful interests in the military opposed, and continue to oppose, more centralized management of intelligence activities," concluded an executive-branch review board in 1971.[31]

The DIA, though, continues to have the backing of the secretary of defense, who enjoys higher official status in the government than the director of the CIA (though this status apparently failed to impress the NSA during the Iran-contra episode). For this reason alone, the DIA remains a formidable bureaucratic rival to other intelligence agencies—and sometimes an incorrigible underling to its titular civilian overseer, the DCI. Recent improvements in DIA analytic skills, plus experience, have enhanced the "clout" of the agency in intracommunity councils—even though a congressional investigative panel in 1976 called for the abolition of the DIA as an unnecessary, redundant agency.[32]

The Department of the Treasury

The Treasury Department, an enormous and incongruous construct of Greek architectural motifs and iron-barred windows, stands on the east side of the White House. Among its several subdivisions are the Internal Revenue Service (IRS), the Secret Service, and the Customs Bureau. The IRS, working in cooperation with the FBI, has conducted numerous investigations into the political activities of organizations to identify contributors and check their tax status. "Some of these organizations may be a threat to the security of the United States. . . . One of our principal functions will be to determine the sources of their funds," stated a memorandum written by the director of the IRS intelligence arm (called the "special services staff") shortly after its creation in 1969. Within five years, the staff had launched investigations into 2,873 groups and 8,585 individuals.[33]

The Secret Service is responsible for the protection of high-level government officials, including the president. Within its files are retained the names and backgrounds of 47,000 "persons of interest," that is, potential disrupters or assassins of a president. The Customs Bureau also operates a small intelligence service to help prevent the import of contraband goods; high among its priorities is the interdiction of illegal narcotics (in conjunction with the DEA). The problem with these lists and investigations, the Church committee found, was that many innocent people and organizations (like the *Rolling Stone* magazine) became subjects for agency harassment because of their legitimate political views (or, in some cases, their criticisms of one of the intelligence agencies), not their genuine danger as "subversives" to the United States.[34]

The Department of Energy

The Energy Department is housed in several government buildings in Washington. The legislation creating the new department in 1977 gave it jurisdiction over nuclear research and development, as well as the job of gathering nuclear intelligence (previously the task of the Energy Research and Development Administration, ERDA, and before it the Atomic Energy Commission). The International Security Affairs division within the Energy Department monitors nuclear testing and the international transfer of nuclear material. It also produces intelligence reports on the nuclear capabilities of other nations, both with regard to weapons production and the development of peaceful reactors. As nuclear energy grows in importance, so will this intelligence on international reactor and bomb-building expertise and the worldwide flow of uranium, plutonium, heavy water, and advanced nuclear technology.

Here, then, is a brief portrait of the American intelligence colossus. Its work is augmented by various satellite agencies such as the Intelligence Division of the Post Office, the Immigration and Naturalization Service, and the Passport Division of the State Department—not to mention untold numbers of CIA and military intelligence "proprietries" scattered around the globe.[35] Binding this constellation together are a few interagency committees and a devotion to the protection of U.S. national security interests—little else.

The job of the DCI is to overcome this dispersed authority: to produce for the president reliable intelligence drawn from the various agencies while at the same time trying to assure that they stay within the confines of law and propriety. "This is not an easy task," concludes one scholar, with obvious understatement.[36] Indeed it is not, especially in a community dominated by agencies whose basic loyalties reside within the Department of Defense and who remain suspicious of an official wanting to "coordinate" their programs at the same time he represents the rival CIA (not counting the exceptional response of the NSA to the secretary of defense during the Iran-contra case).

An incident involving the "discovery" in 1979 of a Soviet brigade in Cuba is instructive in this regard. The DCI at the time, Adm. Stansfield Turner, recalls the total lack of control he had over the National Security Agency (which he refers to as "a loner organization"[37]) during the diplomatic flap. Without coordinating its findings with the DCI, the CIA, or any other intelligence agency, the NSA evidently informed the White House that Soviet training units in Cuba had reached "combat" brigade strength. This evaluation in itself went far beyond the mandate of the NSA for intelligence collection. In Turner's opinion, the NSA "habitually stretches its authority to process data and goes into full-scale interpretation of it."

Neither the DCI nor the secretary of defense, continues Turner, "exercises firm control over grey areas such as whether what the NSA is doing is interpretation or collection. The NSA understands this ambiguity and is skilled in taking advantage of it to do what it wants." The NSA sought to impress the White House with an intelligence scoop, Turner suggests, so it went ahead without regard to interagency coordination. (The scoop turned out to be false, when months later it became known from other intelligence records—slow in emerging—that the Kennedy administration had agreed in 1962 to allow the brigade to remain in Cuba as part of the negotiations over the removal of Soviet intermediate-ranged ballistic missiles.) Turner concludes, "[O]ne of the lessons of the Soviet brigade fiasco is that the Director of Central Intelligence could be given enough additional authority over the NSA to curb the degree of independence it shows with regard to doing analysis and distributing its product."

The DCI has four major opportunities to exercise management control over the NSA and the other members of the intelligence community, chiefly through his Intelligence Community Staff (ICS, usually called the "IC Staff"): providing direction on the "tasking" (assignment) of collection priorities, organizing long-range planning sessions, coordinating CI efforts, and, most important from his perspective, determining (in cooperation with the Office of Management and Budget) the priorities for budget allocation within the community. The DCI is unable to dictate on these matters, but his position does make him at least first among equals on program recommendations. Within the National Foreign Intelligence Program, each operation remains under the supervision of the separate departments and agencies, but the DCI does establish a unified NFIP budget [through extensive interagency consulting at the Senior Interagency Group (SIG) and ICS levels] and defends these spending proposals before the Congress in a mammoth annual document called the Congressional Budget Justification Book or CBJB. The DCI's leverage over intelligence policy remains limited, nonetheless, and the job—like the president's—is usually characterized more by opportunities for persuasion and cajoling than for direct command.

The unavoidable conclusion about the design of American intelligence is this: the leader of the community, the DCI, has found the task of coordinating intelligence—the mandate of the 1947 National Security Act—staggering. Successes have been limited, and tribalism continues to be the community's hallmark. The number, size, and complexity of these institutions alone would assure hardships of coordination; added, however, are a host of other problems—addressed in the next portion of this book—that further entangle these snarls and snags into a Gordian knot.

II

PROBLEMS OF STRATEGIC INTELLIGENCE

FOUR

Seven Sins of Strategic Intelligence

Magna est Veritas

Part II of this volume begins with an introduction to seven important controversies that have plagued the course of intelligence policy in the United States. Behind each controversy lies a question of dubious conduct—"sin" for short. These sins involve distortions in the reporting and the receiving of information (Sins Nos. 1 and 2); the indiscriminate collection of information (Sin No. 3), the indiscriminate use of covert action (Sin No. 4), inadequate cover abroad (Sin No. 5), improper use of intelligence within the United States (Sin No. 6), and inadequate accountability (Sin No. 7). This chapter provides a short introduction to each sin, and subsequent chapters explore in greater detail the most controversial ones: information distortions (chap. 5), excessive covert action (chap. 6), improper domestic intelligence (chaps. 7, 8, and 9) and inadequate accountability (chaps. 10 and 11).

The most valuable contribution the CIA can make to American democracy—indeed, its raison d'être—is to seek and report the truth to policymakers, so they may better serve the people. Two major weaknesses (Sins Nos. 1 and 2) have interfered with this essential mission.

1. Failure to Provide Policymakers with Objective, Uninhibited Intelligence

As I emphasized at the beginning of this book, nothing so persuaded the president and the Congress of the necessity for improved intelligence after World War II as the memory of Pearl Harbor. The intelligence portions of the 1947 National Security Act sought above all to improve the country's protection against surprise attack. The best shield seemed to be information about possible dangers that was timely, well coordinated, and accurate. Of utmost importance was the requirement of impartiality: hard evidence free of emotion, political calculation, or other distorting biases—"neither exaggerated nor minimized," as former CIA director William Colby has put it.[1]

The intelligence agencies, naturally, would have failures. The affairs of people and nations too often resemble a state of Brownian disorder for precise predictions. A senior CIA analyst, reflecting on events preceding the Iranian revolution in 1979 remembers:

> We knew the Shah was widely unpopular, and we knew there would be mass demonstrations, even riots. But how many shopkeepers would resort to violence, and how long would Army officers remain loyal to the Shah? Perhaps the Army would shoot down 10,000 rioters, maybe 20,000. If the ranks of the insurgents swelled further, though, how far would the Army be willing to go before it decided the Shah was a losing proposition? All this we duly reported; but no one could predict with confidence the number of dissidents who would actually take up arms, or the "tipping point" for Army loyalty.[2]

This is certainly not to say that U.S. intelligence efforts in Iran during this period were distinguished. Rather than jeopardize American relations with the Shah (who, among other things, provided access to useful SIGINT collection sites close to the Soviet Union), policymakers consented to curb all but minimal intelligence collection against his political opponents.[3] It is to say, though, that no one has a crystal ball.

Intelligence distortions resulting from policy bias or inhibition are a different matter, however. While apparently infrequent, transgressions against objectivity do occur. Adm. Elmo R. Zumwalt, Jr., former chief of naval operations, points to "the disinclination on the part of senior managers in the intelligence system to give unpleasant information to the President."[4] Thomas L. Hughes, former director of INR, remembers that "especially on Vietnam, INR's role as objective analyst and interpreter had to find its place in an environment where others were seeking and supplying intelligence to please."[5]

No doubt the most well known intelligence controversy from the Vietnam War era involved Gen. William C. "Westy" Westmoreland, the eagle-visaged U.S. military commander there from 1964 to 1968. In a television documentary, CBS news claimed that during the war General Westmoreland had falsified intelligence reports on the strength of the enemy (known as "order-of-battle," or OB, estimates). His purpose, allegedly, was to lower the enemy's actual troop counts to make the chances of an American victory seem more plausible to civilian and military leaders in Washington and, thereby, increase their support for the war. The documentary, entitled "The Uncounted Enemy: A Vietnam Deception" and televised in 1982, charged Westmoreland with participation in a "conspiracy" at the "highest levels" of military intelligence to "suppress and alter critical intelligence on the enemy in the months preceding the Tet offensive of January 1968."[6] This powerful enemy offensive, though turned back in a disasterous military defeat for the enemy, did nevertheless call into serious doubt the original OB estimates. Tet demonstrated that the enemy could strike at the very heart of the American presence in Vietnam, the U.S. embassy in Saigon.

At the core of the dispute was the question of what to count in the OB calculations. Robert W. Komer, deputy to Westmoreland in Vietnam, argues:

> It was particulary difficult to assess the Viet Cong–North Vietnamese Army order of battle, since conventional order of battle analysis simply didn't work too well with the highly unconventional VC-NVA lineup. So estimating its size became a highly arcane and complex process bound to engender debate, especially over irregular or auxiliary groups that did not appear in organized units. Vietnamese intelligence and our own had a much better fix on the enemy's regular units than on the guerrillas and other auxiliaries.[7]

Westmoreland chose to report the firmer figures on regular units, setting aside such entities as the Self-Defense (SD) and the Secret Self-Defense (SSD) units, for the most part older men and women ("mama-sans") in civilian dress who comprised "self-defense" cadres—part-time "nebulous" guerrillas, according to Komar. Westmoreland deleted these units (even though they may have numbered from 120,000 to 150,000 people) from the OB counts on the grounds that they represented no offensive military threat. Critics countered that a grenade tossed by a mama-san was just as deadly as one thrown by an NVA soldier.

Dean Rusk, secretary of state at the time, supports Westmoreland's position. "Order-of-battle figures in a war like that are inherently unreliable, in the first place," he says. "Moreover, for us to count political cadres would have been as far-fetched as for the North Vietnamese to count [Sen.] Frank Church or [the outspoken actress] Jane Fonda [American critics of the Vietnam War] in their order-of-battle calculations."[8] Robert S. McNamara, who was secretary of defense at the time, and his deputy secretary, Paul Nitze—the very men Westmoreland was supposedly trying to fool—agree. "It was extremely difficult to be certain about intelligence estimates in a war of that complexity," Nitze has testified. Further, to have added the self-defense forces to the enemy's regular forces was to have added "flies" to "elephants." And "when you aggregate elephants and flies," Nitze concludes, "you get nonsense."[9] From this vantage point, the debate becomes, in Komer's phrase, "an honest disagreement, not an attempt to deceive."[10]

Moreover, other Westmoreland defenders point out that the type of warfare being conducted by the Americans has to be taken into account. The general's approach was to employ large-scale "search-and-destroy" operations: sweep out from U.S. military compounds in huge numbers, find the enemy, destroy him, then return to the compounds. Westmoreland's successor, Gen. Creighton Abrams, changed military tactics, relying on small units of soldiers engaged in discrete actions against specific targets. For Westmoreland's approach, most vital was intelligence on large, regular enemy divisions that he could engage; for Abrams, intelligence on irregulars became important, for in targeting individual villages his units often encountered the enemy in this guise.

The CBS documentary, however, dismissed this dispute over what to count as merely a "tactic," employed to lower the total OB figure and put a good face on the progress of the war of attrition in Vietnam. The driving force behind this view that the military had indeed "cooked up" the numbers was Samuel A. Adams, an analyst on the CIA Vietnamese Affairs staff from 1965 to 1967. He fervently believed that the lower figures advanced by Westmoreland and his aides amounted to "faked intelligence." He resigned from the CIA, protesting its acquiescence to "a monument of deceit" by the military.[11] Adams subsequently found support for his views in the Pike committee,[12] and with the producer of the CBS documentary (whom he served as a paid consultant during production of the film).[13] At least two current National Intelligence Officers (NIOs) at the CIA privately believe that Adams was essentially correct, if indiscreet, in his efforts to publicize the dispute, and Adams's mentor in Vietnam, George W. Allen—a CIA officer with seventeen years of experience in Indochina—estimates that as much as "40 percent of American loses" in Vietnam were inflicted by the self-defense forces.[14] Nonetheless, for Komer, the case remains "one man's vendetta against all who disagreed with him."[15]

In 1984, General Westmoreland sued CBS over its charge of "conspiracy" in the 1982 documentary. In February 1985, he withdrew his suit, apparently sensing that the jury was persuaded more by CBS's view than his own. (His case suffered a major setback when two high-ranking Army officers testified against their former commander.) The outcome was inconclusive enough, however, to exonerate either party fully, and a more definitive judgment will have to come from scholars of the Vietnam War sometime down the road.

The Vietnamese OB issue arose specifically in a high-level Washington meeting of CIA and military intelligence officials on March 14, 1967. On that day, CIA Director Helms convened a session of the Board of National Estimates (BNE), an interagency panel responsible at the time for approving National Intelligence Estimates (NIEs) before they were forwarded to the White House. (An NIE, discussed at greater length in the next chapter, is an appraisal of a foreign country or situation, authorized by the DCI and reflecting the composite opinions of the intelligence community.) Helms knew that the figures on enemy troop strength in Vietnam provided by military intelligence were wrong—or, at any rate, quite different from CIA figures. Yet he signed the estimate without dissent. The apparent reason, according to his biographer, was that "he did not want a fight with the military, supported by [Walt] Rostow at the White House."[16]

In another instance involving Helms and the Vietnam War, a CIA estimate suggested that an American invasion of Cambodia would fail to deter North Vietnamese continuation of the war in South Vietnam. Helms received the estimate thirteen days before the Cambodian "incursion" was launched by the Nixon administration on May 31, 1970, but he neglected to pass this perspective along to the White House. Apparently Helms concluded that it would have been futile, since the invasion plans already seemed set in concrete, as well as fatuous, since the analysts who prepared the study had no knowledge of the impending invasion.[17]

More recently, in 1983, staff aides on the congressional intelligence committees grew suspicious of CIA estimates on Central America as further examples of intelligence to please. Reportedly offered (and eagerly received by the Reagan administration) as gospel were two conclusions in the estimates that committee staffers found debatable: that the Sandinista government of Nicaragua was dedicated to the extermination of Moskito Indians in the northern provinces, and that the Catholic Church of Central America was an agent of Marxist revolution.[18]

2. *The Disregard of Objective Intelligence by Policymakers*

Just as distortions of reality may occur as intelligence is transmitted up the chain-of-command, so may they occur as it is received. Indeed, no shortcoming of strategic intelligence is more often cited than the self-delusion of policymakers who brush aside—or bend—facts that fail to conform to their *Weltanschauungen.*

Of Kaiser Wilhelm, historian Barbara Tuchman has observed: "[He] was interested in gold-plated news only and disliked above all else those tiresome visits from ministers with their reports of inconvenient facts that did not fit in with his schemes." One of Hitler's close associates, Albert Speer, recalls how the Nazi leader "gladly sought advice from persons who saw the situation even more optimistically and delusively than he himself."[19]

This is not a limited Teutonic disorder. "It has been my experience over the years," writes Robert M. Gates, the second in command of the CIA (DDCIA) and a seasoned analyst, "that the usual response of a policymaker to intelligence with which he disagrees or which he finds unpalatable is to ignore it."[20] Another experienced intelligence analyst, Thomas Hughes, regrets that during the critical years from 1964–1965 in the Vietnam War, U.S. policymakers "did not better exercise their own power to listen."[21] The evidence against a quick American victory in Indochina was available, compelling—and ignored. Former CIA official Ray S. Cline remembers that in 1966 American policymakers began to "lose interest in an objective description of the outside world and were beginning to scramble for evidence that they were going to win the war in Vietnam." By 1969 this pathology had reached disease proportions, lasting through 1974, when, Cline continues, "there was almost total dissent from the real world around us. . . ."[22]

The obstacles between prepared estimates and the president's desk can be formidable. One recent CIA estimate (1983) concluded that the controversial Soviet oil pipeline to Western Europe would fail to make America's allies vulnerable in any significant way to Soviet pressures, as argued by the Reagan administration. A senior staffer on the NSC telephoned the CIA, complaining that "it is not helpful to have an NIE suggesting disagreement with White House policy." The thinly veiled implication was: bury the estimate and start over. When told of the call, then DDCIA John McMahon reportedly responded with searing scatological advice for the NSC staffer. On this occasion the NIE stood.[23]

The CIA has sometimes been less steadfast when confronted by more imposing officials. Two blocked the door of the White House against an estimate in 1969–70: Kissinger, at the time President Nixon's advisor for national security affairs, and Melvin Laird, the secretary of defense. The issue was whether or not the Soviets had MIRVs (multiple independently targeted reentry vehicles) on their SS-9 missiles. If so, it made the missiles much more threatening as first-strike weapons. The CIA estimate concluded no; Kissinger and Laird preferred to believe yes.

According to CIA testimony before the Church committee, Kissinger's request for a new draft of the estimate to provide additional evidence was viewed by the Agency "as a subtle and indirect effort to alter the DCI's national intelligence judgment."[24] At first the CIA refused to budge. It simply bolstered its original conclusion with further evidence. Then, three months later, DCI Helms deleted a paragraph from the estimate after, according to the Church committee, "an assistant to [Laird] informed Helms that the statement contradicted the public position of the Secretary."[25]

Efforts by policymakers to keep their cocoons snugly free of disturbing facts or ideas can be seen in their frequent banishment of experts from high councils. Orders from the White House to Helms not to tell the BNE about the Cambodian invasion plans amounted to a rejection of expert opinion on the wisdom of the invasion. A comparable isolation of key intelligence analysts occurred during preparations for an earlier invasion, at the Bay of Pigs in 1961. This time, too, the BNE was kept in the dark, and as in 1970, CIA analysts had marshaled persuasive evidence against what was to them a hypothetical invasion. Contrary to the view advanced at the White House by the head of CIA covert action, Richard Bissell, top CIA analysts were strongly skeptical that Castro could be easily overthrown. The Cuban premier was "likely to grow stronger rather than weaker as time goes by," concluded the BNE chair in a secret (since declassified) memo to the CIA Director. Contrary to visions of elated Cubans joining an invasion force in the toppling of their government, the memo warned that Castro "now has established a formidable structure of control over the daily lives of the Cuban people."[26]

President Kennedy apparently neither saw this memorandum nor spoke with a single CIA analyst regarding the invasion plans.[27] Ambitious, cerebral, persuasive, a member of Kennedy's social milieu, Bissell assured the president that the Cuban irritant could be removed by a covert action—happy news for the moment, but soon to be proven disastrously wrong. A trained analyst from the CIA (or from the Department of State, whose Cuban specialists were also excluded[28]) presumably would have warned the planners—at minimum—that the contingency escape route to the Escambray mountains was blocked by the impenetrable marshlands of the Zapata Swamp.

3. *Indiscriminate Collection of Intelligence*

The intelligence community is often criticized for a lack of restraint in the gathering of information—a "more is better" mentality, regardless of cost-benefit ratios. This "sin" against democracy—essentially, a waste of the people's resources—seems to be overstated but, given its widespread acceptance as a genuine problem, it deserves mention.

The vast collection systems of the NSA produce a daily rush of information (Noel Gayler's "firehose"), much of it stored in warehouses to be analyzed in the distant future (when time and the deciphering of codes allow, if at all). A former NSA official whom I interviewed in 1988 estimated that only about 20 percent of the information gathered by the NSA is ever used. This imbalance between collection and analysis affects most intelligence agencies. Thomas Hughes remembers a former American ambassador lamenting the excessive intelligence gathering by U.S. military attaches assigned to his embassy, despite his best efforts to rein them in.[29] Richard Betts notes how, throughout the intelligence community, gathered intelligence has outraced the ability of linguists to translate it.[30]

In a discussion of Soviet weapons systems, Colby puts his finger on one reason why collection may receive more attention than analysis: "[S]ince the easiest things to count are tangible forces and weapons and the hardest are military readiness, effectiveness, discipline and will to fight, the tendency is to rely more on the former than the latter."[31] An executive branch report written in 1971 pointed to a distinctly bureaucratic influence. As summarized by the Church committee, "each department or agency sees the maintenance and expansion of collection capabilities as the route to survival and strength within the [intelligence] community."[32]

A fascination with new technology has contributed to the burgeoning collection efforts. The funding for sophisticated hardware must continue at reasonable levels, of course, for the value of technology in this realm has proven itself; but some argue that the expenditure of huge sums of money for marginal improvements in detection is questionable. Money spent on machines means money taken away from classic human espionage. While HUMINT may be notoriously inefficient from an accountant's point of view, a single successful spy out of a hundred might be worth the weight in gold of the other ninety-nine.

Though the gravamen in this debate remains the volume of data collected, questions arise, too, over whether the agencies pursue the right kind of information. The United States has an abundance of facts on the half dozen hard targets perceived to be America's major adversaries; but the government's knowledge proved to be deficient, for instance, with regard to the Indian nuclear detonation in 1964, the shock of oil-price increases in 1973, and, also in 1973, the ease with which Egyptian troops crossed the Suez Canal during the Yom Kippur War. So, in a world where budgetary trade-offs are inevitable, critics look with dismay as the balance in the intelligence community shifts toward technical "gadgetry" designed to suck in information out of the ether like so many goldplated vacuum cleaners, and away from well-trained officers and agents pinpointed against carefully selected targets of political, economic, and military importance around the globe.

An ethical consideration enters into this discussion of collection methods as well: To what extent should American officials ally themselves with foreign officials of an unsavory character? Granted that few saints inhabit this world, should the CIA nurture ties with an individual like Panamanian strongman General Manuel Antonio Noriega—a known drug dealer and suspected murderer—because he has access to useful intelligence information? Should the CIA's philosophy be: collection above all, regardless how perverted the source?

For the U.S. government from the Nixon through the Reagan administrations, the answer with respect to Noriega seems to have been yes. Noriega met with George Bush in 1976 when he was CIA director and again in 1983 when he was vice president; and the Panamanian reportedly enjoyed an even closer relationship with CIA Director William J. Casey during the Reagan years. In return for his cooperation with the CIA and U.S. military agencies—such as providing intelligence on Fidel Castro, assisting with supply operations to the contras, and allowing U.S. intelligence and (ironically) limited anti-drug operations in his own country—American officials reportedly overlooked Noriega's criminal activities. Since 1972, the predecessor to the Drug Enforcement Administration evidently had "hard information" of Noriega's dark side, but the Panamanian had made himself increasingly valuable to the U.S. intelligence community in equal measure with his deepening corruption. By 1986, the Reagan administration seemed to have concluded that his assistance with the contra effort outweighed the evil of his personal avarice, until, in 1987, this greed—and the degree to which Vice President (and presidential candidate) Bush and other administration officials had averted their eyes from his drug-trafficking—became a subject of national controversy in the United States.[33]

Professional intelligence officers argue persuasively, though, for a continued expansive collection of intelligence. Former INR chief Hughes notes, for example, that "once we break the codes, a warehouse of Soviet information may be useful."[34] Seconds a high-level CIA official: "You never know when you might need something on 'Falklandia.' " Another emphasizes how the "warehouse problem" has diminished, since "the microchip has enhanced the storage and retrieval of data."[35]

4. *Indiscriminate Use of Covert Action*

In the flood of new books and articles on intelligence, perhaps no topic has emptied more inkwells that that of covert action. As outlined in chapter 2, its statutory authority is thin, stemming from the catchall phrase in the national Security Act of 1947 that allows the CIA "to perform such other functions and duties related to intelligence affecting the national security" as the NSC found necessary.

Former CIA director William Colby points to two primary successes of covert action: postwar Western European resistance to Communist political subversion and Latin American rejection of Cuban-stimulated insurgency.[36] Certainly among the most conspicuous early successes (at least over the short term) were the covert actions waged in 1953 and 1954 that bought to power pro-American leaders in Iran and Guatemala, respectively. Hardly a shot was fired in either coup; the operations seemed to flow with the ease of a silk handkerchief from a magician's sleeve. Presto, the Communists were out, the CIA boys were in.

Coming as they did on top of earlier good fortune in Greece and elsewhere in Europe, these coups encouraged the view that the CIA could orchestrate events throughout the world, remaking its image more in America's own likeness. Such quick and unobtrusive results gained through the use of this co-called quiet option held strong appeal over the frustration of diplomacy and the dangers of overt warfare. Here was a way to engage the enemy while still keeping the noise and tension levels low. As the Church committee put it, covert action "held the promise of frustrating Soviet ambitions without provoking open conflict." In its global chess game with the USSR, the United States had discovered a wonderful third option, one—so it seemed—that worked.

The national security establishment began to rely upon covert action as a panacea to cure Marxist infections practically wherever and whenever they occurred (see chap. 6). Similar pressure came from the CIA's worldwide network of foreign agents known as the "infrastructure" or, before Watergate gave the term a bad odor, the "plumbing." These individuals, admits an Operations Directorate insider, are not beyond concocting various schemes "to make themselves appear busy and worth their keep."[37]

The short list of early CA successes grew into a long list of failures: the Bay of Pigs, Indonesia, Vietnam, among others, as well as bungled assassination plots against Castro and Patrice Lumumba (see chap. 2). Several of the PM operations grew too large to be covert, yet were too small to succeed. Experts remained ambivalent over the "success" of some operations. In Laos during the 1960s, for example, the United States secretly fielded about four hundred CIA officers to support Meo tribesman in a war against a rising tide of Communist guerrillas supplied by the North Vietnamese (whose numbers swelled from seven thousand to some seventy thousand). Overt assistance would have violated an understanding with the Soviets and perhaps drawn them overtly into the conflict. The resulting stand-off in this "secret" war (which in reality was widely reported on throughout the world) is viewed as another success by Colby, among others. Critics, though, point to the eventual decimation of the Meo people once the United States withdrew, and the eventual consolidation of North Vietnamese power in the region.

Some of the CIA's schemes seem to have been written for the theatre of the absurd: the plan to incite rebellion against Castro ("the anti-Christ") with fireworks shot from submarines off the coast of Havana, accompanied by leaflets proclaiming the Second Coming dropped simultaneously from airplanes;[38] the plot to rob Castro of his charismatic beard, with an application of the depilatory thallium salts;[39] and a recommendation from a CIA consultant, initially approved and only at the last minute squelched by Director Allen Dulles, that American scientific journals be laced with false research findings to fool Soviet scientists (and only "inconvenience" America's own scientific community).[40]

Though failures and absurdities abound, the CIA also claims some recent CA successes. Among them: political support to moderates in Portugal during the Ford administration, which helped prevent a Marxist electoral victory; and PM assistance to the Afghan resistence (the mujahedeen) during the Reagan administration, which contributed significantly to the Soviet decision in 1988 to withdraw its troops from Afghanistan—especially after the CIA provided shipments in 1986 of Stinger missiles capable of shooting down Soviet military aircraft.

Despite these claims, critics contend that in an obsession to preserve the global status quo, the United States has made a grievious error in turning its intelligence apparatus more toward political intervention abroad than toward its original purpose of gathering and assessing information. Senator Church declared in 1976 that covert actions were increasingly directed against "leaders of small, weak countries that could not possible threaten the United States. . . . [N]o country was too small, no foreign leader too trifling, to escape our attention."[41] Coups were directed even toward leaders who had been democratically elected by the people of their country, as with Arbenz of Guatemala (1954) and Allende of Chile (1970–73).

To criticize covert action, though, is to miss the point altogether, retort practitioners. Turning aside hostile questioning from the chairman during a 1975 briefing on covert action before the Church committee, CIA director Colby concluded simply: "What we are really talking here is policy, not covert action."[42] Since the beginning of the Vietnam War, Church had opposed executive branch policies that encouraged "compulsive interventionism" abroad. Covert action was just one form—the most invisible—of this intrusion. If Church could persuade presidents to adopt a less interventionist stance, was Colby's implication, then a decline in covert action would follow; but, as long as the U.S. government sought to influence events abroad, covert action would remain an arrow in its quiver. "What does everybody think we've really been doing all these years?" asks a CIA man with a ready answer: "*Fighting the Cold War!*"[43] Here was the Procrustean bed of American foreign policy that all DCIs were expected to fit.

Thirty years ago, this meant adopting tough measures "in the back alleys of the world" (former secretary of state Dean Rusk's phrase[44]). Despite the controversy these measures have stirred up over the years, CIA officialdom continues to advocate covert action in support of policy. As one high Agency official put it in 1978:

> I am not thinking of returning to those days when covert action was a large percentage of the CIA budget; but I do feel that the present very small percentage [less than 5 percent] is below mimimum. It is an assignment that is a lawful part of the Agency's duties. As soon as we can improve the understanding of it and clear up the controversies that surround it, I think this area should be given more resources if we are to carry out effectively a mission given to us by the president.[45]

Though some CIA officials may remain enthusiastic about the quiet option, critics inside and outside the Agency have posed serious reservations about its value. For some observers, covert actions have amounted to rather limited harassments of perceived foes, a little assistance here and there for a few (chiefly Third World) friends, and some support efforts to help curb terrorism and narcotics trafficking. Overall, covert actions since 1947 seem most often to have been modest (even trivial); or, when more ambitious (like the Bay of Pigs), they have been unlikely to succeed—or even to remain covert. As Gregory F. Treverton notes, "Rarely is it possible to achieve large foreign policy purposes, to wage war on governments or decisively alter their prospects, through secret political actions without the world knowing."[46] That the United States could have gotten by without modest covert actions is plausible; that this nation would have been better off without the failures is self-evident. The persistent failures have been costly to the nation's treasury and—more important still—to the reputation of a people who claim to be more honorable than their foes.[47]

Of all the CA operations since 1947, propaganda no doubt has been the most steadfastly used. Skeptics wonder—as always in an evaluation of covert action—whether such operations are really worth the risk or the money. Yet proponents of secret propaganda maintain that if a Soviet citizen can have a chance to read contraband Solzhenitsyn or *Time,* if a few West European newspapers covertly help the United States sell a policy designed to strengthen NATO—all to the good.

Opponents of CIA propaganda are not so easily persuaded though. They observe that if America's policies are insufficiently sound to garner support by their merits in the open marketplace of ideas, they are unlikely to be rescued by a smattering of covert media placements around the globe. A few editors, here and there, cannot turn a pumpkin into a carriage. Moreover, continue the critics, the dangers are real. First, there is the problem of blow back or replay, where false information planted by the CIA overseas finds its way back to American shores only to mislead American citizens. Second, media assets "bought" by the United States are often for hire to U.S. adversaries the next week (notoriously so in some parts of the world, especially the Middle East); some, refusing to be bought, expose in their columns efforts by the CIA to recruit them. Then there is the moral argument that the United States—of all nations—should be loathe to engage in practices that undermine the free press of those few countries in the world so blessed as to have them (this argument is examined more closely in chap. 9).

Debate like this over the use of media assets is reasonably tame compared to the more passionate opinions on political, economic, and especially paramilitary covert action. While operations like those conducted by the Reagan administration in Nicaragua clearly meet the criterion of "important" in terms of resources and risks, critics doubt their ultimate worth and suggest that they mainly interfere with diplomatic initiatives to achieve the peaceful settlement of disputes in Central America.[48]

Whatever one's view on covert action in Central America, Colby is correct: the lines of argument return to policy—the containment doctrine, the domino theory, and the other postulates advanced to reinforce the anticommunist theme that has dominated American foreign policy since 1945. If the use of covert action has been indiscriminate, then this is because the reaction of the United States toward tumult and revolution in other lands has been indiscriminate.

Arguably, some covert actions have been useful and should be maintained. The best case can be made for selected propaganda operations to spread the truth where America's adversaries would sow lies, for paramilitary and political assistance to popular factions resisting an external invasion (like the Afghan mujahedeen), and for covert actions designed to combat genuine terrorism and the narcotics trade. More compelling still is the argument that the United States must have a CA capability for extreme circumstances, say, to thwart a terrorist nuclear attack again this country. The critics of covert action are persuasive, though, in their central point: to resort to covert action not as a last or a penultimate option (before the use of overt force), but rather whenever someone, somewhere, becomes a nuisance to the United States, runs counter to the nation's democratic belief in fair play and, fairness aside, usually ends in a tragic waste of lives, money, and esteem.

5. *Inadequate Protection of Officers and Agents Abroad*

The loss of lives as a result of poor cover abroad or inadequate efforts to protect foreign agents in the secret service has been a particularly tragic side of U.S. intelligence.[49] Just how important this problem is can be seen in the origins of the Iran-contra scandal of 1987–88. The motivation for the dubious sale of arms to Iran seems to have been the gaining of leverage over terrorist factions in Lebanon for the release of U.S. hostages—most urgently William Buckley, the COS in Beirut, who was reportedly being tortured by his captors and forced to reveal the identities of CIA officers and agents throughout the Middle East.

Poor cover appears to have played a role in the most well-known murder of a CIA officer, Richard S. Welch, the COS in Athens, Greece, gunned down by terrorists in December 1975.[50] Welch's identification as a CIA officer was published on various occasions, including about a month before his death by the *Athens Daily News* (an English language daily published in Greece); yet, he continued to reside in suburban Psychicho in a villa reportedly widely known as the COS residence in Athens.[51] The accurate—and, more often, inaccurate—publication of CIA identities by foreign media is beyond U.S. control, but critics are convincing in their insistence that the government do more to conceal its officers and agents abroad.

Increased attention to better cover is one answer. The problem has been exacerbated by what Colby has referred to as a "melting ice floe of adequate cover."[52] Various groups in the United States have balked at providing cover for the CIA under the banner of their professions. Journalists do not want CIA officers feigning to be foreign correspondents, since if even one is discovered a pale of suspicion is cast across the entire corps of overseas journalists, possibly closing their access to news sources or even endangering their lives. Academics, missionaries, and diplomats argue similarly (see chaps. 8 and 9).

A solution offered by some observers is the extended use of business proprietaries—CIA business fronts, which, if blown, do not endanger non-CIA personnel or tarnish the reputation of reporters, professors, clergy, diplomats, or established businesses.[53] Such "deep" or nonofficial cover (NOC), however, carries its own perils for CIA officers acting in this guise. As an experienced American ambassador points out, nonofficial cover is the "only answer . . . [but it] entails great inconveniences and danger. . . . Outside if you are exposed, and if you appear to be an ordinary citizen . . . people don't always feel they have to be as careful about you, about doing away with you. It is dangerous, but it is the only effective thing. I don't consider the cover we provide today very useful."[54] Others conclude that some government agencies will also have to be more cooperative, though certain ones—like the Peace Corps and the United States Information Agency—have a strong, legitimate claim to maintain their strict neutrality.

Other measures that deserve exploration include tightening security procedures (for example by requiring body guards for "fingered" CIA personnel and having CIA officers drive foreign rather than conspicuous American cars), using less prominent housing (even rotating CIA homes every few years), and transferring to Headquarters those who becomes too visible in the local press. Not even these modest suggestions are easy to implement: for one thing, they run into bureaucratic red tape. The government is geared bureaucratically to use American-made automobiles abroad and to buy attractive homes. Moreover, CIA officials and their families understandably refuse to be hermits, hidden away in caves somewhere. They enjoy the status of fine housing and want to lead reasonably normal, open lives. Perfect solutions to protect U.S. personnel abroad do not exist, but some improvements can be made.

The obligation to protect those in the U.S. secret service abroad extends beyond just officers. Espionage and covert action depend upon the help of indigenous citizens motivated to work for the United States. Yet too often America has abandoned these friends—sometimes whole armies without warning, leaving them to fare alone against a once-common foe. One might expect this behavior from an authoritarian or totalitarian regime, but not from a high-minded democracy.

The results have been disastrous for those who have been deserted: the Ukrainian emigres left to await their deaths in Carpathian caves; the few surviving Meo tribesmen of Laos eking out an existence in Thai refugee encampments; South Vietnamese components of U.S. intelligence operations during the Vietnam War, left behind (along with personnel files and dossiers that would identify them) to face North Vietnamese interrogators; the Khambas in Tibet; the Nationalist Chinese in Burma; the Bay of Pigs invaders; the Kurds—all, in the words of one critic, "so many causes and peoples briefly taken up by the CIA and then tossed aside like broken toys. . . ."[55]

Here are unhappy chapters in the annals of American foreign policy, even if, as one CIA official insists, "these people wanted to fight and would have anyway—we only helped them do what they wanted to do."[56] Asks a critic of the covert action in Nicaragua: "Are we now encouraging the Moskitos to fight the Sandinistas? If so, who is worrying—and doing something—about our obligation to them if they are overrun?"[57]

6. *Improper Use of Intelligence Within the United States*

This sin is twofold. First, the management of counterintelligence in the government has been in a state of chaos over the past decade, and the failure to correct this problem administratively remains a major weakness of the U.S. intelligence community. Second, and more alarming still, has been the use of the CIA (and other intelligence agencies) against American citizens, for counterintelligence, intelligence collection, and even political purposes (covert action). The most notorious illustration is the so-called Huston Plan (the focus of chap. 7).

MUDDLED COUNTERINTELLIGENCE "The Director of Central Intelligence *did away with* counterintelligence," concluded a staffer on the Senate Intelligence Committee in 1980, with reference to Colby.[58] While this overstates the case, changes in counterintelligence wrought by Colby in 1974 did create a festering sore of recriminations within the heart of the CIA. The feud has been over fundamental administrative practices, and at its center has stood the shadowy figure of James Angleton, chief of CIA Counterintelligence for two decades until he was fired by Colby on December 17, 1974. The main tenet of Angleton's philosophy was the belief that counterintelligence must be centralized not only because it would improve security but also because only a limited number of specialists have the requisite temperament and skill to be trained in the convoluted intricacies of this discipline.[59]

While the argument for a centralized CI mission is compelling from the vantage point of security, it has thoughtful professional detractors. They speak of the need for wider sharing of CI information, an increased sense of responsibility for counterintelligence in every division of the CIA, and wider contact among CI officials throughout the intelligence community.[60] In contradistinction to Angleton's elite corps of CI officers, for example, one specialist in the field argues: "Frequent and routine rotation of personnel into counterintelligence and then out into operation units is a most important method of insuring that a wide body of counterintelligence understanding, awareness, and favorable recognition exists outside the counterintelligence organization itself."[61]

Behind such arguments, though, always hovered the main grievance: a distrust of Angleton and his staff for holding counterintelligence too tightly to their vests, for operating too independently—maybe even freewheelingly (for example by allegedly conducting operations in foreign countries without the knowledge of the resident COS). Angleton was widely perceived as aloof, elitist, threatening even, in the bureaucratic sense of controlling too much information with the implied power this might give him over others. Deservedly or not, Angleton had become an object of fear and dislike for key factions within the CIA; their critique of his CI management, while with some merit, often had more to do with Angleton himself than with the proper modus operandi for counterintelligence. Like J. Edgar Hoover, say critics, he had served too long. Norman Smith, once on the CIA's CI Staff, alludes to the problem: "The centralized unit can profit from its highly experienced and specialized personnel for only so long before the organization becomes inbred (and its personnel over the hill) with a resultant isolation from the rest of the organization."[62]

Centralization or decentralization; corporate memory or cross-fertilization; security or shared responsibilities; elitism or the democratization of counterintelligence. Tough trade-offs, but the weight of the arguments falls on the side of a specialized CI corps, relatively centralized but subservient to the COS in the field, valuing continuity but avoiding inbreeding through gradual recruitment of new personnel. Above all, as David Ignatius sums up, "you need a wise judge on top who can decide how long is too long for someone to stay in one job."[63] Or, perhaps, a set term of office for all senior intelligence administrators, say ten years (presently, FBI directors may serve only one ten-year term).

INTELLIGENCE COMES HOME This question of proper administrative practices for counterintelligence pales in significance, however, when compared to a second issue deserving of separate and lengthier treatment in subsequent chapters: the menace to American civil liberties posed by an overzealous use of the CIA and other secret agencies. The CIA's greatest sin is to turn its capabilities against the very society it is meant to protect. Those who find such a prospect remote in a democracy as robust as America's need only reflect upon the facts of the Huston spy plan, approved by President Nixon in 1970. Chapter 7 presents a case study of the Huston Plan, which would have chilled the bones of George Orwell himself. Chapters 8 and 9 examine the CIA's ties to two important American institutions, the media and the academic community. These ties have been less shocking than the horrors of the Huston Plan, but they raise cause for concern nonetheless.

7. *Inadequate Accountability in the Intelligence Chain of Command*

The CIA might be "a rogue elephant rampaging out of control," opined Senator Church in 1975.[64] Contentious diction aside, one can hardly blame him for wondering about the question of accountability in the U.S. intelligence service. The first three subjects examined by his committee in 1975 were the improper sequestering of shellfish toxin, the Huston Plan, and alleged assassination plots. Each exhibited evidence of Agency behavior beyond the control of the presidency (and, certainly the Congress). President Nixon had ordered that the shellfish toxin (and other poisons held by the CIA) be destroyed five years earlier, yet an Agency scientist had disobeyed the order. The episode was hardly earthshaking, but in Church's view it did illustrate "how elusive the chain of command can be in the intelligence community."[65]

The Church committee's most extensive research, on the assassination schemes, raised a host of serious questions about accountability. Presidents, secretaries of State and Defense, attorneys general—indeed, no one outside the CIA seemed to know about the plots. Even inside, CIA Director John McCone was never told about them.[66] The CIA officer in charge of the Mafia-aided attempts against Castro, for example, kept the plans (according to Richard Helms) "pretty much in his back pocket."[67] Nor did testimony before the committee from various "wise men" offer much comfort, as when Clark Clifford told the senators about the slippage in covert action "from point A to point B. . . . When point B is reached, the persons in charge feel it is necessary to go to point C, and they assume that the original authorization gives them such a right. From point C, they go to D and possibly E, and even further. . . ."[68] The "rogue elephant" hypothesis, then, did not materialize from thin air.

The Iran-contra affair, revealed in 1986, and the questionable operations of the secretive U.S. Army Intelligence Support Activity, which may have been involved in covert action in Laos and Central America recently without reporting to Congress, stress the need for constant vigilance by the congressional intelligence committees.[69] This is doubly true for covert action, where, writes former DCI Turner, "one of the risks . . . is that it may get out of control." This risk seems particularly high in the field (though as the Iran-contra affair reminded us, the CIA's top can spin out of control, too). In the field, Turner continues, "the people working for us gain sufficient momentum of their own at some point to go on without us if necessary."[70] All too frequently, acknowledges one high CIA official, the people at this end of the policy process are "goof balls—unreliable and uncontrollable."[71] Adds another intelligence official knowledgable about counterintelligence in Nicaragua: "Control? No way is there control. They [CIA assets in the field] are not just interdicting weapons bound for El Salvador insurgents [as the Reagan administration claimed in 1983]. They're blowing up electrical power plants!"[72]

This looseness in the chain of command was twice perfectly illustrated in 1984, in the Wilson case and with the controversial Nicaraguan guerrilla manual. Edwin P. Wilson was an officer in the CIA from 1955 to 1971. In the latter stages of his Agency career, he ran a proprietary. As the House Intelligence Committee reported: "Ed Wilson soon discovered that CIA's control over his activities and the uses to which he put the propriety firms were very loose."[73] Wilson went into business for himself, eventually selling military items to Libya (an illegal act) and earning millions of dollars. Though let go from the CIA in 1971, Agency officials continued to have contact with him even after high-level internal orders were issued to employees that all ties were to be severed. "The desire to share [in Wilson's Libyan] wealth apparently led a number of intelligence officers to violate not only their own professional code of conduct but also the laws of this country," concluded the House Intelligence Committee following its investigation.[74] With the guerrilla manual, the House Intelligence Committee concluded that it had been produced and used with inadequate supervision. "The incident of the manual illustrates once again," stated a committee majority, ". . . that the CIA did not have adequate command and control of the entire Nicaraguan covert action."[75]

At higher levels, though, the CIA is now probably the most closely watched agency in the national security establishment; the FBI, NSA, and others have escaped comparable attention (see chaps. 10 and 11). Yet even with this close supervision, officials on the NSC staff (its director, Vice Adm. John Poindexter, and aide Lt. Col. Oliver L. North) worked with DCI Casey to sell arms covertly to Iran in 1985 without the knowledge of the Congress, as required by the 1980 Intelligence Accountability Act. North also assisted the contras beyond statutory limitations (see chap. 6). So while lack of accountability may be more likely in the field, the Reagan administration—not to mention the example of Tom Charles Huston (chap. 7)—demonstrates that "rogue elephants" can appear even at the highest elevations, where wayward behavior is apt to have even more somber repercussions.

For some observers, though, the most significant problem of strategic intelligence is not the lack of accountability, but too much accountability. "The real Seventh Sin," says one former CIA official, "is micromanagement—excessive involvement of Congress in intelligence matters. The end result is paralysis."[76] Yet a former senior staffer on the Church committee offers a countervailing perspective: "Congress should continue to monitor the way a covert action plays itself out. It could get off the tracks. It could get into the hands of irresponsible individuals. It could develop as the famous Operation Mongoose [code name for several covert actions directed against Castro] in which perhaps a good purpose at the beginning went into bizarre and stupid behavior."[77]

So, in still another of the many debates over strategic intelligence, the nation continues to ponder the appropriate balance between "micromanagement" and proper democratic control, a subject explored at greater length in later chapters of this study. First, however, I turn to a more thorough examination of Sins Nos. 1 and 2: deficiencies in the CIA's most important responsibility, its search for truth to guide policymakers in the service of the American people.

FIVE

Pathologies of the Intelligence Cycle

The Intelligence Process

The collection, analysis, and coordination of information useful to the United States is the primary mission of the Central Intelligence Agency. At the heart of this mission lies the so-called intelligence cycle. The CIA defines the cycle as "the process by which information is acquired, converted into intelligence, and made available to policymakers."[1] The cycle has five phases: planning and direction, collection, processing, production and analysis, and dissemination (see fig. 5.1), though, as a former CIA analyst notes, the "cycle" is really less a series of discrete phases leading from one to another than a matrix of steady interaction between producers and consumers of intelligence, with multiple feedback loops.[2] This complex matrix is treated here, however, as a discrete sequence of events, both for analytic purposes and in conformity with the common practice among intelligence officials.

Planning and Direction

The first phase of the intelligence cycle entails the identification of what kinds of data need to be gathered, and the assignment of specific agencies to accomplish the task—that is, the management of the mission. The chief responsibility for collection and analysis falls upon the CIA's Directorate of Intelligence, headed by the deputy director for intelligence (DDI). The DDI is, in essence, the top analyst within the CIA, and he more than anyone deserves the credit—or the blame—for the quality of the completed ("finished") intelligence offered to policymakers.

The whole purpose of the cycle is to provide useful knowledge to U.S. poli-

FIGURE 5.1 The Intelligence Cycle

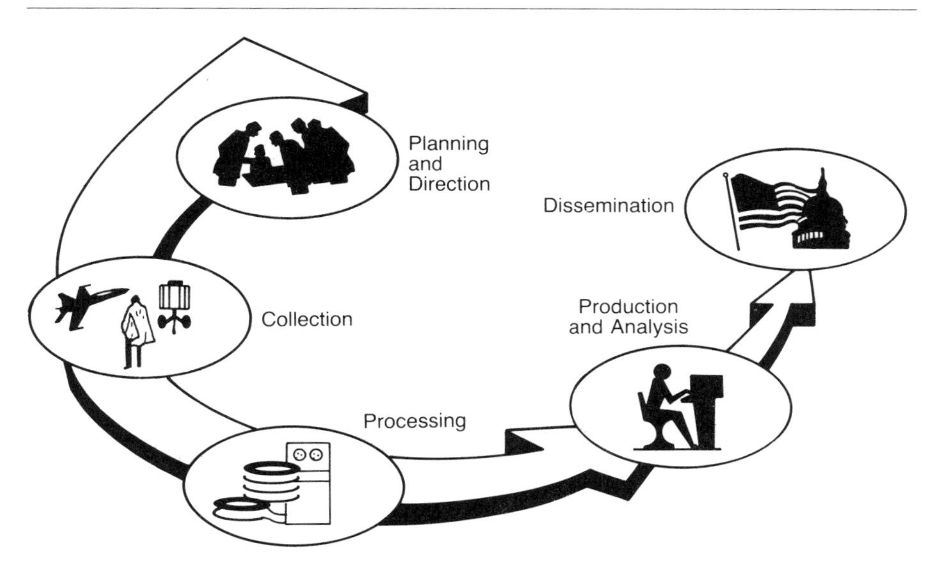

From *Fact Book on Intelligence,* Office of Public Affairs, Central Intelligence Agency, April 1983, p. 16.

cymakers in advance of their decisions. As the CIA puts it in *The Fact Book on Intelligence,* "Intelligence is knowledge and foreknowledge of the world around us—the prelude to Presidential decision and action" (p. 17). Sound foreign-policy decisions are less likely to occur if America's leaders act in ignorance of the capabilities and intentions of other nations.

Major difficulties arise, however, in this seemingly straightforward relationship between the CIA (the producer of knowledge) and the policymakers (the consumers). The informational needs of policymakers are sometimes never made known, or never made clear, to the producers; the needs may extend beyond the capabilities of the CIA and other intelligence agencies, or the producer may disbelieve or ignore information provided. So the management problems can become complex, as the DDI and his assistants attempt, on the one hand, to obtain the best information available, and, on the other hand, to make sure the right people know about it at the right time. What may appear at first glance to be a simple and smoothly flowing process in figure 5.1, in which essential and accurate data is swept along from spy ("agent" or "asset") to analyst to consumer, is instead a complicated series of interactions among men and women and their organizations that often result in the loss or distortion of vital information.

At each stage in the intelligence cycle, the director of the CIA (DCIA), his deputy director (DDCIA), the DDI, and their assistants confront the problem of conflict and disagreement inherent in human relationships. As intelligence moves on the precarious journey from its sources in the field, through the CIA and other intelligence agencies, and on to the ultimate policy councils, it is buffeted along the way by personal biases, group conflict, and bureaucratic pressures. Between

FIGURE 5.2 The Intelligence Cycle as Funnel of Causality

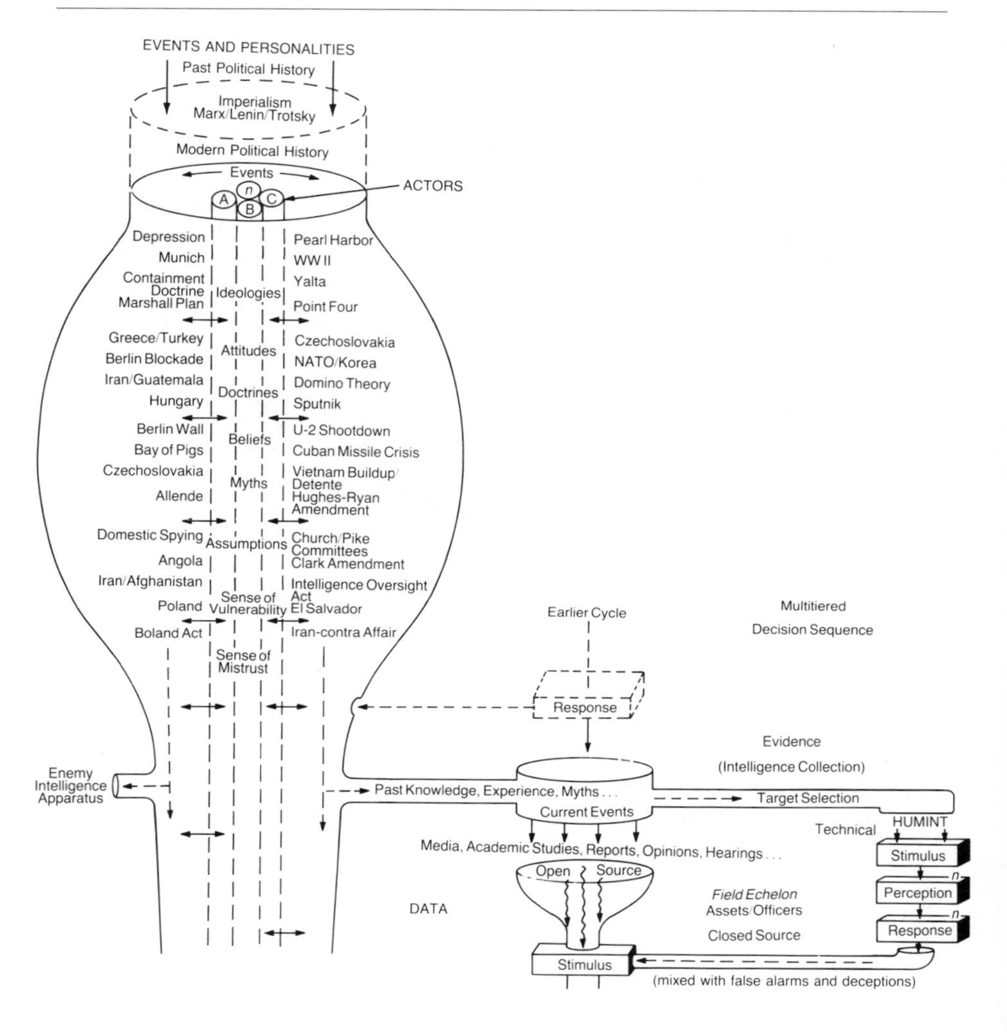

this reality and the ideal of fully reliable data for decisionmakers, based on carefully weighed assumptions, meticulous analysis, total impartiality, and common sense, falls the shadow of human behavior.

So the cycle is anything but smooth; often bumpy and disjointed, it sometimes collapses altogether. Even when working approximately as the sweeping arrow optimistically envisions in figure 5.1 (or in the alternative "stimulus-response" depiction of figure 5.2), the process requires sensitive supervision and

FIGURE 5.2 (continued)

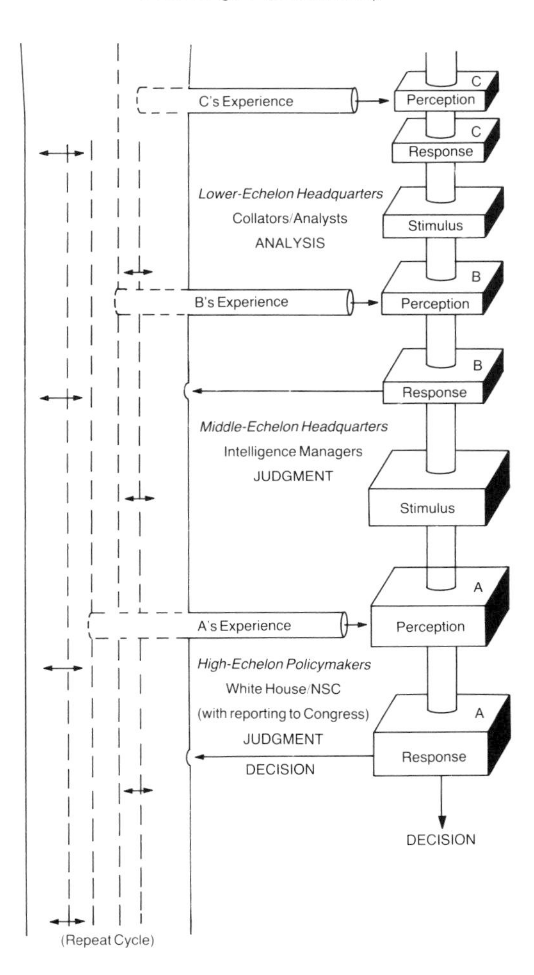

extensive dialogue among collectors, analysts, managers, and policymakers along the way—beginning with the establishment of collection requirements.

Collection

From a cosmological perspective, this planet may be but a tiny speck in the universe, but its dimensions are vast from the viewpoint of a government agency trying to keep up with events in some 184 nations. No agency can be omniscient, so priorities must be set. The managers in the intelligence cycle must decide, in consultation with policymakers, which countries to target, what kind of information about the targets is most important, and what means of gathering the information will be employed.

TARGETING According to former DDI (later DDCIA) Robert M. Gates, the Soviet Union has been the target of most American intelligence collection operations since the establishment of the CIA.[3] This is hardly surprising, given the commitment of the United States to the Truman Doctrine and subsequent expressions advocating the containment of Soviet communism behind the Iron Curtain.[4] American defense spokesmen have all echoed the centrality of the USSR in U.S. national security planning. "The most obvious and most significant [threat to the United States] is the global challenge posed by the only nation that rivals us in military power—the Soviet Union," stated Harold Brown, secretary of defense during the Carter administration.[5] Discussing the American defense budget, Casper Weinberger, secretary of defense in the Reagan administration, observed, "It's the threat that makes the budget. You've got to build your budget on the Russian budget."[6] Similarly, the Soviet Union has been the target that has attracted U.S. intelligence resources for the most part.

In the past decade, however, an evolution has taken place in CIA targeting priorities, as the following in-house Agency memorandum suggests:

> The traditional idea of intelligence is the spy who provides the enemy's war plans. Actually, intelligence is concerned not only with war plans, but with all the external concerns of our government. It must deal with the pricing debates of OPEC [Organization of Petroleum Exporting Countries] and the size of this year's Soviet crop (and here our foreknowledge comes from CIA's pioneering in new analysis techniques). It is concerned with Soviet strength along the Sino-Soviet border, with the intricacies of Chinese politics, with the water supply in the Middle East, with the quality of Soviet computers and its impact on our own export controls, with the narcotics trade in Southeast Asia, even with the struggle for control of Portuguese Timor.[7]

The Soviet Union still attracts one-half of U.S. intelligence budget resources for collection, according to Gates, but the remaining half is broadly distributed.[8] The nations of Eastern Europe, North Korea, the People's Republic of China (PRC), and Cuba continue to be high on the list of so-called hard targets. Now, though, a host of other targets—less countries than topics—crowd in line for limited budget resources. Among them are debt financing in the developing countries and other economic questions, human rights, international energy shortages, science and technology, international terrorism, narcotics, illicit arms sales, agriculture, natural resources, immigration flows, global water and food supplies, population projections, and arms control.[9]

As the Carter administration turned beyond the traditional issues of East-West confrontation to address issues of North-South reconciliation between rich and poor nations, the policymakers' list of new intelligence needs grew rapidly. The policy community became increasingly dependent on CIA data for a wide range of subjects, with arms control a conspicuous example. The Strategic Arms Limitation Talks (SALT) were based almost entirely on a CIA data base. In arms-control negotiations, verifying compliance with agreements has become an extremely important problem and is dependent upon monitoring capabilities based on intelligence surveillance. As a result, CIA and other intelligence experts have even been drawn into the negotiations with the Soviets on this problem. This can be tricky. Monitoring—that is, finding out what in fact is going on, especially in circumstances when the Soviets are suspected of cheating—is a legitimate intelligence job; but ultimate judgments on verification—that is, deciding whether or not Soviet behavior violates an agreement—involves political and legal assessments arguably beyond the proper range of an intelligence agency.

On the question of technology transfers, too, policymakers have relied more and more on the intelligence community to verify compliance and to make judgments regarding what technology might be inadvisable to sell to potential adversaries. In such matters the CIA has tried to remove itself from a policy-making role, but clearly its support role is less ancillary in these two areas (as well as in some others) than has traditionally been the case. "The CIA is now a more important player," concludes Gates.[10]

The selection of intelligence targets ("setting requirements" or "tasking") is a job involving many hands, sometimes including the president himself. "When I became President I was concerned, during the first few months," said President Jimmy Carter in a press conference, "that quite often the intelligence community, itself, set its own priorities as a supplier of intelligence formation. I felt that the customers—the ones who receive the intelligence information, including the Defense Department, myself and others—ought to be the ones to say this is what we consider to be most important."[11]

One reason why the intelligence agencies have frequently set their own priorities stems from the difficulty of ascertaining what the policymakers want. Most policymakers understand little about the intelligence process; this fact, according to one experience CIA analyst, prevents them from articulating the proper questions. " 'What do I do about Nicaragua?' is not a good intelligence question," he observes.[12]

Various methods have been tried by the CIA to elicit more specific requirements from consumers. One initiated by CIA Director William E. Colby (1973–76) was the establishment of Key Intelligence Questions, or KIQs (pronounced "kicks"), a list of around 70 subjects commonly of interest to policymakers over the years.[13] The list was circulated to consumers, who were asked to check off or add items of current interest. This procedure proved less than satisfactory, however; policymakers and producers alike found it cumbersome and vague. A check next to "Cuba" or "Jamaica" was little better than "What do I do about Nicaragua?" The KIQs system was replaced during William J. Casey's tenure (1981–86) by a similar document called the National Intelligence Topics, this time listing 185 issues. Not surprisingly the NITs have run into the same criticisms leveled at the KIQs.

Occasionally, as President Carter's comment indicates, desired targets may be missing altogether from the CIA's shopping list. When the Reagan administration came to Washington, it too was unhappy with certain omissions. Casey, this administration's first CIA director, once recalled that the intelligence community was excessively country-oriented. Reports were available on Nicaragua, Honduras, and El Salvador, but "no one was looking at the regional interplay," and, "no one was concentrating on the economic component of these situations."[14] Moreover, in twenty years the CIA had produced only five major evaluations of the Soviet economy. Casey took steps to correct these perceived shortcomings and, in addition, created two new analysis centers on subjects of high interest to the Reagan administration: technology transfer and "insurgency and instability."

The tasking priorities of the intelligence community, then, continue to concentrate on the potential threat of the Soviet Union and other communist powers; but below these top targets the list is subject to frequent changes according to the fluctuating views of policymakers on what information is critical to the protection of U.S. national security. Zaire, for instance, is of somewhat less interest now to decision makers than it once was because the advent of plastic pipe has diminished U.S. dependence on copper tubing.

The wellspring of suggestions for what to target never seems to run dry as consumers from throughout the executive branch continue to add to it. "Often the requirement has more to do with the last item that popped into the policymaker's in-box or some harebrained idea he concocted while shaving that morning than with any well conceived intelligence problem," DDI Gates once remarked.[15] Even members of Congress have their recommendations, but here the intelligence community draws the line. "We don't seek input from the Hill on collection requirements, but sometimes we get it anyway," Gates added ruefully. "We definitely don't want 535 new taskers; serving the executive branch is tough enough."[16] Often the issue of debate between consumer and producer is less about the proper targets of intelligence operations than the types of intelligence to be produced from the collected data.

TYPES OF INTELLIGENCE Ideally, policymakers might like to know what has occurred in, say, Monrovia in the past two hours, or what is apt to occur in Nicaragua over the next year. In the first instance, the policymaker desires current intelligence—up-to-the-minute information often vital in times of crisis; in the second instance, he or she would want more thoroughly researched and analyzed, longer-range predictions—known as "estimates" in the intelligence business. And, in a third instance, the policymaker might require a truly in-depth study on a specific subject, perhaps a definitive examination of Soviet oil reserves. Respectively, these three categories are often referred to as *current, predictive,* and *research* (sometimes called *basic*) intelligence. They differ from one another in a number of ways, including closeness in time to actual events, degree of speculation, and richness of detail. The current ("daily" might be a more accurate term) intelligence reports obviously place a premium on providing a window for policymakers to see—as instantaneously as possible—what America's foreign adversaries are doing. Especially important is "indications and warning" (I and W), that is, intelligence that detects and provides an early warning of impending threats. While this quest for dynamic, "real time" information is hardly simple, more difficult still is the challenge of long-range prediction—what these adversaries intend to do down the road. "What will the military in El Salvador be like in ten years?" is an example of one predictive intelligence project prepared recently.

Intellectually most satisfying to the intelligence analyst is the full-blown research project, for it allows more room for tracing nuances—the analyst's delight (though seldom what the policymaker wants to wade through). Often the press of time or the waning attention span of policymakers relegates this art form to the status of second-class citizenship as current intelligence and estimates move to the fore.

Under the Reagan administration, the Intelligence Directorate established a research program that emphasized two objectives. First, each of the ten offices within the directorate (see fig. 3.2) was expected to redouble its efforts to produce solid research papers. In one recent nine-month period, over nine hundred such reports were in progress, including one on the state of computer technology in the PRC and another on the beliefs of various Muslim factions in the Middle East. Second, all the offices were expected to contribute toward five major research endeavors of special interest to policymakers in the Reagan administration: the Soviet military-industrial complex, penetration of the Third World by the PRC, East-West technology trade, foreign intelligence threats to the United States, and the dangers of nuclear proliferation. The Intelligence Directorate is typically the most boastful of its research intelligence; this form of reporting is the most reliable and, claims a DDI, "the CIA is the only place where long-range research on national security issues is being done—two, five, ten, twenty years down the road. The Department of State Policy Planning Staff conducts this kind of research only sporadically."[17]

While current, predictive, and research findings are the most common packages of intelligence sought by consumers, others exist. Sometimes a policymaker will simply ask for a brief response to a query in the form of a typed memorandum. In one page, what can you tell us about the Chinese coal industry? About the West German Chancellor's agenda for his approaching visit to the United States? On other occasions, the consumer will prefer a quick oral briefing, perhaps on the run between meetings or in a limousine on the way to the airport (practices discouraged by security personnel for fear of eavesdropping by foreign intelligence services). "Policymakers will accept fifteen-minute briefings over the investment of five minutes to read a report," notes a recent DDI.[18]

Policymakers frequently come from a political background. They are often less comfortable reading than talking and listening; after all, it was these oral skills that led to their success in the political milieu. Regardless of what form they may want the collected intelligence to assume, the policymakers usually want it *now,* want it perfectly accurate, and want it without too many nuances—demands that often defy human capabilities and protective instincts, as well as technological feasibility.

GATHERING INTELLIGENCE The intelligence community relies upon three approaches to the gathering of information vital to America's national security. The first is overt collection, and the other two, already familiar to the reader from earlier chapters, are forms of covert collection: classical espionage using spies (HUMINT) and modern technical surveillance (TECHINT). Harry Howe Ransom estimates that "80 per cent or more of intelligence material in peacetime is overtly collected from nonsecret sources such as newspapers, libraries, radio broadcasts, business and industrial reports, or from accredited foreign service officers. . . ."[19] To supplement these "open sources" of information, the intelligence agencies attempt to gain access to ("penetrate") Kremlin safes, Beijing high-council meetings, secret sessions of the Palestine Liberation Organization (PLO), OPEC negotiations, and a hundred other inner sanctums around the world where the eyes and ears of U.S. officials are unwelcome.

To assist the efforts of analysts as they sift through open sources of intelligence, the CIA maintains vast library holdings. The Agency library contains practically every publication openly available from around the world. These newspapers, reports, studies, periodicals, and other documents are augmented by extensive computer printouts available on the personalities of political and military figures from every nation, along with many other computerized reference services. (Reportedly half the floor space at CIA headquarters is now occupied by computers.[20]) The Agency's Foreign Broadcast Information Service (FBIS) monitors global radio and television emissions, which are translated and circulated to the offices of country analysts. The problem posed by open sources is less one of having too little data than of sorting out what is truly useful from the blizzard of information that descends daily.

To speak of the covert collection of intelligence is to broach the controversial debate over HUMINT versus TECHINT. While overt collection provides the bulk of information gather by the CIA (and the KGB), covert HUMINT and TECHINT often unearth the most important knowledge for decision making. Soviet HUMINT directed against the British reaped a rich harvest of intelligence as a result of the successful KGB recruitment of highly placed British intelligence officers, including Guy Burgess, Donald Maclean, Harold "Kim" Philby, Anthony Blunt, and others.[21] The British and Americans have had their HUMINT successes, too, as with Col. Oleg Penkovsky of Soviet military intelligence (GRU), who provided the West with thousands of documents on Soviet missilry.[22]

The controversy enters with attempts to weigh the cost-effectiveness of HUMINT. For each Penkovsky, the CIA may have hundreds of agents who produce little more than snippets of gossip. Often HUMINT reports are simply unreliable. Prior to the Cuban missile crisis of 1962, intelligence from CIA agents in Cuba proved notoriously inaccurate; of 200 reports of missile sightings, only 6 proved accurate.[23] Then there is always the possibility that one's agent is actually a double agent through whom a deception operation is being conducted by the adversary—the "wilderness-of-mirrors" problem confronted by CI specialists.

Yet what if the United States had been able to recruit a Japanese embassy official in Washington, D.C., in November 1941? What if the United States had worked harder at placing agents in the religious movement led by the Ayatollah Ruhollah Khomeini before the revolution of 1979 in Iran? What if someone burrowed deep within the Kremlin (a "mole") were secretly working for the CIA?

Such "what ifs" assure that monies will continue to be available for HUM-

INT, though how much funding to provide remains a topic of heated debate within the intelligence community. A mole in the Kremlin would be handy (though unlikely), but how much money should the United States devote to HUMINT in Africa? Grenada? A cell of the terrorist group, the Red Brigade, in Italy?

The costliest phase of the intelligence cycle is collection (especially technical), and budgeteers from the various intelligence agencies engage constantly in a tug-of-war over the distribution of these large dollars. The end result of the budget battles over the years has been for HUMINT to come up with the short straw when compared to TECHINT (with all of its expensive hardware) by a ratio of about one to seven.[24] "Is this the right balance?" asks one experienced intelligence analyst rhetorically. "We don't know for sure; we're not that smart."[25] Nobody is. HUMINT defies evaluation by normal budgetary criteria, for one Penkovsky makes the deficits column—however lengthy—suddenly appear insignificant.

As the one-to-seven ratio indicates, however, the Penkovsky argument is difficult to sell in budget councils. As researcher after researcher has discovered, the people who make decisions about national security priorities prefer "to concentrate on things that can be 'scientifically' measured."[26] Just as counting missiles and—the favorite—warheads becomes the fascination of strategic planners, so are intelligence planners drawn to the counting of mechanical devices for surveillance, with their glittering, awesome technology. A reconnaisance spy plane is something tangible—it takes pictures, and its appurtenances can be demonstrated with slides before legislators on the appropriations subcommittees in closed session. Moreover, as William E. Burrows has noted, "No reconnaisance camera has ever lied for purposes of expediency or because it worked for the opposition, had a lapse of memory, or became confused."[27] In a word, the hardware is impressive and promises immediate, less ambiguous results (cloud cover willing).

In contrast, the nameless spy, who for security reasons can never be discussed explicitly in budget meetings (indeed, his true identity will be know only to his CIA supervisor—case officer—and perhaps one or two others at Headquarters), may, or may never, obtain information of great significance. Moreover, the giant aerospace corporations and laboratories persistently lobby for surveillance-hardware contracts, with skills finely honed through experience in weapons contracting with the Department of Defense. HUMINT has no such external advocate and, despite occasional coups like Penkovsky, remains the stepchild of the collection process.

Adm. Stansfield Turner, DCI during the Carter administration, found himself in 1978 at the center of the debate over the proper balance between HUMINT and TECHINT—and, for the first time ever, with the public involvement of a president. The impetus was the Iranian crisis. In mid-August of 1978, the CIA reported to President Carter, "Iran is not in a revolutionary or even prerevolutionary situation."[28] That appraisal soon proved about as accurate as the 1941 declaration of William Borah (R, Idaho), chairman of the Senate Foreign Relation Committee, that world war would be forestalled. The Shah of Iran came tumbling down, and the president wished to know why. At a November 30 press conference, he pondered whether the deficiency resulted from insufficient attention to HUMINT: "I have been concerned that the trend that was established fifteen years ago to get intelligence from electronic means might have been overemphasized."[29]

Turner, once the president's classmate at the Naval Academy, was—on the surface—vulnerable to the charge because one of his first acts as director was to relieve from duty 805 officers in the Operations Directorate, home of HUMINT in the Agency. This move, which he made with little tact, created sharp acrimony within the CIA. The view was widely held within Washington that Turner was a hardware man, dedicated to technological spying over more ancient forms.

Turner, however, denied these charges, claiming that the Operations Directorate had been "pared down and streamlined" only because it was "overstaffed." He sought "efficiency" and would allow "no meaningful reduction in overseas strength or activities."[30] Stressing the importance of HUMINT, Turner continued: "Only human collectors can gain access to motives, to intentions, to thoughts and plans. They will always be vital to our country's security." Privately, before the congressional oversight committees, he made a similar case for HUMINT; but, like his predecessors, he advocated the usual imbalance of spending between the two covert means of collection. Further, his occasional briefings on collection focused far more on the latest technological achievements in aerial surveillance—some of it quite astounding in its technical proficiency—than on the less flashy (and less discussable) matter of HUMINT placements abroad.[31]

The attraction to intelligence hardware goes far beyond ease of counting, measurability of success, or industry lobbying. The overmastering fact is that this hardware has revolutionized the collection task—and has made the world an infinitely safer place to live. According to intelligence experts, the fact that the United States can now monitor SALT through aerial and electronic surveillance, though enormously expensive, has alone been well worth the cost of the new technology. These advances in remote surveillance allow this country to know, much more immediately than before, the capabilities and movements of its adversaries. The equipment cannot see through roofs or fathom intentions, for which HUMINT will remain indispensable, but it can spot military movements and missile installations from afar.[32] Continued access to this invaluable data strongly encourages the lopsided spending ratio favoring technical collection.

The actual methods of covert collection involve, for HUMINT, the case officer and his stable of agents and, for TECHINT, the scientist and technical operator with their equipment. As the reader knows from chapter 3, within the CIA are two types of subdivision, geographic (as in the European subdivision of the Operations and the Intelligence Directorates), and functional (as in the Office of Scientific and Weapons Research in the Intelligence Directorate.) Among the Operation Directorate's tasks (covert action being the most controversial) is the responsibility to conduct HUMINT abroad through each of its geographic subdivisions in order to gather information for geographic and functional analysts in the Intelligence Directorate. Overseas, then, are CIA officers (case officers) whose job it is to recruit local nationals to spy on their own country (just as other nations have this goal within the United States).

This process of recruiting agents is, as one can imagine, difficult and delicate.

Its primary phases, as introduced briefly in chapter 2, include spotting and assessing, gaining access, recruiting, and training. This sequence can be time-consuming, even lasting several years and then often failing; but sometimes it may lead to the recruitment of a sitting or potential prime minister.

Spotting and assessing involves finding a would-be agent, someone who might defect and who has good information—or ties to those who do. One West European nation has almost five hundred Soviet and PRC personnel—one big happy hunting ground for double agents and defectors. Gaining access in order to attempt a recruitment, say, of a young KGB officer in Bonn ("Boris") can be awkward. With some Asian diplomatic delegations, access may be close to impossible because they move from place to place in groups and are rarely alone. Perhaps "John" (the CIA officer who is pretending, let us say, to be an American businessman) can join Boris's club and ask him for a game of tennis, which may lead to a drink in the clubhouse, an agreement for another game next weekend, and subsequent meetings.

One day, John begins an approach to Boris, as subtly as possible at first. ("My wife and I have been thinking about taking a vacation next month. Do you have any brochures about holidays on the Black Sea?") Later (perhaps months, even a year or more after further incubation of the relationship) comes "the pitch," the actual espionage recruitment. ("Listen, Boris, old man, can you obtain documents for me about troop placements in Poland? I can make it worth your while.") Here is the moment of truth—the "publish-or-perish" challenge of the case officer. If he has done his job and is lucky, the answer may be yes. Inducements might include money, ideology, sex, and other human incentives. Boris may have fallen in love with an American women and will spy for the CIA for two years in exchange for a new life under a fresh identity in the United States. Perhaps Boris's child is ill and needs treatment at the Washington Children's Hospital. Possibly Boris wants his daughter educated at a prestigious American university. Maybe he was denied membership in the Communist party and is now bitter and contemplates revenge.[33]

Though hard figures are unavailable, interviews with former CIA case officers suggest that most of the time the pitch results in a flat rejection by the potential recruit. Sometimes, though, the answer is yes. Then Boris must be trained in the arts of acquiring what the CIA wants and communicating this information secretly—dangerous endeavors requiring much skill. Enter here the tradecraft of invisible inks, microdots, disguises, electronic burst transmissions, concealment devises, dead drops, minature cameras (one CIA camera is as small as the last joint on your thumb and can shoot ninety frames), and the rest—the preserve of the Directorate of Science and Technology (DS&T), which provides the high-tech communications gadgetry and other paraphernalia now a part of modern espionage.

The DS&T works closely with scientists in the private sector to devise equipment for covert technical collection. Here is the world of high-speed, high-altitude reconnaissance planes, sophisticated satellites, long-range sensor devices, and a plethora of other dazzling machines that look and listen without being seen or heard.

A major technological breakthrough for intelligence came in the 1950s, with the development of improved film, panoramic cameras, sky-to-earth lenses, computerized grinding of lenses for enhanced resolution, and—the centerpiece—the development of the U-2 reconnaissance aircraft. The CIA initiated the U-2 program when it became clear that the United States could no longer easily penetrate Soviet air defenses. The design began to take shape in November 1954 and, remarkably, the aircraft took its initial flight nine months later. In July 1956 it made its first reconnaissance over the Soviet Union. Thanks to the U-2, the United States knew of the Soviet missiles in Cuba before they became operational, and learned also that the "bomber gap" between the United States and the Soviet Union was nonexistent. Here were new eyes to see the truth where once speculation and fear had reigned.

In May 1960, Soviet air defenses shot down U-2 pilot Francis Gary Powers, ending such flights over the USSR. America had lost its past method for monitoring the development of the Soviet intercontinental ballistic missile (ICBM) program. The CIA was forced to base its projections of Soviet missile advances on hunches about their capabilities rather than on hard photographic evidence. To be on the safe side, the hunches were high, and thus was born the "missile gap."[34]

By 1960, however, the CIA and the Air Force together had placed in the sky the first successful photographic satellite; by 1962 a fully operational model circled the earth, providing reliable information about the Soviet ICBM program.[35] American intelligence had achieved a new level of confidence and accuracy. "From that time onward," notes a CIA briefing paper, "every new Soviet ICBM system has been detected, and its characteristics determined, well before it was deployed. The number of missiles operational are well known within a narrow range."[36]

To supplement U.S. surveillance capabilities, the CIA had also begun work on an aircraft able to fly higher than the U-2, for it was already clear before 1960 that Soviet surface-to-air missiles (SAM-2s) would soon pose a threat. Such a plane came off the assembly line a decade before anticipated and was to the U-2 as the U-2 had been to the Piper Cub. It successfully flew missions over North Vietnam beyond the range of SAM defenses and at speeds of Mach 3.2. The current version of this surveillance craft is the SR-71. Soon to be airborne is a new spy plane, transatmospheric and hypersonic, with a speed that can carry it from Los Angeles to New York in twelve minutes![37]

The CIA has probed the depths of the earth's seas as well as its outer rim. The *Glomer Explorer,* a deep-sea mining vessel constructed by eccentric billionaire Howard Hughes, was redesigned under CIA direction in the mid-1970s for the purpose of recovering a sunken Soviet submarine. Under this mining cover, part of the submarine was retrieved from the incredible depths of 16,000 feet—over three miles, an engineering first. Although the submarine was old, it provided a benchmark for projecting what the state of Soviet technology may now be like (see also chap. 9).[38]

Satellites and aircraft have their limits (half the world remains covered with clouds at any moment and, of course, darkness is also a hindrance.), and *Glomar* missions are unlikely to succeed often. Coupled with HUMINT collection, however, America's capabilities for knowing about the world have increased dramatically. Surprises will continue to occur, for neither Americans nor Russians are all-knowing; still, surprises are now fewer.

Processing

This next step in the intelligence cycle is the least disputatious. Here the collected information undergoes various refinements for closer study by analysts. For example, coded data are decrypted, foreign languages translated, and photographic material interpreted to make the information readable and understandable for the analyst. "The major problem during the processing stage is information overload," states a CIA official. "Fortunately, the use of computers offsets this to a large degree."[39]

The processing procedures can nevertheless be time-consuming. A professional photointerpreter may require four hours to decipher fully a single frame of satellite photography. The art of cratetology, interpreting the often obscure markings on crates aboard ocean freighters, can also require hours of careful study by experts. In November 1984, the cratelogists at the CIA failed to agree over what the markings indicated on boxes unloaded from a Soviet cargo vessel in Nicaragua: the presence of MIG 21s, antiaircraft missiles, attack helicopters, or something more benign. A reliable HUMINT asset on board ship would have been valuable.

After processing comes the central purpose of the entire cycle: the conversion of raw information into usable intelligence. Here is the marriage of data, overtly and covertly gained, with thoughtful assessment—in a word, analysis.

Production and Analysis

While some policymakers, like Henry Kissinger when he was secretary of state, often prefer to have "raw" (unevaluated) information brought to them directly so they can perform their own analysis, most want the intelligence experts to make some sense of the data, to spell out its possible meanings. They want "finished" intelligence. After examining the finished evaluation, if time allows, policymakers may wish to see the raw supporting evidence to check the reasonableness of the analytic judgments on important matters.

The individuals responsible for the conversion of raw information into intelligence are the analysts—highly educated men and women who are experts on a single country such as Zaire, or on broad topics, such as the flow of petrodollars, or on quite narrow topics, such as the efficiency of Soviet rocket fuels. Within the CIA are several hundred specialized analysts with these skills. Each, ideally, has the characteristics advocated by Sherman Kent, for years the chief analyst in the Agency: "the best in professional training, the highest intellectual integrity, and a very large amount of worldly wisdom."[40]

The forms into which analysts craft their finished products are protean, depending upon whether policymakers prefer written reports or oral briefings and whether they are interested in current, predictive, or research intelligence. President Carter, for example, was a voluminous reader; Lyndon Johnson preferred conversation. Analysts must be equipped to present their judgments through either format. Sometimes they are asked to travel with policymakers, presenting peripatetic evaluations along the way. (I examine these approaches more fully later, when I address the dissemination phase of the intelligence cycle.)

While some policymakers thrive on the freshness of an oral briefing—short-circuiting the delay inherent in written preparations—most finished intelligence comes from the producer to the consumer in written form. The essence of the production-and-analysis phase is to lay out the usually fragmentary and inconclusive evidence gathered by the collectors and processors, study it, and write it up in short reports or long studies that meaningfully synthesize and interpret the findings.

For current intelligence, the CIA produces (in coordination with the wider intelligence community) a number of daily documents, among which the President's Daily Brief (PDB) is preeminent. The PDB is the most tightly held document in Washington; nothing has ever leaked from it, according to senior intelligence officials.[41] It arrives early in the morning, via CIA courier, and is placed on the desktops of five individuals: the president, the vice president, the secretaries of state and defense, and the president's assistant for national security affairs (who directs the NSC staff). This top-secret "newspaper" briefs these officials on major political developments throughout the world that have occurred in the previous twenty-four hours; it often sets the agenda for early morning discussions among themselves and with aides.

According to a top NSC aide during the Carter years, Zbigniew Brzezinski, Carter's assistant for national security affairs, not only used the PDB "as a checklist to discuss possible trouble spots with Carter, but would often use that document as a way to engage the President in discussion about an issue, like a development in the Mideast." From this discussion might come a decision to call a meeting of the NSC (or, more likely, one of its subcommittees) to review the policy issue further and form a recommendation for the president. "So the PDB," continues the NSC staffer, "was not just an end in itself, but often the catalyst for further action."[42]

Under the Ford presidency, the PDB was quite lengthy. President Carter had it reduced in size to a maximum of about fifteen pages, though occasionally DCI Turner included longer pieces tracing specific global trends at the end of the daily report (say, on the status of OPEC oil pricing). Carter ordered Turner to make the PDB a kind of inventory of those items the DCI thought the president should know each morning before he started his day. Carter often wrote in the margins of the PDB, asking Turner for more information. As the NSC staffer cited above recalls: "It kept the CIA boys hopping, but—most importantly—it let them know what was of interest at any given time to the President."

The PDB is supplemented with other daily reports, similarly designed for limited distribution. The Situation Room in the White House, the president's communications center, pulls together up-to-the-minute intelligence materials (mostly cables and reports from the NSA), summarizes them, and forwards them to the national security adviser (as the assistant for national security affairs is less formally known) for possible presentation to the president. Usually a few of these items make it to the Oval Office. Also, the Department of State's morning intelligence report (the Morning Summary), prepared for the secretary of state and his top assistants by INR, goes to the national security adviser and, almost always, on to the president.

A document similar in format to the PDB, resembling an offset tabloid newspaper and stamped "top-secret" (though with some of the ultrasensitive PDB material excised), is the National Intelligence Daily (NID), with a distribution of about 500. Of this number, 125 are recepients outside the CIA: assistant secretaries and above for the most part.[43] Until January 1976, two committees of Congress also received the NID: the Senate Foreign Relations and Armed Services committees, which kept copies of the document securely ensconed in their respective safes during the day, with limited access to members and only two staffers; at day's end, the document went back to the CIA.

This security system failed to impress Colby, who was DCI at the time. Citing "massive disclosures of classified information by Congress" in a letter to the chairman of the Foreign Relations Committee, Colby tried to justify the termination of the NID product to Congress—even though the alleged disclosures were about covert action, not the data analysis found in the NID.[44] Protests from members of the committee reinstated the NID distribution, though the incident seems to have been something of a tempest in a teapot since, according to a senior Senate official, "with rare exception, none of the members were ever interested in reading the Daily."[45] The NID now comes to the two new intelligence committees on Capitol Hill; again, the members show little direct interest, but do expect key staffers on the committees to monitor the document on their behalf.[46]

The Pentagon also publishes a daily intelligence brief of limited distribution, called the Defense Intelligence Summary (DINSUM), prepared by the Defense Intelligence Agency (DIA). A host of weekly and monthly reports are issued, as well, to feed the appetites of policymakers (or at least their staffs) for current intelligence, prepared by sundry functional and geographic divisions of agencies throughout the intelligence community. Among them is the *National Intelligence Survey* (NIS), containing detailed information on practically every country in the world—from political party structures to transportation networks. An abbreviated form of the NIS is the *National Basic Intelligence Factbook,* prepared in classified and unclassified forms. Often senior policymakers request, on an ad hoc basis, brief responses to specific topics. Here the analysis takes the form of the typescript memorandum—informal "quickies" on everything from the health of Ferdinand Marcos to rescheduling Brazil's international debt.

Harder to prepare in a satisfying way are the longer-range reports that go beyond current intelligence to cover predictive and research intelligence. The CIA regularly receives high marks for quickly assembling data during a crisis ("crisis management") and for having extensive maps, economic data, and personality profiles on tap; but its marks have been lower for forecasting or "estimating" what is apt to happen down the road. Until the invention of a crystal ball, this disparity will remain inevitable. Given the absence of reliable human clairvoyance, the

CIA's record in the area of long-range intelligence has been, on balance, remarkably good (as discussed later).

The key documents for this production task have been, for short-range projections, the Interagency Intelligence Memoranda (IIM) and for mid- to long-range projections, the showpiece National Intelligence Estimates (NIEs), prepared routinely on different parts of the world, and the Special National Intelligence Estimate (SNIEs), prepared in response to specific subjects requested by senior policymakers. Lyman Kirkpatrick, former executive director of the CIA, defines the NIE as "a statement of what is going to happen in any country, in any area, in any given situation, and as far as possible into the future."[47]

The NIE, then, is a vital report that assesses the current situation in some part of the world and often makes a prediction about future developments. An NIE can be an indispensable link to reality for a president—"the single most influential document in national security policy making, potentially at least," concludes Ransom.[48]

An NIE attempts to set down, and often rank, possible outcomes that might threaten or present opportunities for the United States. The principle purpose is to assess trends—what is presently going on in Patagonia—and then, if possible, to predict what will happen (the rebellion will occur at 2:00 A.M. next Tuesday, Patagonian time). In an NIE, emphasis is placed on the most likely outcome, but in recent years, this best shrewd guess has been augmented with a listing of several possible alternative hypotheses. Moreover, an NIE is expected to suggest explicitly what opportunities may lied down the road for American foreign policy. The best NIEs are accurate, timely, and dispassionate—the final result of careful combing and evaluating data drawn from throughout the intelligence community.

The estimating process begins when a senior policymaker—from the president down to a ranking departmental official—makes a formal request, usually through the NSC, for an appraisal and prognosis of events in another nation. The request is examined by a panel of intelligence experts within the CIA (the name and composition of which have undergone several permutations), in consultation with intelligence analysts throughout the community as well as with other senior policymakers. (An exception to this rule occurs with the special "net estimate," a highly classified, closely held paper comparing, say, U.S. and USSR military capabilities and intentions [not just presenting capabilities]; such estimates are handled directly by NSC and [chiefly] Pentagon staffers.[49]) The panel decides whether to proceed on a course of action and, if so, devises a plan on how to proceed. The communitywide offices that would be involved are designated at this stage, and they are provided with an outline of objectives expected of them. These contributions would subsequently come back to the CIA, where they would be fashioned into a draft NIE prepared by a group of senior analysts in the Agency known, since 1973, as National Intelligence Officers (NIOs).

When this system was first instituted, the eleven NIOs were given one assistant and a secretary and charged with the responsibility for working with interagency intelligence specialists to hammer out NIE drafts. Though the emphasis changes now and then, the set of NIOs in 1975 had these responsibilities:

1. USSR and Eastern Europe
2. Western Europe
3. China
4. Japan and the Pacific
5. South and Southeast Asia
6. Middle East
7. Latin America
8. Strategic Programs
9. Conventional Forces
10. Economics and Energy
11. Special Activities

Since 1980, the NIO team (now sixteen strong) has been expanded to include more generalists—some from outside the CIA—and more assistants. The use of experts outside the government, employed in a limited fashion since the beginning of the NIO system, has also been expanded in recent years. This opening up of the analytic side of intelligence to citizen-experts represents a democratizing influence on the CIA, allowing at least a limited circulation of ideas from the broader society into these hidden hallways.

The NIOs make the final judgments on the appropriateness of all the findings and conclusions presented in each NIE, from the soundness of data and hypotheses to the wisdom of the forecast, but the actual initial drafting rests largely on the shoulders of the various analysts picked by the NIOs to work on the project. According to one NIO, he and his colleagues were expected to work closely with the NSC and other consumers, "scrubbing information honestly and adapting to the working style of the policymakers receiving the NIE," and to keep in touch with the various intelligence entities whose analysts were indispensable contributors to NIE production. This NIO considers the NIE an "art form" that "requires a corps of floating linebackers, flexible and easily collapsible, to charge an intelligence problem—that is our job."[50]

That job has hardly been trouble-free, even beyond the difficulty of predicting the future of the world. As Stephen J. Flanagan notes, the role of the NIO as liaison between the intelligence community and the policymaker has led to some politicizing of the analytic process as NIOs become too involved personally with policy positions. The NIE has also evolved into too much of a CIA effort, without sufficient attention to interagency coordination. Further the NIOs have failed to work as a team, becoming more free safeties than supporting linebackers. They have tended to concentrate more on responses to crises than to long-range research, and, most important, the NIEs under this system have proved to be fewer in number and slower in coming (at times taking years!), and have sometimes been of questionable quality. In Flanagan's view, "the net effect of these production problems was a further erosion of the influence of NIEs, IIMs, and other art forms; particularly short typescript memoranda prepared by the NIOs in response to specific requests from policymakers, proliferated as substitutes for more traditional reporting."[51] Directors Turner and Casey took useful steps to improve the

NIE process (Casey was especially instrumental in increasing the production of NIEs; only five went to the Carter White House, but some fifty-five went to Reagan by 1988), but many of those problems have stubbornly resisted correction.

The proof of the pudding, it is said, lies in the taste. For estimates, the taste has been inconsistent, but more often sweet than sour. Among the sour, or mistaken, estimates: the failure to predict, prior to 1962, the placement of Soviet offensive missiles in Cuba (although CIA Director John McCone disagreed with the faulty NIE and told the White house so[52]); the misconceived missile gap in the early 1960s; and, among other examples, the failures to forecast Soviet invasions into Hungary in 1956 and Czechoslovakia in 1968, the Arab-Israeli war in 1973, and the fall of the Shah in 1979. Sen. Jesse Helms (R, North Carolina) has been especially critical of CIA analytic shortcomings and has offered his lengthy public list, which includes "underestimation of Soviet submarine capabilities."[53]

The CIA, though, has had some impressive successes. The list its officials like to display includes predictions of the following:[54]

- the Soviet Sputnik (1957)
- the Sino-Soviet split (1962)
- the Chinese A-bomb (1964)
- new Soviet strategic weapons systems over the years (though precise numbers have been off somewhat)
- Vietnam War developments (1966–75)
- the Arab-Israeli War (1967)
- the Soviet Antiballistic Missile (ABM) system (1968)
- the India-Pakistan War (1971)
- the Turkish invasion of Cyprus (1974)
- the Chinese invasion of Vietnam (1978)
- the mass exodus from Cuba (1978)
- the Soviet invasion of Afghanistan (1979)

Moreover, the CIA points to its ability to monitor the SALT accords, the flow of petrodollars, OPEC investment strategies, and the rise of political leaders around the globe. Some predictions will forever remain impossible to verify or refute. The CIA, for instance, publicly projected a sharp decline in Soviet oil production; though this decline failed to happen, the reason may be a concerted effort by the Soviets to prove the estimate wrong by changing their production plans.

One conclusion, nonetheless, is certain: the intelligence community—like other organizations—will never escape failure. As Richard Betts has emphasized in his research on the constraints of situational ambiguity and personal misperception in the intelligence process, "some incidence of failure [is] inevitable." He urges a higher "tolerance for disaster."[55] In 1975, Sen. Frank Church summed up his inquiry into the estimates process with these words: "The CIA Directorate of Science and Technology has not yet developed a crystal ball. Predicting the future must remain probabilistic. Though the CIA did give an exact warning of the date when Turkey would invade Cyprus, such precision will be rare. Simply too many unpredictable factors enter into most situations. The intrinsic element of caprice in the affairs of men and nations is the hair shirt of the intelligence estimator."[56]

The production-and-analysis phase of the intelligence cycle, then, is more replete with pitfalls than might seem to be the case at first glance. One might think that the final phase, dissemination, would be child's play once the challenge of analysis has been met. On the contrary, the dissemination problem has proven to be among the most intractable.

Dissemination

"Being right isn't enough," stresses Donald Gregg, a CIA aide on the NSC staff and a long-time intelligence professional. "You have to inject that 'rightness' into the policy process. Unless a particular concern is actually raised at the denouncement in one of the key meetings, say, the NSC, the good analysis done by the intelligence community is often lost. The chip must be put in the pot at this stage—and forcefully."[57]

As Gregg's observation reminds us, it is neither preordained nor automatic that good intelligence will be heard and accepted in high places. Recall that a central explanation for the successful Japanese surprise at Pearl Harbor was the faulty dissemination of available intelligence regarding the possibility of an attack. Recall as well the examples from chapter 4 of top officials refusing to accept CIA intelligence reports, or, occasionally, the DCI himself failing to transmit intelligence.

The key to success for the CIA during the dissemination phase of the intelligence cycle may be summarized, according to experienced analysts and managers, in one word: dialogue. (Indeed, this interaction between producer and consumer is crucial throughout the cycle for, to reemphasize the distinction made at the beginning of this chapter between a matrix and a cycle, "collection never stops, analysis never stops; intelligence is not neat or finite, but more of a continuum wherein incremental changes take place based on a series of feedback loops between analysts and policymakers."[58]) Senior intelligence officers and senior policymakers must talk to each other if this transfer of intelligence is to work: "Not just every six months," points out DDCIA Gates, "but much more constantly, every two weeks at least, and usually more often."[59] One goal of this interaction, from the CIA viewpoint, is to find out—beyond NITs and other formal procedures—what the policymakers' priorities are by actually working with them.

This producer-consumer relationship can be as fragile and unpredictable as it is important. It can sour faster than cheap wine. For one thing, intelligence frequently challenges policy views. Policymakers—often political appointees fresh from national presidential campaigns—may have a strong sense of mission, an overabundance of zeal, perhaps an ideological axe to grind. Usually they are imbued with an undue optimism about what their administration can do to change the world. The hard, cold facts of intelligence and the nuances of the analytic estimate can be as welcome as the proverbial skunk at a lawn party. Indeed, according to Gates, the toughest assignment for the CIA is "to give honest assessments and still keep the policymaker reading."[60]

A former senior CIA analyst draws this conclusion: "Many of the consumers are ideologues. It is hard to work with them—especially when we're usually dealing with highly ambiguous data. When consumers criticize our product, it is often on grounds that it fails to support their suppositions."[61] Facts rarely speak for themselves; they have to be interpreted. "Intelligence may be defined as the subjective evaluation of ambiguous data," suggests this same analyst.[62] Even in the case of the Cuban missile crisis, Secretary of Defense Robert S. McNamara had the same facts as other key policymakers, but he argued that the Soviet missiles did not matter that much; the United States was already under the threat of Soviet ICBMs.[63] His colleagues disagreed; the closeness of these medium-range ballistic missiles (MRBMs) did matter to them, if only psychologically. Interpretation, especially when colored by ideology, can lead to sharp conflicts between analyst and consumer.

Referring to the involvement of the Reagan administration in Lebanon in 1984, which cost the lives of almost three hundred Marines, one seasoned CIA officer, Donald Gregg, recalls: "The policymakers screwed up. They wanted to counter Soviet influence in the region. We told them that Syria was not a puppet of the Soviet Union, but they refused to listen. Our country was lured up a side alley and got mugged."[64] Another analyst remembers being told by Secretary of State George P. Shultz that "the product on Nicaragua is trash." It was the analyst's impression that Shultz felt this way because the interpretation of the data failed to agree with the secretary's worldview of Nicaragua as a Soviet puppet and aggressive exporter of arms to Marxist guerrillas throughout Central America.[65]

Sometimes the policymaker wants more than the evidence will support. "Analysts don't tell more than they know," observes one old hand within the Intelligence Directorate, "whereas the policymakers often want unfounded speculation—if it agrees with them."[66] "The Reagan Administration wants to believe that the assassination plot in 1984 against Pope John Paul II was concocted in Moscow and directed from Sofia," illustrates Gregg, "but the intelligence we have pieced together just doesn't support this."[67]

On other occasions, the policymaker may like the intelligence provided by the CIA because it agrees with his predispositions or political hopes. Yet according to Gates, this happy situation can be short-lived: "We always get new information, and we may have to change our original assessments."[68] In the meantime, the policymaker may have already delivered a widely reported speech or press comment based on the first evaluation. President Carter spoke strongly in favor of pulling U.S. troops out of South Korea in the early days of his administration; he had to reverse his position, Gates remembers, when the intelligence community began to report increases in the number of North Korean fighting forces.

Part of the potential for irritation between producer and consumer comes from the latter's inevitably filled—overflowing—plate. The policymaker has enough problems already, thank you, yet here comes the intelligence officer with another pail full. Moreover, Congress is now very much in the dissemination phase of the cycle, and this has exacerbated relations between analysts and policymakers within the executive branch. In the CIA's carefree days from 1947 to 1974, when it was relatively free from congressional monitoring, legislators asked for little information and directors offered little. The "intelligence wars"[69] between the branches from 1975 to 1980 changed all that; now CIA directors face stiff reporting requirements from Congress, whose members demand a greater share of the intelligence product.

"Few intelligence documents went to the Hill before 1976," Gates observes. "Today only a tiny percentage of intelligence that goes to the executive doesn't go also to the two Intelligence Committees of Congress."[70] In 1983, the Intelligence Directorate alone gave five hundred briefings on Capitol Hill,[71] not to mention the many times the staff of the Operations Directorate and senior CIA managers responded to individual legislative inquiries. All this provides the Congress—recently flexing its muscles in foreign affairs—with ammunition for the critique of executive-branch programs, a favor unlikely to endear the CIA to administration policymakers. These days during hearings one is likely to find a staffer whispering to a member of Congress: "That testimony is wrong, and you can cite CIA document such-and-such."

And, of course, the CIA and other agencies are sometimes just plain wrong in their reports to policymakers. Their current intelligence or, more likely, their estimates prove to be misguided. In the instance of reports on North Korean troop strength during the Carter years, the CIA apparently made a mistake in its counting that was subsequently discovered and corrected by two DIA analysts.[72] Mistakes, naturally, make the policymaker truly fond of the analyst, especially when the analyst enjoys the luxury of returning to his office, leaving the policymaker to face the press.

In light of these and other formidable barriers,[73] one may wonder if a harmonious relationship can ever exist between the consumer and the producer. It can, and does, as both try to make the adjustments and accommodations usually necessary for individuals to work together successfully. Intelligence officials have become quite sophisticated in ways of nurturing the relationship.

"We use to throw things over the transom," remembers a senior analyst. "Now we *market*. We still try to give an honest account, but we've become more adept about how to keep people reading."[74] One way, apparently, is through a variant of Chinese water torture. The CIA will write a sufficient number of analytic papers on a topic and send them to a policymaker so that, eventually, he may become interested—sometimes whetting his appetite with a brief teaser or two in the PDB, NID, or some other current-intelligence publication. "The more provocative we are, the more likely we are to be read," is Gates's rule of thumb.[75] Another CIA tactic is to use the typescript memorandum as a door opener. One senior analyst takes the time to respond to a specific policymaker on a specific question. "I make the investment to entice the principal to read subsequently longer, more formal documents. The method is expensive but, for me," he concludes, "it has yielded the highest payoffs."[76]

The chief manager of current intelligence in the CIA in 1984, Douglas MacEachin, Jr., emphasizes the importance of packaging, timing, and building rapport. "Good packaging is vital," he observes. "You must focus the policymaker's attention. They are busy. They like pictures and graphs. Videotapes have been a big hit with the Reagan team, especially in portraying international personalities like [the late Soviet Premier] Brezhnev."[77] A videotape on Col. Muammar Qaddafi was popular with the Reagan White House, too.

Timing is crucial. "You have to get [your data and analysis] in early," MacEachin further notes. "This is risky, because you can be wrong, especially on current intelligence—just as the best golden-gloves shortstop is going to make more errors than the right fielder. But if you don't get in early, the *Washington Post* or the wire services will scoop you. This means hitting the consumer before his second cup of coffee, prior to the 8:15 staff meeting." Nor is it simply a matter of being there first in the morning; the analyst also has to hit the right day. "You have to be there when the options are being formed," stresses MacEachin.[78]

The producers have learned—the hard way—that their primary masters, the deputy assistant secretary and above, will not look up much on their own accord, will not turn back the page in a report, and will often slough off reports to staff. Typically, when an intelligence report is prepared for a policymaker, his staff aide (the "gatekeeper") will read the "executive summary" at the beginning; if it seems worthwhile, the aide will then read the document itself, perhaps underlining a key paragraph on a single page. The principal will then look at the paragraph and maybe request an oral briefing from the analyst. Once in the policymaker's office, the skillful analyst will then use the opportunity to draw his attention toward other pages in the report, along with a key table or two.

To improve the chances of gaining access to the office in the first place, the CIA now encourages the growth of closer personal ties between analysts and policymakers, which needless to say is easier said than done given the particularly harried existence of the latter. (After working with policymakers for over a decade, one CIA analyst calculates that at best they spend about fifteen minutes a day on intelligence.[79]) This "personal chemistry" may be the most important aspect of the entire intelligence cycle, emphasizes MacEachin.[80] But even when the chemistry is good, complains a senior DIA manager, "getting on the policymaker's calendar is hard."[81]

From this standpoint, the CIA never had a better relationship with the most important policymaker—the president—than under Director William Casey. President Reagan and Casey were close friends, and Casey served as his national presidential campaign manager in 1980. The director often saw the president, in contrast to former directors like Colby and Richard Helms who were rarely invited to the White House. (Helms reportedly saw President Nixon *once* alone.[82]) "It's very nice to have someone who has the ear of the president," states Charles Briggs, CIA executive director, though he further notes an important disadvantage: "This has sometimes led to a perception of the politicizing of intelligence."[83] As the Iranian arms scandal of 1986–87 reveals, it can also hold the danger that a president and a CIA director who are that close will enter into questionable operations based on personal trust, without adhering to established decision procedures involving others in the CIA, the NSC, and the Congress.[84]

"The purpose of the intelligence cycle is to find the best minds available to produce the best one-page report on subject x," concludes MacEachin. "After that, it's all technique."[85] As these remarks on the dissemination phase indicate, the best minds and the most brilliant reports will be of little use without the skillful management of interpersonal relations between producer and consumer during this last phase of the cycle. Without an audience to listen to it, the importance of an analyst's intelligence becomes irrelevant. One may speak endlessly about the intelligence cycle in the abstract, but what it boils down to largely—not just in the intelligence establishment but throughout the government—is the quality and match of people charged with making the machinery work. As always, the success of a democracy depends ultimately on the talent and integrity of the people in office, whether that office is as open as the Department of Health and Human Resources or as closed as the CIA.

SIX

The Quiet Option

Covert action (CA), as defined in chapter 2, is a phrase used to identify the policy of secret intervention by the United States into the affairs of other countries. Few policies within the federal government are as closely held. No intelligence operations—with the important exception of domestic spying—have been so controversial.

These characteristics have made covert action a subject resistant to scholarly analysis. In the face of thick veils of government secrecy, documentation becomes difficult. The researcher is forced to rely heavily on interviews with those who have a practical familiarity with the policy and who, in most cases, demand anonymity in order to protect their identities. The approach must be rooted chiefly in historical description, rather than the more satisfying and modern methods of transhistorical generalization. As a result, the scholar is pushed toward a search for "wisdom" in a region inhospitable to more exacting forms of scientific inquiry—though even in this difficult terrain the obligation remains to bring scholarly rigor to bear wherever possible.

Despite the methodological obstructions that hinder research into secret policymaking (and they are considerable), one conclusion is certain: students of international affairs can scarcely ignore covert action. From the Truman through the Reagan administrations, this option has often been a favorite of presidents and their advisers in the pursuit of foreign-policy objectives. For this reason, further examination of covert action seems warranted even though its elusive quality frustrates a definitive exegesis. The purpose of this chapter is to map the paths of decision making for covert action that have evolved within the executive and legislative branches, and to appraise the recent efforts to tighten its supervision. The findings indicate that despite the presence of a highly formal and comprehensive decision-making process in recent years, covert action has escaped the full accountability envisaged by reformers during the 1970s.

The Incidence and Targeting of Covert Action

Incidence

To what extent has the CIA engaged in covert action, and against which nations? The data necessary to answer these basic questions are spotty, for the complete record remains highly classified. Nonetheless, from interviews, the published recollections of former CA practitioners, and the few existing scholarly works, as well as from congressional documents (especially the findings of the Church committee and its House counterpart, the Pike committee), one can piece together a likely portrait of covert action over the years.[1]

A longitudinal estimate of the percentage of the CIA's budget dedicated to covert action from 1947 to 1986 is presented in figure 6.1. The number of operations started off slowly, though not from zero. The CIA continued selected projects initiated by the OSS (Office of Strategic Services, America's first civilian intelligence agency created during World War II) while it began to develop a global infrastructure—that is, a network of spies (agents or assets) in the field built upon this foundation of OSS contracts. At first, the CIA's interest in covert action resided exclusively in the realm of propaganda operations—psychological warfare (psy war) as it was called at the time. By 1948, according to a report from the Church committee, the CIA "had acquired a radio transmitter for broadcasting behind the Iron Curtain, had established a secret propaganda printing plant in Germany, and had begun assembling a fleet of balloons to drop propaganda materials into Eastern European countries."[2]

From these modest beginnings, the Agency—strongly encouraged by the Policy Planning Staff in the Department of State and with the approval of the NSC—soon turned toward a wider range of CA projects. An interest in influencing the outcome of elections overseas (political covert action) became the main focus, especially the European elections of 1948. And by the summer of that year, the Agency had received further authority through NSC directives to launch operations involving economic and paramilitary (PM) warfare as well. (Paramilitary operations were also referred to as "preventive direct action" at the time.) Propaganda activities remained important, particularly the nurturing of media assets, but high on the list of priorities now were operations against Communist attempts to control labor and refugee groups, assistance to friendly political figures, and the establishment of PM "stay-behind" forces should war resume in Europe (this time against the Soviet bloc).

So within the brief span of a year after its creation, the CIA had been assigned the major responsibility within the American government for the conduct of a secret cold war against—primarily—the Soviet Union, using the full range of CA capabilities at its disposal. The Agency concentrated at first on the Soviet threat to Western Europe, in part because these war-wrecked nations seemed highly vulnerable to Communist intervention (while the Allies were celebrating

D-Day, Communist guerrillas were busy robbing banks in Paris and elsewhere to finance their political aspirations), and in part because the imperious Gen. Douglas MacArthur, commander of U.S. forces in the Far East, had declared that region of the world off-limits to the CIA.

As outlined in NSC directives, the initial intention was to keep the CIA's CA capabilities modest, for use when emergencies arose; but with the outbreak of overt warfare in 1950 against the Communists in North Korea, the NSC abandoned its early restraints. According to the Church committee, as the CIA undertook major PM operations in Korea and China, its CA component (steered at this time by the Agency's Office of Policy Coordination [OPC]) assumed "massive" proportions and its budget "skyrocketed."[3] Moreover, the CIA shifted its geographical attention from Western Europe to the Far East, as PM warfare there absorbed an increasingly larger percentage of the CA budget.

By the time the Eisenhower administration had settled into Washington in 1953, covert action was in full swing as the United States responded to intensified Soviet experiments in global interventionism; as the CIA infrastructure became more elaborate, experienced, and reliable; and as John Foster Dulles (President Eisenhower's secretary of state) and his brother Allen Dulles (the CIA director from 1953 to 1961) reached key positions to advance their own unalloyed anti-Soviet policies. During the Dulles years, political covert action enjoyed the highest priority at Agency Headquarters—above all the care and feeding of pro-West candidates and parties in Western Europe. With the Korean War over, the Agency returned its primary attention to Europe, though the interest of CIA officials in the Far East continued at a high level, followed by interest in Latin America. Among the CIA's political operations, the creation of international organizations—for students, academicians, workers, and many others—favorably disposed toward American foreign-policy objectives became an important undertaking.

With the fascination of the Kennedy administration for "counterinsurgency" (its name for PM operations conducted by the CIA and the military), this form of covert action came to dominant the genre during the 1960s—"surpassing," according to the Church committee, "covert psychological and political action in budgetary allocations by 1967."[4] In the aftermath of the Bay of Pigs disaster (May 1961), President Kennedy scaled back on covert action, but this remission was limited and short-lived. As America became increasingly involved in the Vietnamese civil war, covert action attracted a greater percentage of the Agency's budget than ever before—a reflection of the extensive use of large-scale PM operations by the United States in this part of the world from 1962 through the early 1970s.

For a number of reasons, covert action went into a decline after a peak budgetary allocation in 1967. The expose published in 1967 by *Ramparts* magazine, which revealed CIA ties to the National Student Association and other organizations, forced the Agency to sever its CA support (political and propaganda) to these entities.[5] Moreover, Richard Helms had become the Agency's director in June 1966, and he was skeptical about the effectiveness of covert action; he preferred instead to see additional resources moved toward the clandestine collection of intelligence.[6] He was concerned as well about the increasing risk of exposure

FIGURE 6.1 Covert Action: A Longitudinal Perspective, 1947–86 (estimated)

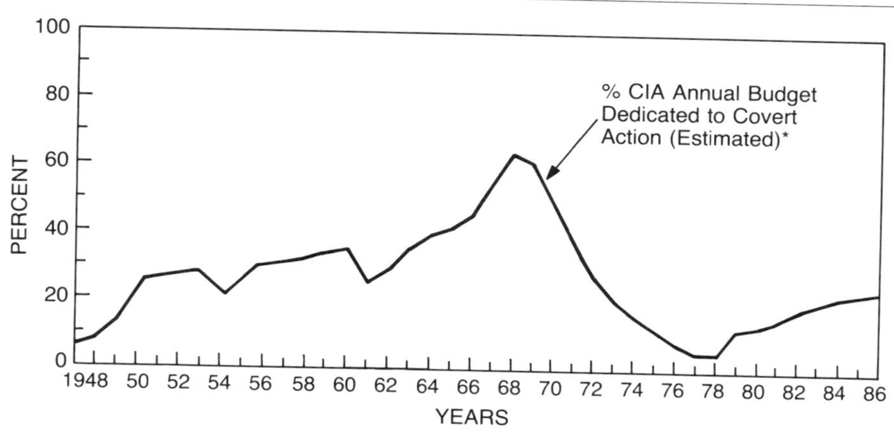

[a]This trend line represents a "best guess" based on the open-source material cited in this study and interviews with CIA and legislative officials from 1979–88. The technique used by the author was to draft a trend line that reflected the occasional references in the open literature to CA funding (e.g., Harrison Salisbury, "Interview with William Colby: The Role of the CIA and that of the Press," *Behind the Lines,* WNET Television, New York, 1975). This estimate was then shown to officials during interviews, and they were asked how they would adjust the line to make it more accurate. The line in this figure represents the approximate mean, or best fit, of these various interview responses—all of which were in high agreement. A total of sixty-four officials commented on all or, more commonly, part of the time span.

for the large PM undertakings in Indochina, especially the covert war in Laos—an operation he was happy to have transferred from the CIA's budget allocations to the Defense Department in 1970.[7] Government-wide budgetary retrenchments in 1967 and 1969 struck the Agency, too, and resulted in significant personnel cutbacks in the CA program.[8] Finally, by this time the Nixon administration had begun to reduce U.S. involvement in the Vietnam War, covert and overt. By 1972, the CIA's PM operations in Indochina were halted altogether.[9] During the Nixon and Ford years, funding for covert action went into a precipitous slide, made all the more slippery by the press exposes in 1974 about CA operations against a democratically elected president in Chile and by the assassination plots revealed in the Church committee proceedings of 1975.[10]

Jimmy Carter defeated President Ford in 1976 and entered office with a strong bias against covert action, which he had expressed many times during his campaign for the presidency. During the election, he came close to selecting Frank Church as his running mate, attracted to his outspoken views against covert action, and he did select Walter Mondale (D, Minnesota), who had also been sharply critical of covert action as a member of the Church committee. The CIA budget allocation for covert action plummeted to less than 5 percent during Carter's early years, a continuation of the trend away from this approach to foreign policy begun in the Nixon years and carried through the Ford administration (despite President Ford's much publicized—and unsuccessful—effort in 1975 to gain congressional support for covert action against Marxist rebels in Angola).

With the Soviet invasion of Afghanistan, however, Carter reversed himself and gave support to an increased use of the quiet option.

When the Reagan administration came to Washington in 1981, with its inflamed cold-war rhetoric and a CIA director (William J. Casey) impressed by secret operations from his days as an OSS officer, the CA budget experienced a further surge upward.[11] This administration also went to the extreme of placing responsibility for key CA operations into the hands of NSC staffers in the White House, outside the normal procedures (see below), and began to raise funds from private Americans and foreign heads of state for "off-the-shelf, self-sustaining, stand-alone" secret operations (an objective attributed to Casey by NSC staffer Lt. Col. Oliver L. North during congressional investigative hearings in 1987). Once revealed in 1986, these activities precipitated major inquiries by Congress, a presidential commission, and a special prosecutor, and stimulated extensive public debate over the proper decision paths and supervision for covert action.

Targeting

Senator Church concluded during his committee's investigation into covert action in 1975 that this instrument of foreign policy had been directed, for the most part, against "small, weak countries."[12] The accuracy of this observation is borne out by interviews with intelligence officials: most covert actions have taken place in the developing nations—not because the United States has invariably viewed these nations as enemies (though some have been), but because the so-called Third World has become the battleground between America and the Soviet Union, the main arena of competition over raw materials, strategic position, and popular support.

While the details remain shrouded in secrecy, the names of those countries targeted for covert action are almost all on the public record. In table 6.1, they are grouped according to geographical region, illustrating the predominant CIA targeting of developing nations from 1947 to 1986 (as measured by the frequency of operations, not the cost)—though Europe and the Soviet bloc have by no means been ignored. East Asia has been the focus over the years of more covert actions than any other region, both in terms of frequency of operations and money spent (in significant part because of the great number and cost of PM operations associated with the war in Vietnam). The high-through-low categories of table 6.1 offer a portrait of U.S. CA priorities aggregated over the years. These priorities fluctuate from time to time, however. If, for example, one were to examine the decade from 1977 to 1987, the regional priorities would shift somewhat (interviews indicate) to Europe, Latin America, the Near East, the Soviet bloc, East Asia, and Africa. If one looked at only the last couple of years, the expanding PM operations in Afghanistan would give the Near East an even higher profile. Covert actions aimed directly at the Soviet bloc have been limited; the necessary assets (that is, indigenous agents working for the CIA) are too hard to come by. Here the United States has had to settle mainly for limited propaganda operations.

TABLE 6.1. Frequency of Covert-Action Targeting, 1947–86 (estimated)[a]

Region[b]	Form of Covert Action				
	Propaganda	*Political*	*Economic*	*Paramilitary*	*Combined*
East Asia	High	High	High	High	High
Europe	High	High	Low	Low	Moderate
Africa	High	Moderate	Low	Moderate	Moderate
Latin America	Moderate	Moderate	Moderate	Moderate	Moderate
Soviet Bloc	High	Moderate	Low	Low	Moderate
Near East	Moderate	Moderate	Low	Low	Moderate-Low

[a]Based on the open-source materials cited in the Bibliography and notes, as well as periodic interviews from 1979 to 1988 with officials of the CIA's Operations Directorate. The categories "High," "Moderate," and "Low" are subjective, stemming from characterizations of CA frequency in the various regions made by experts in the literature and during my interviews. In the interviews, I asked CIA officials to comment on the frequency with which the major forms of covert action were used in the different regions, employing the three broad categories for their estimates (similar to the "thermometer" technique in voting studies, where respondents are asked to comment generally on the extent of warmth which they feel toward a particular party or a candidate).

[b]The regional categories are based on *The World Factbook* (Central Intelligence Agency: April 1984), with Egypt placed in the Near East and East Asia combined with South Asia and Oceania.

Decision and Accountability

Choosing Covert Action

The official approval process for covert actions in the modern era has gone from the simple to the complex. For CA authority in the early days of the CIA, officials drew upon the section of the 1947 National Security Act that directed the agency to carry out secret operations for the protection of the national security as ordered by the NSC. While today CIA attorneys find this language a questionable legal basis for CA operations (they turn instead to the Hughes-Ryan Act of 1974[13]), at the first meeting of the NSC in December 1947 its members relied upon this "authority" to approve NSC Directive 4 (and its annex, 4-A). This document ordered the CIA to engage in covert actions designed to discredit international communism. The quiet option attracted broad support among the NSC principals, for "it held the promise of frustrating Soviet ambitions without provoking open conflict."[14]

This and subsequent NSC directives on covert action were (until December 1974) subject to minimal discussion and supervision.[15] In the early years, the CIA's Office of Policy Coordination (established by the NSC in June 1948) provided a forum for the discussion of CA proposals with officials outside the Agency, namely, representatives from the departments of Defense and State. These meetings, though, were not designed to give approval to the CIA's plans. The sense among OPC participants was that the Agency already had approval for covert operations from NSC-4 and 4-A, along with subsequent supporting directives. The OPC forum was established only to offer "guidance"; the CIA could conceive of and implement covert actions as it saw fit, in harmony with the broad mandate from the NSC to combat communism. As the Church committee reported, "loose understandings rather than specific review formed the basis for CIA's accountability for covert operations," and even within the CIA, covert action soon became a "pocket of privilege."[16] The Controller's Office, for example, which was responsible for tracking Agency spending, found the doors to the Covert Action Staff (CAS) closed to its auditors by order of the CIA director.

As the CIA's covert actions proliferated during the 1950s, government officials concluded that a more systematic review was warranted. The NSC created a subcommittee, the 5412 Committee (as it came to be known after the number on the council directive creating it). The 5412 was composed of designees (staff aides) representing the NSC principals, the council's four statutory members: the president, vice president, secretary of state, and secretary of defense. Until 1959, however, the 5412 rarely met and the CIA enjoyed broad discretion in its conduct of the covert cold war. Even when the 5412 began to meet regularly, its members rarely challenged the CIA proposals. The stunning failure on the beaches of the Bay of Pigs stirred the NSC, though, into a closer look at covert action. Officials in the Kennedy administration upgraded the membership of the "Special Group" (as they preferred to call the 5412 Committee) to encompass the chairman of the Joint Chiefs of Staff (JCS), the CIA director, the under secretary of state, the deputy secretary of defense, and the president's assistant for national security affairs (who served as chair).

In its fresh look at the approval procedures for covert action after the Bay of Pigs shock, the Kennedy administration instituted more frequent meetings of the Special Group and spelled out more clearly the criteria for seeking its approval for covert action. Expense—any project costing over $25,000—was to be the key criterion for triggering outside authority from the Special Group. The administration also established other NSC subcommittees to supplement the work of the Special Group. A Special Group on Counterinsurgency supervised large-scale PM activities, and a Special Group (Augmented) supervised Operation MONGOOSE, a series of covert actions designed to undermine Fidel Castro's rule in Cuba.

Yet these attempts to ensure greater accountability for covert action stood at odds with a resilient counterphilosophy: the doctrine of "plausible denial." According to this concept, the president and other high officials had to be shielded from clear responsibility for a covert action so that, if the operation were exposed ("blown"), they could deny culpability without—in the words of one former CIA professional—"the government's being caught in a barefaced lie."[17] In this manner, the president (and therefore the country) could presumably save face.

Reconciling accountability and plausible denial presents an obvious dilemma: under a system of plausible denial it often becomes uncertain who really does know about, and has approved of, any given covert action. The lines of accountability wash away like markings in the sand. Moreover, critics contend, it is fatuous to believe that a president can escape blame simply by claiming ignorance; he is, after all, the chief executive and, as such, is ultimately responsible for the acts of his minions. Of great concern is the possibility that the president might not only be unaware of an operation, but could actually be opposed to it if he were aware. This is what the Reagan administration claims was the case in 1986 when the president's assistant for national security affairs, Vice Adm. John M. Poindexter, approved—apparently without President Reagan's knowledge—the diversion of illicit funds to the contras in Nicaragua from a secret arms sale to Iran (both, by the definitions widely accepted by intelligence overseers, PM operations).

"Although I was convinced that we [the NSC staff] could properly do it [divert funds to the contras] and that the President would approve it if asked," testified Poindexter before the Inouye-Hamilton committees in 1987, "I made a very deliberative decision not to ask the President so that I could insulate him from the decision and provide some future deniability for the President if it ever leaked out."[18] Yet subsequently the president said he would never have approved the diversion. This case, evidently a result of Poindexter's attempt to establish plausible denial for the president, stands as a damning commentary on the doctrine, though the cynic might wonder if the president, despite his protestations to the contrary, had simply carried the doctrine to its logical end: sustaining the lie to protect his office—and himself.

Soon after Lyndon Johnson came to the presidency in 1963, the NSC changed the name of the Special Group to the 303 Committee. The NSC's main business, though, took place in weekly lunches at the White House, which soon became known as the "Tuesday Lunches." Over sandwiches, the NSC members discussed the pressing national security issues of the day, often in the company of the DCI, the assistant for national security affairs, and the JCS chairman. These sessions were highly informal and tended to focus on military outcomes in Vietnam; they seldom addressed questions of covert action.[19]

In 1970, the 303 Committee—already with more aliases than a bank robber—took on another name: the 40 Committee (from National Security Decision Memorandum No. 40). The only significant change in CA procedure introduced by the Nixon administration, though, was to require an annual review by the 40 Committee of all extant CA projects. Then during the Ford administration (1974–77) the 40 Committee became the Operations Advisory Group (OAG), but in other respects closely resembled its predecessor. For both the 40 Committee and the OAG, as with their forerunners, the weakness in the decision process remained the failure to refer most CA proposals to the president or even to his top cabinet officials responsible for foreign and defense policy. In 1962, for instance, only about 16 percent of all CA projects received approval from the Special Group, according to a CIA memorandum reported by the Church committee. The committee added that, by its reckoning, only about 14 percent of all covert actions from 1961 to 1975 had been approved by the NSC.[20] In the committee's opinion, "these ambiguous arrangements were intentional, designed to protect the President and to blur accountability."[21]

The Congress Awakes

During the years from the Truman through the Ford administrations, Congress was rarely in the intelligence decision loop. The CIA did not want to tell, and, conveniently, the Congress did not want to know. The political risks might be too high for legislators if an operation went awry and became public; it was safer to remain untutored. One could then claim innocence and pillory the CIA for its poor judgment. As Senator Church once recalled, the legislators on the CIA watchdog committee in the Senate had the attitude: " 'We don't watch the dog. We don't know what's going on, and, furthermore, we don't want to know.' "[22] According to a DCI, William Colby, "the old tradition was that you don't ask. It was a consensus that intelligence was apart from the rules. . . ."[23]

That the NSC was uniformly apprised of CA operations, let alone the Congress, the figures presented earlier show to be a fiction. The tendency seems to have been for the CIA to ask the NSC for broad grants of authority that would then be the sire of large broods of subsidiary operations—many arguably warranting separate and specific approval. One statutory member of the NSC, Dean Rusk, the secretary of state for presidents John Kennedy and Lyndon Johnson, apparently remained in the dark on most CIA operations. He remembers, "I never saw a budget of the CIA, for example. . . ."[24] Nor does Rusk—or any other NSC official—recollect any decisions or briefings on CIA assassination plots; each denied, under oath, knowledge of any presidential authority for this form of PM operation—though clearly they took place.[25]

In December 1974, a post-Watergate Congress—troubled by press revelations alleging domestic spying and unsavory covert actions in Chile—formally confronted this lack of accountability and took the first step since 1947 to rein in the CIA. In a flurry of last-minute legislative activity before the end of the session, its members approved the Hughes-Ryan Act, sponsored by Sen. Harold E. Hughes (D, Iowa) and Rep. Leo J. Ryan (D, California). This legislation required the president himself to approve all important covert actions (the assumption was that he would do so in writing), and established a procedure for informing congress of these decisions. Its provisions required that

> [n]o funds appropriated under the authority of this or any other Act may be expended by or on behalf of the [CIA] for operations in foreign countries, other than activities intended solely for obtaining necessary intelligence, unless and until the President finds that each such operation is important to the national security of the United States and reports, in a timely fashion, a description and scope of such operation to the appropriate committees of the Congress. . . .

The Hughes-Ryan legislation represented nothing less than a bold attempt to replace plausible denial with a clear trail of accountability for covert action—one that led straight to the Oval Office. The law formally forbade all covert actions unless found important and approved by the president himself.

From the verb "finds" in the act came the term "finding," that is, the anticipated written document of approval bearing the president's signature. The "appropriate committees" to whom the finding was to be delivered in a timely fashion (within twenty-four hours came to be the understanding) were initially three in the House of Representatives and three in the Senate: the committees on appropriations, armed services, and foreign affairs. In 1976, legislators added the Senate Intelligence Committee to the list, and, in 1977, the House Intelligence

Committee. Then, in 1980, with the passage of the Intelligence Accountability Act (usually referred to more informally in Washington circles as the Intelligence Oversight Act), Congress trimmed the list back to the two intelligence committees.[26]

This 1980 statute, the most important formal measure taken by legislators to tighten their control over intelligence operations, further had the effect of clarifying that the Congress wanted to be informed of *all* important covert actions, not just those sponsored by the CIA (thus closing a loophole for possible presidential recourse to other agencies—presumably the military, but perhaps any entity from the NSC staff to personnel in the Department of Agriculture—for covert action to avoid the Hughes-Ryan reporting requirements). Moreover, with this legislation the Congress took a strong stand in favor of prior notice on covert actions, not after the fact as "timely fashion" allowed; in emergency situations, the president was allowed to limit his prior notice to eight leaders in Congress (the so-called Gang of Eight). The wording of the statute is quite clear on this point: "[I]f the President determines it is essential to limit prior notice to meet extraordinary circumstances affecting the vital interests of the United States, such notice shall be limited to the chairmen and ranking minority members of the intelligence committees, the Speaker and minority leader of the House of Representatives, and the majority and minority leaders of the Senate. . . ."

Yet in spite of this language, at least one recent DCI, Adm. Turner, rejected the obligation of prior notice;[27] and, as hearings on the Iran-contra scandal made clear, the Reagan administration failed to report to Congress at all on this covert action—indeed, the belatedly signed and retroactive finding for the covert sale of arms to Iran explicitly forbade the CIA from reporting it.

At the 1988 Symposium on the Management of Intelligence at Georgetown University, former senior staff overseers on the congressional intelligence committees—some of whom had helped draft the 1980 Intelligence Oversight Act—argued vociferously against diluting the prior-notice requirement, as proposed in pending legislation. A sponsor of a bill to require a weaker forty-eight-hour provision for reporting (S.1818, introduced on October 27, 1987), Sen. Arlan Specter (R, Pennsylvania), retorted that prior notice had failed to work and that, if the Congress had to go to the court of public opinion with a complaint about inadequate reporting by the CIA, the standard had to be "clear and simple, in black-and-white." The matter continues to provoke divided opinion, as does much of the relationship between the intelligence community and the Congress in the wake of the investigations of 1975, the stringent new oversight procedures of 1976–80, and the collapse of trust between the CIA and the Congress that came with the Iran-contra scandal in 1986.

The Congress was not alone in its fresh insistence on improved controls over the intelligence community. In 1976, Jimmy Carter of Georgia campaigned for the presidency in part on a platform of intelligence reform. This interest was clearly reflected in the drafting of the Democratic party platform that year (a process controlled by Carter delegates), which included for the first time references to intelligence and, specifically, CA reform.[28] Once in office, President Carter tempered his campaign zeal, but nonetheless instituted far-reaching changes in the structures and procedures of the intelligence community. On January 26, 1978, he signed a major Executive Order on United States Intelligence Activities (No. 12036), which, among other things, established a remodeled and much more rigorous decision-cycle for covert action.[29] This chapter concentrates on the Carter years as an illustration of the organization and decision process for covert action in the United States since the passage of the Hughes-Ryan Act—though the procedures used by the Reagan administration (with the large exception of the Iran-contra episode) were quite similar.

The Approval Process for Covert Action in the Carter Administration

Under Carter, the NSC remained the highest organization in the executive branch to review and guide the conduct of all national foreign intelligence and counterintelligence (CI) activities. As replacements for the Ford administration's Operations Advisory Group, two new committees were designed to fit within the NSC structure, each with important intelligence responsibilities: the Policy Review Committee (PRC) and the Special Coordination Committee (SCC). The PRC was created to establish the objectives and priorities for foreign intelligence collection, to review resource allocation for this mission, and to evaluate the intelligence product; the SCC was to be the central entity for the review and approval of all CA proposals before they were forwarded to the president. The SCC was further responsible for the review and approval of especially sensitive intelligence collection and CI programs. Since the SCC had primary control over covert action, its membership warrants listing here: the secretary of state, secretary of defense, attorney general, director of the Office of Management and Budget, assistant to the president for national security affairs, chairman of the Joint Chiefs of Staff, and the director of the CIA.

The SCC, then, stood as the Carter version of—to recapitulate—the 10/12 and 10/13 committees (Truman), the 5412 committee (Eisenhower), the Special Group (Kennedy), the 303 Committee (Johnson), the 40 Committee (Nixon), and the Operations Advisory Group (Ford). The names have changed but the purpose and membership have remained fairly constant (with an upgrading in the Kennedy years)—until the Reagan administration. Although the Carter administration is the focus of this chapter, contrasts with the Reagan administration will be offered from time to time for comparative purposes. President Reagan abolished the interdepartmental consultation that characterized all the earlier CA groups on the NSC, in what was said to be an attempt to tighten the secrecy of covert actions.[30] In place of the military and diplomatic experts on the SCC, Reagan's new organization—called the National Security Planning Group (NSPG)—included more of his own personal advisers. He retained the vice president, the secretaries of state and defense, the national security adviser, and the DCI on the new NSPG, but excluded the JCS chairman and his staff aides. Three political advisers were added: the White House chief of staff and his deputy, plus the president's counselor.

According to a former senior official, the Reagan changes were really made less to preserve secrecy—very few of the total number of covert actions have been disclosed to the public—than to assure that the president had "his senior advisers to focus on the issues themselves and not have people around always saying why things could not be done."[31] The end result seems to have been a loss of expert staff advice and a sharp increase in the budget allocations for covert action during the Reagan administration—some fivefold over the nadir reached during the Carter years.[32]

ORGANIZATION The director of central intelligence (the DCI, who, recall, also serves as CIA director) provides the vital link in the chain of command between the NSC and the CIA. Within the Agency, covert action is housed inside the Operations Directorate, headed by a deputy director for operations (DDO). As the reader will remember from chapter 3, this directorate consists of (among other subdivisions) a unit for political and economic covert action (the Covert Action Staff, or CAS), for paramilitary (PM) covert action (the Special Operations unit), for counterintelligence (the CI staff [CIS]), and for several geographic desks responsible for the collection of foreign intelligence. Like the definitions of the CA forms presented in chapter 2, these compartments are somewhat artificial. A single asset, say, a bartender in an exclusive hotel in Hong Kong, may collect foreign intelligence, carry out counterintelligence (that is, operations against hostile intelligence agencies), and conduct covert action of some sort—though it is unlikely (at least outside the cinema exploits of "007") that the same individual would have the skill and access to accomplish all of these missions.

This functional overlap can lead to tangles in administration and requires careful coordination among staffs. It can also lead to controversy, as in Angola in 1975. Following the Clark amendment, which prohibited covert action in Angola, a disgruntled former CIA officer charged that the Agency had a PM operation underway there despite the new law.[33] William Colby, DCI at the time, countered that these Agency officers in the field were engaged strictly in the collection of intelligence; they were specifically prohibited from participating in the civil war, or even advising the anticommunist guerrillas.[34] Given that the same individual can either collect intelligence or, if he wants, provide PM advice, former CIA officers, legislative overseers, and others may have had some grounds for skepticism; but Colby's assurance could have been, and probably was, perfectly honest. To know for sure, one would really have had to be next to the officer in the jungles of Angola—a rather demanding form of oversight.

As outlined in chapter 3, the Operations Directorate also houses special groups for conducting counterterrorism and counternarcotics, for tracking nuclear proliferation, for monitoring international energy issues, and other tasks. The cadre of experts in the PM group (Special Operations) are highly skilled in weaponry; covert transport of personnel and material by air, sea, and land; guerrilla warfare; the use of explosives; and escape and evasion techniques. They are prepared to respond quickly to a myriad of possible needs, from parachute drops and communications support to assistance with counternarcotics operations and defector exfiltration. For PM tasks and its other responsibilities, the Special Operations staff attempts to recruit assets with the appropriate specialized skills, though the geographic desks remain the principal units involved in the recruitment of Soviet,

East European, and other useful personnel in so-called denied areas. In a nutshell, the CAS and Special Operations provide policy guidance and staff support for the geographic desks and for the deputy director of operations in his planning and implementation of covert actions; in addition, Special Operations provides special air, ground, maritime, and training support for the Agency's intelligence collection operations.

While Headquarters must provide planning and guidance, the field has the tougher job of achieving results. The heart of covert action lies in the activities of the Operations Directorate abroad. The CIA has stations and smaller bases around the globe. Operations Directorate personnel in the field are called, as I indicated in chapter 3, case officers, each responsible for recruiting and managing ("running") local assets—some of whom they will have inherited from a predecessor. The objective is to use the indigenous population for espionage (HUMINT), for counterintelligence, and for covert action. Case officers report to the country chief of station (COS), the CIA equivalent of an ambassador, and all operations conducted in support of approved CA programs are supervised in the field by the COS. The stations in turn report to their county desk back in the Operations Directorate at Agency Headquarters and, if the relations are cordial (see chap. 11), to the American chief of mission in the field (the ambassador).

PROCESS During the Carter administration, the normal line of authority for covert action ran from the CIA through the SCC (which became the NSPG in the Reagan administration) on to the president and back, touching a number of key entities and officials along the way. (See fig. 6.2.) Most proposals for covert action were conceived in the CIA stations (roughly 85 percent[35]), but some came from the Department of Defense, the Department of State, ambassadors in the field, and occasionally from the president himself or his close advisers in the White House.

The Hughes-Ryan Act required that before the CIA conducts any operations abroad (other than intelligence collection), the president must find the activity "important to the national security of the United States." Before the president issues his "finding" (or approval) of a covert action, however, it is reviewed by several policy groups. Typically, after a covert action originates from a station overseas, it advances through the Agency hierarchy, requiring approvals from the appropriate staff components of the CAS and, if a PM proposal, from the Special Operations unit in the Operations Directorate, then the deputy director for operations, other offices within Headquarters, and eventually the DCI himself.

Prior to gaining final approval from the DCI, CA proposals were reviewed by a variety of offices within the CIA during the Carter administration, including the Comptroller, the Office of General Counsel, the Legislative Counsel, and the National Foreign Assessment Center (the name of the Intelligence Directorate during the Carter years, that is, the division of the CIA devoted to the production and analysis of intelligence reports for policymakers). Proposals were also reviewed by two organizations outside the Agency before going to the DCI for his final approval: the Department of State and a SCC working group for covert action (the "special activities working group," after the Carter administration's

FIGURE 6.2 Covert-Action Decision Process

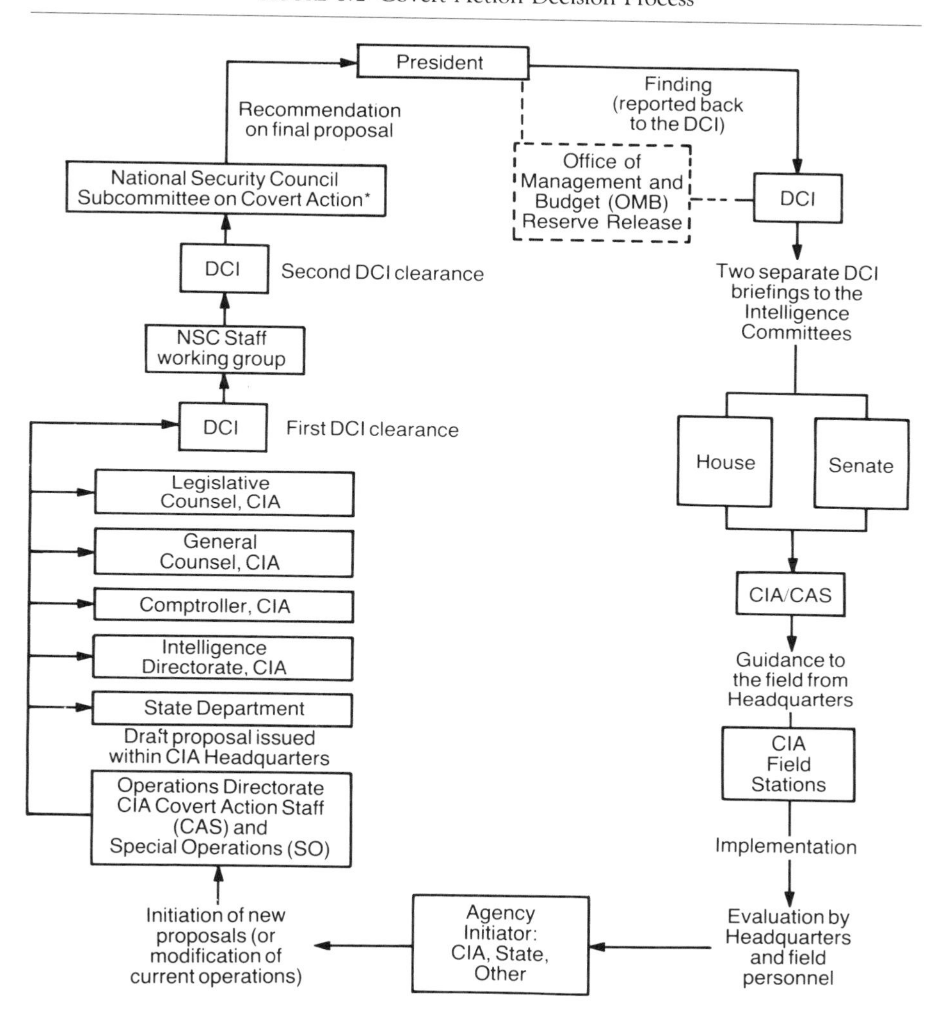

[a]Known as the Special Coordination Committee (SCC) during the Carter administration and the National Security Planning Group (NSPG) during the Reagan administration.

preferred euphemism for covert action). Of all these entities, the SCC working group normally provided the acid bath for CA proposals.

The working group consisted of senior designees from each organization represented on the parent Committee. Here each proposal received close scrutiny. A high percentage were rejected outright at this stage, or, in most cases, at least sent back to the CIA for clarifications or modifications. In preparation for this review before SCC staffers, the CIA had to make sure that its proposal addressed several points: justification of the project, expense, alternatives, risks, prior coordination, past related activites, and importance (that is, whether the proposal warranted presidential review and congressional consultation, as required by Hughes-Ryan). Consideration of CA proposals, then as now, depended greatly upon the attitudes of individual reviewing officers; but risk, compatibility with American foreign policy goals, likelihood of success, value of outcome, cost, and the prevailing political climate all weighed in the balance.

The special activities working group was not a decision-making committee; rather, it acted in a staff advisory relationship to the principals on the SCC and as a conduit back to the initiating agency (usually the CIA). Its job—like that of the staff group attached to its successor, the NSPG—was to ensure that full and complete deliberations, including supporting and dissenting opinions, were reported to the SCC and to the director of the agency that generated the proposal.

If a covert action was supported at the level of the staff working group (typically following one or more revisions) and then by the DCI, the full SCC examined its merits and submitted a policy recommendation to the president, along with comments and dissents (if any) from individual members of the SCC. It was then, as now, expected that an affirming decision by the president (the finding) would be reported to the Congress. This last provision applied only to the CIA until the 1980 Intelligence Oversight Act (at which time it applied to all government agencies), and not at all during times of declared war or when the president was operating under the War Powers Resolution (passed in 1973).

AN ILLUSTRATION To illustrate the approval cycle for a major CA proposal, the resumption of a program to provide Country X with special security equipment offers a useful example.[36] The purpose of the program was to improve the physical protection of the country's highest leaders. Country X requested a resumption of the program, which had been allowed to lapse for a prolonged period. During the spring of one year, a proposal was drafted in the Operations Directorate and reviewed by staff officers throughout the Agency, including individuals in the Comptroller's Office and the Office of General Counsel. The deputy director of the CIA then approved the proposal on behalf of the DCI.

During the summer, the SCC working group reviewed the proposal twice, requested additional data, and then recommended submission of the proposal to the formal SCC. In early fall, the DCI approved the proposal and it was delivered to the SCC. One month later it was approved by the SCC, and the next month it was signed into a finding by the president. Since the program was unexpected and had not been included in the Agency budget, the CIA obtained funding through a reserve release and, as required, notified the congressional intelligence committees of this procedure.[37]

BROAD FINDINGS The Hughes-Ryan Act required a presidential finding for all important covert actions but provided no guidance on how detailed a finding should be. During the Ford administration, according to one report (substantiated by my interviews), the president elected to sign a series of "worldwide" findings that guaranteed authority in advance to the CIA for the conduct of covert actions against terrorism, the narcotics trade, and CI targets.[38] The purpose of the worldwide findings was to allow the CIA some flexibility to carry out routine covert actions, that is, to task its international infrastructure without imposing on the president a requirement that he make a separate finding with respect to each such operation. Subsequently, in response to legislative criticism that the worldwide findings provided too much of a blank check to the CIA, the Carter administration combined the worldwide findings approved by President Ford into a somewhat more detailed overall finding. In his initially limited use of covert action (there were no new ones during his first six months in office), President Carter relied chiefly on this combined finding, which he augmented later with selected propaganda themes and specific findings focused on individual countries.[39]

The congressional intelligence committees remained concerned, nonetheless, that the precedent of a broadly worded finding might encourage President Carter and his successors to endorse proposals that were too open-ended and vague. Carter's DCI, Adm. Stansfield Turner, has commented on the distinction between broad and narrow findings:

> Under a broad finding, an operation can be expanded considerably; with a narrow one, the CIA has to go back to the President to obtain a revised finding if there is any change of scope. The Congress is wary of broad findings; they can easily be abused. The CIA is afraid of narrow findings; they can be a nuisance. What has evolved is a working understanding that whenever the activity being carried out under a finding is widened past the original description to the Congress, the CIA will advise the committees.[40]

PROPAGANDA REVIEWS To help assure that at least one common form of covert action based on broad findings—the use of propaganda—was really in tune with ongoing government policy, a special oversight routine evolved within the executive branch.[41] This review takes the form of a policy paper, or viewpoint, written by the CAS to explain the need for a certain propaganda operation. Specifically, it outlines a suggested theme; two hypothetical examples might be "Pershing II missiles are an indispensable counter to the threat posed to Western Europe by Soviet SS-20s" or "the Strategic Defense Initiative offers the best shield against nuclear war for the NATO alliance." Once drafted, such themes are sent to the Department of State for approval, modification, or disapproval. The intelligence arm of the Department of State, the INR, determines who within the department should receive a proposed theme, forwarding most to the appropriate geographic bureaus where they are reviewed by desk officers and frequently passed on to the deputy assistant secretary level, less frequently to the assistant secretary. The secretary of state is apprised of the propaganda theme as his staff deems necessary and proper.

If blessed by the Department of State (about 25 percent are turned down here), the CIA sends the proposal to the relevant field stations. If a new propaganda scheme is proposed from the field that, in the judgment of the Operations Directorate, falls somewhat outside the guidelines for existing themes, the CIA would go back to the Department of State with a revised proposal—assuming the DDO finds the scheme useful. A widely divergent theme would require a fresh finding. The whole purpose of this cycle is to insure that the CIA and the Department of State are not working at cross-purposes in their propaganda programs.

EXECUTIVE CONTROL The CIA divides covert actions into the "important" and the "routine." Since Hughes-Ryan, the former require a presidential finding and the latter flow from broad findings already approved. The existence of findings—even when broad—reduced somewhat the discretion of DCIs in determining whether a covert action should be considered routine or important enough for presidential decision; however, the judgment of the DCI continues to play a significant role in sorting out this distinction. Some proposals, though, always require NSC review—particularly economic and paramilitary schemes, strategic deception operations, and political covert action or propaganda directed toward prominent foreign leaders or groups.

Nevertheless, the line between the important and the routine can still be fuzzy because "important" remains a subjective word. One high-ranking CIA official suggests that overseers will simply have to trust the DCI to make an accurate judgment on what is important or else enter into the "micromanagement" inspection of each and every proposal.[42] While administratively persuasive, this prescription of trust understandably makes overseers uneasy. The Congress trusted the CIA blindly for almost three decades, only to learn in 1975–76 that the Agency had abused this trust on some occasions (recall the catalogue in chap. 1); in light of the Iran-contra episode, Agency appeals to "trust" will be all the more strained on Capitol Hill.

Occasionally, in midstream, a covert action must undergo a significant alteration. In such cases, steps have been instituted to trigger a review by the NSC of the proposed change. Without presidential approval, the change cannot be made. An alteration is considered significant if it involves substantial political repercussions, if exposed; an increase in the risks of exposure; a marked change in operational direction; a substantial (usually defined as 20 percent) increase in costs; or any request for the release of funds from the CIA Contingency Reserve Fund.[43]

SUBTHRESHOLD PROPOSALS The Carter administration used an additional procedure to initiate a review of less important covert actions that the CIA believed might fall into a gray area; that is, those proposals neither important enough for formal review at the SCC threshold nor sufficiently covered by the generic findings to be administered with full confidence of compliance. This method proved especially useful for what one retired CIA official has described as "lesser tactical CA activities that we felt required special interdepartmental coordination."[44] In this review cycle, the Agency notified SCC members of a "moderately" risky covert action and asked for their comments and recommendations. In the event of a dissenting opinion or serious question by any member, the CIA either tried to resolve the matter informally with the dissenter, withdrew from the operation, or sought adjudication from the SCC staff working group and, if necessary, from the president's assistant for national security affairs.

A counternarcotics operation in Country Y illustrates how this subthreshold procedure worked. The general authority for such operations derived from a broad finding, as well as from propaganda themes approved by the Department of State; but in this case the CIA station in Country Y felt that its idea for a covert action was sufficiently sensitive to warrant additional authority. The COS proposed a dual operation employing covert media assets and agents of influence (that is, assets placed close to decision makers and opinion leaders) to expose and disrupt the activities of a major heroin dealer. Once the dealer was exposed through the media of Country Y, the station wanted to use well-placed assets inside and outside the government to encourage the leader of Country Y to move against the narcotics trafficking.

Under the subthreshold procedure, the CIA referred the proposal to the SCC working group and, in this instance, the group reached a consensus that only the media portion of the operation should be implemented because the government of Country Y, and in particular the position of the chief of state, was at the time too unstable. Informed of this decision, the Country Y station chose to postpone the entire operation until the government had stabilized; the COS did not believe the media exposure alone would meet the objective. Four months later, when conditions in Country Y changed dramatically, the proposal was resubmitted by the station, approved by the SCC working group, and authorized by the president.

Major CA programs passing through the system for NSC review and presidential authority have tended to move at a pace dictated by a sense of urgency, or by the need for more careful and exhaustive policy examination. During the Carter years, proposals to the SCC working group under the subthreshold procedure usually received rapid attention. Each proposal had a deadline, normally five working days, for comment by members of the SCC working group.

Carter and Covert Action

The finding process under the Carter administration, then, was quite exhaustive in its consideration of CA recommendations. Proposals were studied closely by staffers within the CIA and those on the SCC working group, who developed recommendations for the president on each proposal. The president in turn considered each proposal and the recommendations of the SCC, including dissents, before coming to a final decision.[45] While the working group and its parent committee focused primarily on the substance of proposals, the Office of Management and Budget (OMB) was represented in these deliberations and reviewed the funding aspects. The membership of the SCC and its working group was designed to assure that each proposal received detailed consideration by a range of foreign-policy specialists from throughout the executive branch. The attorney general was also always present during SCC deliberations (a first) to provide legal counsel, a practice discontinued by the Reagan administration.

Several methods have been used within the executive branch, in sum, to guide the CIA and its agent network in directions harmonious with approved U.S. policies. They have included presidential findings, which are binding until such time as they are rescinded or replaced by a subsequent finding; State Department–CIA coordination of propaganda themes; subthreshold notifications; NSC background guidelines supporting findings; ongoing guidance from the assistant to the president for national security affairs; and—since the Iran-contra scandal—a more detailed quarterly review of ongoing covert actions by the national security assistant, the CIA, and the two congressional intelligence committees.

Congressional Supervision of Covert Action

Since the passage of the Hughes-Ryan Act in December 1974, the Congress has become a part of the decision process for covert action, though its role continues to be somewhat ill-defined and uncertain—and it was totally ignored during the Iran-contra affair. Once the president "makes a finding," the DCI or his designee (often the deputy director for operations) has the responsibility to notify the appropriate committees—presently the House and Senate intelligence committees (each with broad memberships from throughout the chamber, including some legislators who sit as well on the foreign affairs, armed services, and appropriations committees).[46] When the notification should take place and with what degree of specificity continue to be a matter of dispute between the branches.

Reporting to Congress

The wording of the 1980 Intelligence Oversight Act seems to require "prior notice," and this is the expectation of at least some members of the intelligence committees. Lee Hamilton, chairman of the House Intelligence Committee, stated unequivocably, "The Intelligence Oversight Act of 1980 provides that the Intelligence Committees of Congress should be given prior notice of activities such as those involving the transfer of arms to Iran."[47] In lieu of reporting to the two intelligence committees, the 1980 Oversight Act permits notification in times of "extraordinary circumstances" to the Gang of Eight, the Congress's designed leadership group for emergency CA reporting—a procedure used only once so far (a Reagan administration operation involving the establishment of antiterrorist strike teams in the Middle East).

Recent DCIs have hedged on this point, however. Admiral Turner claims that the law does not always require prior notice. A former White House legal adviser in the Carter administration has noted further that if President Carter had insisted on reporting to Congress in advance, Canada would have refused to cooperate with the United States in a covert action to help rescue a separate group of American diplomats from Tehran, who otherwise would have been rounded up as hostages in December 1979 like their less fortunate colleagues; the Canadians feared a leak that might endanger their own personnel involved in the operation. In this instance, Carter chose to ignore the prior-notice provision and waited until after the successful completion of the operation (over three months) before he reported to the intelligence committees—one of three instances when the Carter administration failed to abide by the prior-notice stipulation (see chapter 10).[48] Turner's successor, William Casey, did his best to avoid a clear statement on this dispute over prior notice—and most other subjects.

Once the committees learn of a finding (its invariably succinct wording bearing the president's signature is hand-carried to the committees), they must decide if and when they wish to receive details on the covert action. If either committee wishes to learn more, the practice has been for the panel members to listen to an oral briefing, without a written report (beyond the finding itself). The quality of these briefings has varied, usually depending upon how interested the legislators are in asking questions and how well informed these questions (and follow-ups) prove to be.

At the very first briefing on covert action presented to the House Intelligence Committee (officially known as the House Permanent Select Committee on Intelligence), Admiral Turner attempted—almost successfully—to prohibit a verbatim recording of the briefing by the panel for its files and subsequent study. The DCI said that such a record would present an unnecessary security risk. Rep. Les Aspin (D, Wisconsin) strenuously objected and persuaded the committee members that they and the staff could hardly carry out their oversight work thoroughly and responsibly without a clear record; relying on memory alone could lead to misunderstandings, elusiveness, and poor accountability. This "hard copy" proved invaluable to the committee subsequently when it needed to compare new proposals against those the CIA already had underway. The Senate Intelligence Committee (officially known as the Senate Select Committee on Intelligence) also demanded a verbatim transcript.[49]

Sometimes the CA briefings to the congressional committees can last hours, spilling over to a second or third meeting; sometimes they take only minutes, when the operation is self-explanatory and noncontroversial, or when the legislators have not had the time or inclination to focus on the proposal. Thorough briefings are more often the rule, for they take place infrequently enough so as not to become too great a surcharge on the time of busy legislators; moreover, covert actions have a certain aura of mystery about them, usually sufficient to capture the interest of committee members and stimulate a series of questions (not always on the mark, to be sure). On one occasion in 1978, an annoyed chairman of the House Intelligence Committee, Edward P. Boland (D, Massachusetts), sent DCI Turner back to his Agency to prepare a briefing on a finding with more detail than he seemed ready to deliver at the time; upon his return, the newly contrite director offered a full-blown description of the fine points and implications of the project—*sans* the names of agents, which are quite properly known only by their case officers and a few senior DDO officials.

The most egregious example of a poor briefing (and evidently poor questioning by committee members, too) occurred in April 1984 (until the Iran-contra scandal, the lowest point in Senate-CIA relations since the intelligence investigations in 1975; see chap. 10). In a briefing to the Senate Committee, DCI Casey slid over the fact that the Agency had escalated its covert actions in Nicaragua to include the mining of harbors. The president had approved this finding in February, but some members of the Senate Committee remained unaware of the risky operation (which endangered not only Nicaraguan vessels but international shipping, including that of American allies). The CIA had briefed the House Committee fairly well (though only because of persistent questioning by members on the details of covert action in Nicaragua) and some individual members of the Senate Committee; but when the full Senate Committee met (after several delays caused by legislative floor business, not the CIA), only one sentence—in a briefing lasting over two hours—dealt with the mining and was mumbled in Casey's inimitable fashion.[50]

In the days that followed, the DCI tried to deflect questions on the subject by flatly denying that *harbors* had been mined. Only later did senators discover this was merely a subterfuge. Casey had relied on a technical distinction: the CIA had mined the *piers* within the harbors.[51] This trickery outraged some committee members and caused even Casey's supporters to blanch. Not even the committee's chairman, Barry Goldwater (R, Arizona)—a reliable advocate on most occasions for each and every CIA project—heard and understood the initial reference to mining. Later, when he learned of the project from a member of his committee who had requested an additional, individual briefing, he was furious and sent a letter to Casey. "It gets down to one, little, simple phrase," he wrote, "I am pissed off!"[52]

Casey had managed a difficult feat: the alienation of one of the CIA's most reliable and powerful allies in the Congress. It is possible, of course, as the CIA argues, that some of the committee members and staff had been insufficiently alert during the briefing.[53] With his succinct reference to the mining, it may be true that Casey had honored the letter of the law, but "hardly the intent," writes his predecessor, Admiral Turner. "The CIA did go through the motions of informing, but it wasn't speaking very loudly."[54]

Then with the secret arms sale to Iran in 1985 (and the funneling of the profits to the contras in Nicaragua), the CIA failed to speak at all. Despite the 1980 Intelligence Oversight Act, the Congress received no report—indeed, by presidential order.[55] The president evidently hoped to guarantee the secrecy of the arms deal by taking refuge in the ambiguous escape hatch found in the preamble to the 1980 act, with its fleeting reference—among other statutory ambiguities—to the right of the president to remain "consistent" with his constitutional authorities and duties.[56] The refusal to report was doubly disconcerting for the Senate Intelligence Committee, because after the mining flap Casey had entered into a series of "understandings" with the committee (communicated through an exchange of still-classified letters) that henceforth he would report promptly, fully, and even audibly on all new covert actions and any notable changes in those already underway.[57]

The Iranian arms deal was a shock to legislative overseers in another important respect: for the first time, as far as we know, a finding was based on an oral approval by the president with no written documentation—precisely the slippery accountability Hughes-Ryan had been designed to overcome.[58] "The oversight process has been fractured," concluded the vice-chairman of the Senate Intelligence Committee, Patrick Leahy (D, Vermont).[59]

Legislative Prerogatives

The Agency's CA briefings are provided (by law) only when a "new" covert action is approved, or (by custom) when a "significant change" to an existing program is made. Following a briefing, committee members face a decision: how to react to the CA project already approved by the president and on the brink of implementation—if in fact the operation did not already begin the moment the ink on the president's signature dried. Under current law, the committees are not required to approve or disapprove a finding. The CIA must only report the finding to the committees and, by custom, then inform the White House of any dissent. Failure to dissent, however, is widely regarded within the executive branch as tacit approval of the finding by the Congress. The briefing on a finding, therefore, takes on greater significance than other CIA briefings; the committees, meeting separately, must decide whether to place their tacit authority and good reputation behind each covert action.

If members of the intelligence committees are displeased with the covert action, they have a number of possible responses. One or more members may voice reservations during the briefing and ask the CIA to make these objections known to the president. Obviously the greater the number of those objecting, the more seriously the negative response is taken in the White House. Exactly who is objecting can be important, too. If it is the chairman of the committee or, worse yet for the executive branch, the chairman and the ranking minority member, here is a force to be reckoned with. To emphasize the seriousness of their opposition, one or more members of the committees might formally write or even visit the president to stress why the project is ill-advised. A committee may decide to take a formal vote on a finding to register its feelings clearly, as the Senate Committee has done on a few occasions. At least twice a negative formal vote by the Senate Committee has caused a president to rescind his approval of a covert action (one instance reportedly involved the renewal of funding for the Christian Democratic Party in Italy[60]).

So even though the Congress has no legal role in the approval of findings, it clearly has an opportunity to critique a project. A formal negative committee vote—or even a tacit one—may be ignored by the president, but he may have to pay a heavy political price. He must work with these committees in the future, and the CIA must come to them for budgetary requests as well as other items of business now that the Congress has insisted on greater involvement in the making of intelligence policy. The members of the intelligence committees, especially the senior ones, are not good enemies for the White House and the DCI to make. The power of the purse held by Congress remains a potent corrective to uncooperative behavior in the executive branch. In 1978, for example, the Senate Intelligence Committee terminated one CA project during the panel's review of the CIA's annual budget request by simply striking the monies designated for it.[61] Commenting at a Tufts University Symposium on "Secrecy and U.S. Foreign Policy" (February 27, 1988), former DCI William Colby has underscored the importance of the annual budget authorizations controlled by the congressional intelligence committees: "In order to persuade the CIA to abandon a proposed covert action, a Committee chairman needs only to say to the DCI at the end of the briefing: 'Write down in your notebook $100 million, because—if you go ahead—that is what is coming out of your CIA budget next year.' "

Either of the committees may also take its opposition to the full house, which by the rules can meet in secret session to hear the case—though it is unlikely the project would remain "covert" with so many people hearing about it. Explicit resolutions before the parent chamber to terminate a covert action suffer the same problem: they lead to public exposure of a supposedly secret operation. Still, as the Clark amendment on Angola (1975) and the various Boland amendments to limit covert action in Nicaragua (1982–1986) illustrate, Congress occasionally reaches a point of sufficient frustration over some policies that a majority of its members are prepared to hold a debate in the chamber and vote on a "covert" action. On occasion, the Reagan administration seemed to encourage publicity on selected "covert" actions (creating a new phenomenon, the "overt-covert operation") as a way of generating pressure on legislators to fall in line behind a popular president.[62]

Ultimately any committee member who hears a briefing (or one of the few staffers who have access) could return to his office and telephone a reporter about the covert action. This "leak item-veto" could put a stop to the whole project. It has been shown time and time again, however, that few leaks come from the Congress; most come from the executive branch, where far more individuals are aware of any given intelligence operation (some 400,000 to 500,000 individuals in the executive branch had top-secret clearance in the 1970s!), and where bureaucratic rivalries and slips of the tongue by presidents and their aides have been responsible for many a disclosure.[63]

In sum, the Congress has no direct authority over the approval of covert action, but the very requirement of reporting on these operations serves as a strong deterrent against madcap proposals like those that surfaced within the intelligence bureaucracy more easily in the past. Yet as the Iran-contra case illustrates, stringent reporting requirements can produce an unintended effect: an incentive to bypass the established decision process. In most instances, however, a major force of congressional influence over covert action has been the law "of anticipated reactions" that political scientist Carl Friedrich knew to be so important in all executive-legislative relations; the potential for a negative reaction on Capitol Hill can have a sobering effect on bureaucrats who must obtain annual funding from the Congress—even if the occasional disregard of statutory prohibitions (like the Iran-contra affair) proves that this "law" is by no means made of iron.[64]

So, while the two intelligence committees do not have the formal authority to approve or disapprove of covert actions, they do hear about almost all of the important ones—the Iran-contra operation stands as the most important and troubling exception—and they may argue and even symbolically vote against them in closed meetings with CIA representatives. The covert action may be short-lived if the executive branch senses strong congressional opposition. Committee members have no veto, but they have the opportunity to persuade. If a sufficient number on an intelligence committee (presumably a majority, or at least an intense minority) objects to a particular operation—particularly on grounds that it is illegal—a prudent chief executive might well have second thoughts about what he believed was going to be a "quiet option."

Potential for Abuse

Although invisible to the general public and to most elected representatives as well, the decision process for covert action has nonetheless matured since 1974 into a complex matrix of checkpoints and overseers—too much so from the perspective of most intelligence professionals. "What we have is covert action by national consensus!" complained one exasperated deputy director for operations.[65]

Covert-action decisions can be time-consuming, exhausting, and nerve-wracking. One recent project was conceived in May of one year but not carried out until February of the next. In another instance involving a terrorist hijacking of an airplane, a counterterrorist team in a NATO country requested help from the CIA. The team sought expert advice from the United States on how to blow out the door of the airliner without harming passengers inside, a skill the CIA and one other Western intelligence agency had developed to a high degree. The CIA station in the NATO country cabled Headquarters for permission to help in this covert action. Hours passes; the hours turned into days. Finally, two days later, with still no decision from the United States, the NATO nation turned to the other Western intelligence agency, which responded affirmatively over the telephone on the first call. Within a few hours, PM commandos from this agency were enroute to the hijacked plane and soon successfully blew off the door for the counterterrorist team.[66]

For intelligence professionals, the conclusion to be drawn is self-evident: U.S. intelligence has been paralyzed, or at least maimed, by oppressive layers of decision makers and overseers brought on by congressional inquiries and investigative journalism run wild. The new circumspection has also produced, according to intelligence professionals, an avalanche of required paperwork for even the most minor operations, and has spread a profound sense of caution throughout the intelligence hierarchy (overcome only partially by President Reagan's efforts to "unleash" the intelligence community). No intelligence professional wants to be the focus of the next congressional inquiry. This attitude of restraint is, from one vantage point, laudable; excesses of the past might be less likely. The nation, however, can ill afford to have a crippled intelligence service at times when it might need one to act boldly and quickly.

The trick—not only for intelligence policy, but for democracy—is to achieve the proper balance between too many checks that stifle the reasonable initiatives of public servants and too few checks that lead to their abuse of power.[67] This balance is not achieved once and for all by the careful derivation of a scientific formula, but rather as a result of constant effort as the branches of government make readjustments in their positions to accommodate new evidence, pressures, and beliefs. Indeed, that such a debate can take place peacefully reveals the strength of American democracy.

The bureaucrats in the intelligence agencies make a strong case against micro-management—that is, too many executive and legislative policymakers enmeshed in the small wheels of covert action and other delicate operations best left to the pros. A case can also be made, however, that too much discretionary authority has been given to some parts of the chain of command. In 1975, Senator Church concluded at one point in his investigation that the CIA had lost control of its clandestine operations; it had become, in his words, "a rogue elephant"; and in 1983 Representative Boland observed similarly, after chairing the House Intelligence Committee for six years, that the CIA was "almost like a rogue elephant, doing what it wanted to."[68]

This harsh judgment comes from a deep sense of frustration over what these key legislators found to be dangerous CIA excesses, in the instance of Church the assassination plots with their Mafia connections, domestic spying (Operation CHAOS), and, among other abuses, shellfish toxins that were sequestered despite a presidential order to destroy them; and, in the instance of Boland—well before he learned of the Iran-contra end run—the CIA efforts to circumvent his amendments limiting PM operations in Nicaragua.

The past and present validity of the "rogue elephant" hypothesis remains a subject of lively debate, stirred up anew by the Iran-contra scandal. As the siphoning of funds to the contras outside the view of legislative overseers demonstrates, opportunities clearly exist to encourage CA projects even when they have been expressly prohibited by the Congress. To achieve its goal of overthrowing the Sandinista regime—despite the Boland amendments to limit further spending on PM operations in Nicaragua—the Reagan administration urged Israel, Saudi Arabia, South Africa, and Brunei, among other nations, to supply weapons and funding to the contras; lobbied wealthy American civilians to fund the contras; "borrowed" weapons from the Pentagon for the operation (since the CIA was prohibited from supplying weapons under the Boland law); gave monies to the contras beyond the legislatively imposed limits; and assigned a NSC staffer experienced in guerrilla warfare, Lt. Col. Oliver L. North, to provide guidance for the operations, while denying his involvement to legislative overseers.[69]

Although the 1980 Oversight Act explicitly covers any "entity" in the government, the Reagan administration attempted to argue that the NSC was excluded from intelligence reporting requirements, raising the possibility that to elude the Congress the executive branch might call upon any number of other unexpected agencies (say, the Drug Enforcement Agency, which in fact did become involved in the Iran arms-for-hostages scheme) to carry out covert action—or might bypass the government altogether by establishing an outside intelligence capability funded not through the appropriations process but by wealthy foreigners and private U.S. citizens (as occurred during the Iran-contra affair)—the privatization of covert action.[70] Were these extraordinary methods allowed to continue, it would spell the end of constitutional controls over American intelligence.

Sometimes, too, ambiguities over whether a field agent is a CA operative or an intelligence collector can cause confusion about proper authority (especially since an agent is often both, depending upon the opportunities presented to him). In the recent past, a liaison relationship with Country Z was first viewed as a covert action warranting a finding but was later continued without such authorization because the CIA decided the operation was essentially intelligence collection. In another case, the NSC gave the CIA official authority for an operation through a broad finding, though the project had been in existence for three decades already under the guise of intelligence collection. In a third case, alluded to earlier, the CIA had radio operators and intelligence collectors in Angola during the civil war in 1975. Just how far their activities went beyond these assignments is still at issue, but it is obvious that individuals in those positions could advise pro-West rebels in one way or another, perhaps during the heat of battle.

"Routine" covert actions present another hazard. They do not require a finding and, therefore, are neither reported to the Congress nor (in most cases) to the president. During the Carter administration, the CIA would seek outside authority for covert actions that seemed "moderately" risky, but this involved only the subthreshold notification procedure discussed earlier, in which the proposal was circulated to members of the SCC to see if anyone objected to it—in a loop that excluded Congress. A major new initiative against terrorists, for example, required a finding; but, short of that, the CIA had considerable discretion to counter this threat and only had to obtain "guidelines" from the Department of State. The problem with these "routine" operations is that in the tangle of notifications, guidelines, and other caveats and exceptions, the chain of command can become so loosened that the original policy intent changes—outside the vision of the White House at times and certainly the Congress. As McGeorge Bundy, who chaired the 303 Committee during the Johnson years, has observed: "It can happen and I think it has happened that an operation is presented in one way to a committee [the 303] and executed in a way that is different from what the committee thought it had authorized."[71]

Within the CIA itself, instances have occurred in which units other than the CAS have planned and implemented CA projects without the approval or even the awareness of authorities outside the Agency (or even the CAS). In one widely reported example, James Angleton, as chief of the CIA Counterintelligence Staff, is said to have doctored the famous "secret speech" delivered by Nikita Krushchev following the death of Joseph Stalin. By adding deceptive paragraphs to the speech and circulating it in Eastern Europe, Angleton apparently hoped to stimulate uprisings against the Soviet regime by painting an even more venal portrait of the Stalinist era than did the unadulterated speech itself.[72] Whether or not Angleton ultimately carried out this deception operation remains a matter of conjecture, but that it would have amounted to covert action (secret propaganda) under the guise of counterintelligence is self-evident. The arcane arts of counterintelligence thus represent yet another avenue to CA projects that could be initiated without the benefit of NSC or congressional review.

A constant quandary in the CIA—as in every other bureaucracy—has been how to control overzealousness or plain unlawfulness at lower echelons that are less visible to the public, to the Congress, and to the bureau chiefs themselves. Individuals working for the Agency may simply ignore properly authorized policies, guidelines, directives, orders, and laws promulgated by the Congress, the president, and the DCI. The Nicaraguan assassination manual is only a recent example.[73] This difficulty is magnified in the intelligence community where agencies from top to bottom are virtually invisible to outside eyes. Commenting on the problem of internal CIA controls during the Carter years, a senior DDO official said with some anguish and with particular reference to his PM cadre: "What do you do with the firehorses when there's no fire!"[74] Ruthless former Somosa National Guardsmen now in the contra army in Nicaragua, Mafia hitmen, and ideologically fervent Cuban exiles have numbered among the CIA assets who have proved difficult—if not impossible—to control (not to mention a few CIA officers, like Edwin P. Wilson, who have sold weapons as well as secrets to terrorist groups and enemy nations).[75]

Further, career incentives and other bureaucratic pressures can lead to CA schemes merely to justify the existence of PM cadres, CA case officers, and their global infrastructure. One former senior CIA administrator points to the "bureaucratization" of this mission as an unfortunate development within the Agency: "In the early years it was a bunch of ex-OSS officers who were, in effect, freelancers. But when it became institutionalized and [the CIA] began recruiting junior people from the Ivy League, training them, setting them upon career patterns, one began to create incentives and bureaucratic pressure for routine covert-action operations. . . . it was a great incentive, because of the bureaucratization of covert action, to create a 'Communist menace'. . . ."[76] Similarly, Senator Church noted in 1976 that the "CIA is a bureaucracy which feeds on itself, and those involved are constantly sitting around thinking up schemes for foreign intervention which will win them promotions and justify further additions to the staff." He continued, "In this way, the CIA has grown in the way most bureaucracies do. And it self-generates interventions that otherwise never would be thought of, let alone authorized."[77]

Successful control over covert action seems to depend heavily upon the attitudes of men and women in the national security establishment and the leeway they are given by the public and its representatives. The interplay between the two can be complicated. Sometimes the public seems willing to allow wide discretion in the use of America's intelligence agencies. In an attempt to achieve leeway, the White House can try to stimulate a sense of crisis and affect the public's perception of external events. The preoccupation of the Reagan administration with Soviet threats, U.S. vulnerabilities, and tumult in the Third World seems to have created to some degree a public uneasiness about America's security and an increased tolerance toward the use of covert and overt force abroad. Sometimes public anxiety springs directly from media reports about events that present danger to the United States, as with the bombing of Pearl Harbor; in such instances, questions of accountability and civil liberties are often displaced by concern for order and security. The internment of American citizens of Japanese descent following the attack on Pearl Harbor is a classic illustration.

In contrast, the public may express dissatisfaction with an unchecked use of intelligence agencies. When the intelligence community (or any other government organization) has been shown to break the law without compelling justification, or in some other way to have transgressed acceptable boundaries, a period of public criticism and resulting bureaucratic retrenchment sets in. During the Carter administration, intelligence officers were frequently inclined to comment that "in the prevailing atmosphere" strong pressures existed both within the CIA and outside to tighten command and control and reduce as much as possible any risk of further controversy. When asked if the CIA would scrupulously follow Hughes-Ryan reporting requirements, the DDO in 1978 responded with a resounding: "Yes! I'm not going to jail!"[78] Within the NSC, President Carter's assistant for national security affairs, Zbigniew Brzezinski, recalls that withholding information about important covert actions from the president "simply didn't occur to us. There was, if you will, a legalistic, an ethical mind-set which simply precluded that as a possibility. . . ."[79]

Ransom has argued persuasively that when a foreign-policy consensus exists, as with the staunch anti-Soviet mood of the 1950s, the intelligence agencies are given greater discretion.[80] Recently, such consensus has been lacking (despite efforts by the Reagan administration to promote the dangers of Soviet global machinations), and the agencies have felt the close eye of reformers—doubly so since the Iran-contra revelations. In a word, real or contrived threats from aboard can lead to public tolerance for giving discretion to intelligence agencies; in contrast, an absence of perceived threat, or the presence of a scandal involving the intelligence agencies, can result in a call for tighter controls.

"With today's supervision, and with the command structure trying to keep things straight, the people in CIA know what they should do and what they should not do—as distinct from the 1950s, in which there were no particular rules," concludes former DCI Colby. "If CIA people today are told to violate their limits, or if they are tempted to violate those limits, one of the junior officers will surely raise that question and tell the command structure, and, if not satisfied there, he will tell the Congress, and, if not satisfied there, he will tell the press, and that is the way you control it."[81] The Iran-contra disclosures demonstrate that this control system is far from perfect, but the checks now in place—and the added warnings to future administrations that have come from the 1987 executive and congressional investigations, coupled with the criminal indictments handed down in 1988 to the principals in the Iran-contra affair—amount to a vastly more serious effort to cope with the supervision of covert action than was even remotely attempted before 1975.[82]

Conclusions

From 1947 to 1974, intelligence agencies decided upon and conducted covert actions with only limited accountability, but in December 1974, Congress insisted upon a much more formal decision process and included several new actors to authorize and monitor this policy option. From 1975–1980, Congress and the Carter administration further tightened controls over covert action, culminating in the Intelligence Oversight Act. For better or for worse, covert action had been democratized, with representatives of the people in Congress now more closely involved in its supervision.

Examples of insufficient accountability continue to occur even with the rigor of the new procedures, as illustrated by the incidents of harbor mining and the assassination manuals in Nicaragua, as well as—most conspicuously—by the Iran-contra affair. Short of further clarifying the congressional expectations with regard to reporting and instituting legal sanctions against those who lie to the intelligence committees (or withhold the truth, which amounts to the same thing), the existing oversight procedures for covert action seem to have struck an appropriate balance between control and discretion. The larger problem lies in the unwillingness of some executive officials to honor the procedures, and, as the Inouye-Hamilton committees put it, to "deal in a spirit of good faith with the Congress."[83] When asked by congressional investigators on these panels why he had withheld information from the intelligence committees about the secret sale of arms to Iran, Vice Admiral Poindexter responded: "I simply didn't want any outside interference."

While the new procedures have worked for the overwhelming majority of covert actions, they have been most often ignored by the executive branch in its planning and implementation of PM operations—as the Iran-contra affair vividly illustrates. In their investigation of this scandal, the Inouye-Hamilton committees and the Tower commission revealed weak links throughout the CA chain of command. President Reagan delayed reporting to the Congress on the Iran arms sale far beyond the intent of the 1980 Intelligence Oversight Act; DCI Casey, the president's close friend and political ally, went along with this (perhaps advocated it) and seems to have encouraged the questionable exploits of Lieutenant Colonel North as well as efforts to establish private funding for covert action; top CIA personnel in the Operations Directorate participated in the contra-fund diversion, even though they realized that this represented a violation of the Boland amendment; several officials, including Operations Directorate personnel and the Department of State's assistant secretary for Latin America, lied to the Congress about the existence of these operations; and the COS in Costa Rica aided the contra army, again in violation of the Boland law. A decade earlier, the Church committee revealed other startling examples of CIA case officers exceeding proper limits—one of whom, the reader will recall, went so far as to hire the Mafia hitmen who were supposed to kill Castro.[84]

Irregularities at the top of this chain are the most alarming, for here are people with great power who can do much mischief. This is the link that deserves the closest scrutiny by overseers. Yet, consistently the weakest link—that is, the place where the new rules of accountability are most likely to be ignored—seems to be at the bottom, beneath the case officers at the level of agents in the field. Here is where rogue elephants are most apt to roam, far away from the rules and regulations of Headquarters and the Washington community; here are the often unsavory and unpredictable soldiers of fortune—like the Mafia hitmen, the factions within the CIA-funded Afghan mujahadeen accused of selling U.S. missiles to Iran[85], or the former Somosa national guardsmen among the contras—who are inclined to have their own agendas that may or may not be congruent with the details of NSC directives.

Like original sin, covert action is unlikely to disappear. Presidents, regardless of their ideological persuasions, are apt to find the quiet option an attractive alternative to the overt use of force. In light of this reality, the objective must not be to ban covert action but—in so far as possible—to bring it within a democratic framework. The remedies are clear, though difficult: at every level of the chain, accountability will continue to depend upon clear guidelines, timely reporting, honest officials, and dedicated overseers willing to invest the necessary time—through hearings, audits, inspections, and less formal discussions with intelligence officers.

Beyond these efforts to find a proper balance between micromanagement and insufficient management lies a broader policy question from which the use of covert action directly flows: To what extent should the United States intervene in the affairs of other countries—especially through hidden means that hold limited opportunity for open debate? Covert action is only one instrument of American foreign policy, and its use tends to reflect broader policy goals pursued by the nation. As a consequence, how one views the quiet option is apt to be a function of one's general approach to the question of how the United States ought to achieve its international objectives. On this question, Dr. Ray Cline, formerly a high CIA official, is representative of the "realist" school. According to him, "we are already engaged in a protracted secret war against the Soviet Union." As a result, the United States needs to get on with the business of winning that war, using covert action wherever it may aid that objective. "The United States is faced with a situation in which the major world power opposing our system of government is trying to expand its power by using covert methods of warfare," Cline continues, asking rhetorically: "Must the United States respond like a man in a barroom brawl who will fight only according to Marquis of Queensberry rules?"[86]

In contrast, George W. Ball, under secretary of state during the Kennedy and Johnson administrations, is representative of the "idealist" school, with its emphasis on world public opinion, image, and morality. "In principle I think we ought to discourage the idea of fighting secret wars or even initiating most covert operations," he argues. "When the United States violates those principles—when we mine harbors in Nicaragua—we fuzz the difference between ourselves and the Soviet Union. We act out of character, which no great power can do without diminishing itself. . . . When we yield to what is, in my judgment, a childish temptation to fight the Russians on their own terms and in their own gutter, we make a major mistake and throw away one of our great assets."[87]

The ongoing debate over the appropriate decision paths and degree of oversight for covert action is important. The proper use of the quiet option, however, must be considered within the broader framework suggested by the dispute between realists and idealists. Which of these two impulses do the people of American support, and to what extent do they wish this country to intervene in matters beyond its shores? While Americans continue to wrestle with these fundamental questions, upon which the use of covert action rests, friends of democracy may enjoy one comfort: judgments about covert action are now more likely to be made with the participation of elected representatives in Congress—and not by the president and his clandestine services alone.

III

THE CIA AND THE RIGHTS OF AMERICANS

SEVEN

The Huston Plan

Mr. Huston and Colonel North

In the aftermath of the Iran-contra affair (1985–86), Lt. Col. Oliver L. North of the Reagan administration received widespread notoriety as the "action-officer" who guided the controversial operations. North served as an aide to President Reagan's assistant for national security affairs, Vice Adm. John M. Poindexter, on the NSC staff. In this capacity, he became deeply involved in the White House scheme to sell arms secretly to Iran in exchange for that country's help in gaining the release of American hostages held in the Middle East. Joining forces with high-ranking officers in the CIA—including its director, William J. Casey—and private Americans with contacts in the Middle East, North stood at the center of a series of top-secret intelligence operations (the "Enterprise" in his codename) that would spread a dark stain of scandal across the administration in its last two years of power. When revealed by a Middle East magazine in 1986, the arms-sale caper drew extensive criticism in the Congress and the media; Iran itself often harbored or otherwise encouraged terrorist groups and hardly seemed an appropriate country for the president of the United States to negotiate with.

Equally controversial was the second phase of the scandal—evidently carried out without presidential approval—in which profits from the Iranian arms sale were secretly diverted through a Swiss bank account to the CIA-backed contras in Nicaragua, who were engaged in covert warfare designed to overthrow the pro-Marxist Sandinista regime. The diversion of funds to the contras took place without reports to the congressional intelligence committees (legally required under the provisions of the 1980 Intelligence Accountability Act). Moreover, congressional investigators on the Inouye-Hamilton committees charged in 1987 that the diversion had violated the Boland amendment, which placed strict limits on the sale of lethal military equipment to the contras. Committee investigators also discovered that Lieutenant Colonel North and his associates had bypassed the normal appropriations process by obtaining funding for the contras from wealthy American citizens and foreign leaders—a privatization of U.S. foreign policy that threatened to undermine Congress's constitutionally mandated check on foreign policy-making through the power of the purse.

As the spotlight of public attention focused on the Iran-contra affair, forgotten was another case of high-level intrigue that only two decades earlier had involved a young White House aide who was similarly all too ready to cross the bounds of law and propriety in the zealous pursuit of administration objectives. In 1970, White House staffer Tom Charles Huston of the Nixon administration aggressively shaped major national security initiatives with little authority from the president or other members of the NSC. As the coordinator of a top-secret proposal for intelligence collection that now bears his name, Huston's machinations struck deeply at the roots of civil liberties at home. The so-called Huston Plan recommended the secret removal of legal restraints on the CIA, the FBI, the NSA, and other U.S. intelligence agencies so they could spy on American citizens. North and Huston: two young men with little intelligence experience, suddenly at the center of power and given key roles in intelligence operations, both determined to advance their vision of what was right for America—regardless of what "outsiders" might think, regardless of what laws might stand in the way.

In terms of physical appearance, the two men had little in common beyond their youth. North, the rugged Marine and decorated combat veteran, had been the boxing champion in his class at the U.S. Naval Academy. At the time of the Iran-contra investigations, he still looked fit enough to go twelve rounds. In contrast, Huston, a former Army intelligence officer, was a slightly built man with a finely boned face, spectacles, and thinning hair, more at home behind a book than a bazooka. The Warrior and the Intellectual. Yet appearances aside, the two men had attributes that closely converged. Both were junior members of their administrations; both were sincere—even zealous—in their convictions about the "rightness" of their cause; and both were driven by a bugbear—for North the evil of the Communists in Nicaragua, for Huston the tatterdemalion Vietnam War demonstrators marching through the streets of America.

Both men serve as reminders of the ease with which the American secret service can be misused, and of the dangerous consequences that can arise when power is delegated to individuals with little regard for the law or for Congress. What follows is an account of how Tom Charles Huston used an ambiguous grant of presidential authority to turn the CIA and other intelligence agencies against the American people. This case and the Iran-contra affair stand as poignant illustrations that America's democracy is fragile, that—as the nation's founders warned—power is easily abused, and that the best laws are of little use if the men and women in high office are prepared to disregard them.

Counterintelligence Comes Home

Counterintelligence (CI) is "that phase of intelligence activity devoted to countering the effectiveness of hostile foreign intelligence operations."[1] Its purpose is to learn about "the enemy" in order to oppose whatever plans he many have to injure the nation or its intelligence services. In this sense, the drafting of a master spy plan in June 1970—now known as the Huston Plan—was an act of counterintelligence. The men who participated in drafting it were primarily CI specialists, only this time the enemy consisted of antiwar protest groups in the United States—and foreign powers suspected of supporting them. (See fig. 7.1.)

The Huston Plan as Counterintelligence

Early in the Nixon administration, key White House officials reached the conclusion that the greatest threat they faced was not the Soviet Union or China but domestic radicals. As the political chronicler Theodore H. White has observed, the domestic unrest sweeping the nation shook President Richard M. Nixon: "[P]erplexed by a street madness which seemed beyond the control of either his staff, his own efforts, or the F.B.I., he groped for solutions."[2] The White House staff presented the Huston Plan to the president as one possible solution. Counterintelligence experts were asked to review existing intelligence collection procedures and to underscore those restraints which, if removed by the president, would result in better information about the schemes of domestic radicals.

The story of the Huston Plan reveals a president frustrated by domestic upheaval; a young White House staffer with ambition and a strong ideological disdain for Vietnam War protesters; a group of insurgents within the intelligence community devoted to the elimination of restraints on foreign intelligence collection; an aged FBI director determined to protect the Bureau's—and thus his own—image; and, as in the Iran-contra affair, a host of high officials prepared to set aside laws, regulations, and a presidential order in pursuit of their own intelligence objectives.

The episode raises vital questions about the American intelligence community: Who rightfully is "the enemy," both within the United States and abroad? Who should determine which kinds of operations are acceptable to thwart hostile foreign intelligence services—that is, who is to set, approve, and oversee CI objectives? What intelligence collection methods are legitimate in a democracy, and under what conditions? What degree of control should the White House and the Congress have over CI operations?

To these questions the Huston Plan represented one set of answers offered by a group of senior CI specialists during the summer of 1970. At best, these answers serve as provocative stimuli for debate on the scope and methods of American intelligence; at worst, they represent a dangerous descent from the worthy objectives for the intelligence community that President Truman and the drafters of the 1947 National Security Act aspired to. Perhaps above all the Huston Plan stands as a reminder of what can happen—even within a vibrant democracy—

FIGURE 7.1 The Huston Plan as Counterintelligence

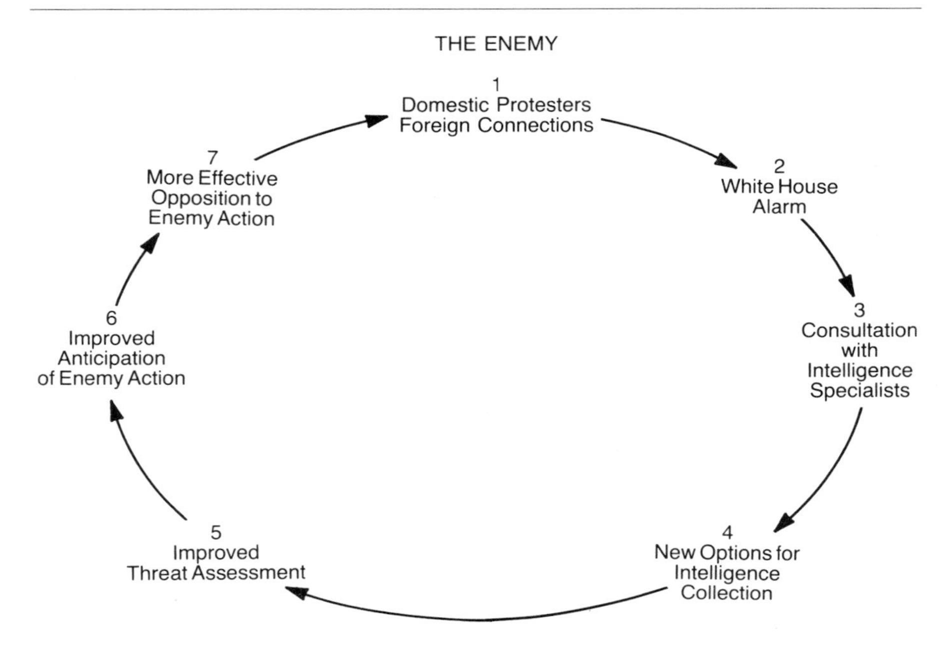

when secret agencies succumb to improper White House commands and their own overzealous devotion to intelligence objectives.

A Time of Turbulence

Richard M. Nixon won his first presidential election in 1968 by less than 1 percent of the total popular vote. The election was a cliffhanger, marked by the most violent and bloody street demonstrations in the history of modern American elections. His first year in office provided the president with ample further evidence of a rebellious mood in the country. In March and April 1969, student riots erupted in San Francisco, Cambridge, and Ithaca, and in Chicago, ghetto blacks battled police in the streets. By October and November, the antiwar movement had become sufficiently well organized to bring to Washington, D.C., the largest mass demonstrations ever witnessed in the United States. Just as the country was obsessed by Vietnam, so, too, did the White House become obsessed with the waves of domestic protests.

In April 1969, the president asked a senior aide, John Ehrlichman, to prepare a report on foreign Communist support of campus disorders. The evidence provided to Ehrlichman by the intelligence community revealed that foreign connections were virtually nonexistent. The president and his aide were dissatisfied with this finding, however, believing it to be inconclusive. Two months later, Ehrlichman assigned a twenty-nine-year-old White House counsel on Pat Buchanan's research and speech-writing staff to prepare a second and more thorough report on this subject. Fresh to the White House staff following a tour of duty as an

Army intelligence officer assigned to the Pentagon (he had worked in the Nixon campaign during his off-duty hours), Tom Charles Huston of Indiana drew the assignment chiefly because he seemed to know more about radical ("New Left") student politics than anyone else in the White House. At Indiana University he had been a rightist student leader opposed to the antiwar dissenters on campus; he now suddenly found himself in a commanding position to carry on the battle, which he took up with the zeal one might have anticipated from a former president of the outspoken conservative youth group, Young Americans for Freedom.

On June 19, 1969, Huston paid his first visit to William C. Sullivan of the FBI. Since 1961, Sullivan had served as the Bureau's assistant director for domestic intelligence, which included responsibility for counterintelligence. Huston related to Sullivan the substance of a recent meeting he had had with the president. Concerned about revolutionary activities by the New Left, Nixon wanted to know the details on the radical movement, "especially," Sullivan remembers Huston emphasizing, "all information possible relating to foreign influences and the financing of the New Left."[3] Sullivan informed Huston that he would have to put his request in writing to Mr. J. Edgar Hoover, the legendary director of the FBI.

The next day, Huston prepared the request. With the FBI's earlier report for Ehrlichman in mind, he brashly informed one of the most powerful men in Washington—and certainly the most heralded government authority on the Community Threat—that U.S. intelligence on the influence of Communists over the antiwar dissenters was "inadequate." On behalf of the president, Huston wrote that he wanted to know what intelligence gaps existed on this subject and what steps could be taken to provide the maximum possible intelligence coverage of domestic radicals. Huston gave the same assignment to the CIA, the NSA, and the DIA.

Each agency submitted a report to Huston before his June 30 deadline. Sullivan, who was in charge of preparing the FBI's response, argued that increased electronic coverage (that is, telephone wiretaps and other listening devices, or "bugs") would be necessary to obtain the data sought by the White House. The Bureau's CI experts warned further that "it appears there will be increasingly closer links between [the New Left and black extremist movements] and foreign communists in the future."[4] The quality of these reports failed to satisfy either Ehrlichman or Huston, and a cloud of disaffection toward the intelligence community settled over the White House.

The Huston-Sullivan Alliance

Throughout the rest of 1969, Huston was responsible for receiving and disseminating FBI intelligence reports sent to the White House. The White House staff expressed a growing contempt for these reports. Huston, however, began to moderate his negative views on Bureau intelligence as he became more acquainted with Sullivan. Listening to the CI specialist made Huston sympathetic to the difficulties of intelligence collection under the restraints imposed by Mr. Hoover.

From June 1969 through the following year, the contacts between Huston and

Sullivan deepened into a working alliance devoted to the lowering of the intelligence collection barriers within the FBI. Though far different in temperament, age, and experience—the quick-tempered Sullivan was old enough to be Huston's father and was a seasoned intelligence professional—the two men found themselves in agreement on several issues. Both viewed the spiraling unrest in the country with grave alarm; both believed in the need for improved coordination among the intelligence agencies throughout the government; both thought the quality of data on domestic radicals was woefully inadequate; and both agreed that most of the CI deficiencies could be remedied if only the intelligence agencies—especially the FBI—could reinstate collection methods that had been common practice "in the good old days."

The New Hoover

Counterintelligence specialists throughout the executive branch were dismayed when Hoover, in 1966, suddenly suspended a number of useful and long-standing undercover FBI operations. The Kennedy administration's emphasis on investigations into organized crime and civil rights violations had already drained personnel from the Bureau's security and intelligence operations. Now Hoover began to terminate specific security programs. In July 1966, for example, he issued a memorandum ordering the termination of most FBI break-ins (known in the trade as "second story" or "black-bag jobs")—burglaries by the Bureau to obtain evidence against domestic "subversives" or information regarding the operations within the United States of foreign intelligence agents. Then, five months later, Hoover added FBI break-ins of foreign embassies in Washington, D.C., to the off-limits list.[5] Tom Charles Huston believed that, with its refusal to use a full array of intelligence-collection methods, the FBI was failing to do its job—a belief shared widely among intelligence professionals. Intelligence chiefs Richard Helms (CIA), Donald Bennett (DIA), and Noel Gayler (NSA) all expressed this view, as did—outside the hearing range of Mr. Hoover—key CI officers within the FBI itself.[6]

Word drifted among the intelligence professionals about "the change" in Hoover. Why was he reluctant now to order what he had allowed before on a regular basis? Some suggested that the wiretap hearings conducted by Sen. Russell Long (D, Louisiana) in 1965 had turned public opinion against the use of certain intelligence-gathering techniques. The shrewd FBI director was merely reading the writing on the wall. Others pointed to the increased risk involved in embassy break-ins because of the presence of the Executive Protection Force, created in 1966 to guard (among other things) Washington's diplomatic buildings. Hoover, according to this theory, was unwilling to engage in past practices when faced with new dangers of being caught.

Most highly placed observers within the intelligence community thought, however, that the director was simply growing old and more wary about preserving his reputation—a wariness nurtured by the protective instincts of his longtime roommate and alter ego, Clyde Tolson, the No. 2 man at the Bureau. Both Dr. Louis Tordella, the top civilian at the NSA, and Sullivan thought that Tolson probably had told Hoover something to the effect that "if these techniques ever backfire, your image and the reputation of the Bureau will be badly damaged." In Tordella's words, Hoover was "like the airplane ace who feels that sooner or later he's going to get shot down."[7] Tolson may have seen himself as an endangered copilot.

Various intelligence professionals screwed up their courage in 1967 and asked Hoover to reinstate the break-ins. Soon after the operations against Washington embassies had been terminated, Tordella and Gen. Marshall Carter (NSA director at the time) paid a call to the Bureau director. A fifteen-minute appointment stretched into two-and-a-half hours as the SIGINT experts endured Hoover's ritual stories about the gangsters Dillinger and Ma Barker, the Communist Threat, and other favorite topics. At last, Tordella and Carter were able to explain to Hoover how much money embassy break-ins had saved the taxpayer by providing valuable coding information that would have taken the NSA thousands of hours and expensive computer time to develop. Hoover seemed to soften. He said their argument was persuasive; he would reconsider the issue.

A few days later, the Bureau liaison to the NSA delivered happy news: Hoover would reinstate the break-in policy. Then, three weeks later and before any embassies had been entered, the liaison returned with another message: the anti-break-in policy was to remain intact after all.

Sullivan, who would have been in charge of the break-in program, called to tell Tordella that "someone got to the old man. It's dead." That someone, "Sullie" (Sullivan's nickname) surmised, was Tolson. Hoover added one sentence to his message, though, that suggested a way to circumvent the impasse: the Bureau would conduct break-ins for the NSA—if the operations were approved, in writing, by either the president or the attorney general. This did not seem like much of a solution to Tordella, since he for one had no desire to plead NSA's case before the attorney general. "I couldn't go to the chief law enforcement figure in the country," he recalls, "and ask him to approve something that was illegal."[8] Tordella seemed to perceive a distinction between requesting Hoover to break the law and asking the attorney general to do so. Moreover, he and Sullivan felt that the president should never even know about the break-in dispute. This was not a subject with which the chief executive should soil his hands—an attitude comparable to that held by Vice Adm. Poindexter of the Reagan NSC staff with respect to the contra fund-diversion.

Tordella, Sullivan, and others in the intelligence world grew steadily more impatient with Hoover's obstinacy on the question of intelligence collection. If they were to counter the enemy, their hands would have to be untied. Yet no one was willing to take on Hoover directly. Tordella and Carter had tried their best shot and failed. Helms and Hoover had little patience for one another. (Hoover distrusted CIA personnel, with their stylish, button-down collars and Ivy League educations: "Ph.D. intelligence" was a derisive term he often used against the Agency.) And Gayler and Bennett, newcomers to the intelligence community, had been warned immediately by their aides not to take on the FBI director on matters related to domestic intelligence. To cross Hoover, the James Bowie of bureaucratic in-fighting, was considered unhealthy. It would take the pressure of events, a strategically placed intelligence greenhorn from Indiana, and strong encouragement from CI officers on the sidelines to mount an assault on Hoover's "New Morality."

The Pressure of Events

Events force decisions. The riots and bombings escalated across the country in the spring of 1970. Within one twenty-four-hour period, four hundred bomb threats occurred in New York City. In March the explosion of a Greenwich Village bomb factory operated by the Weathermen (a New-Left group) initially stunned, then angered the White House. A month later, additional Weathermen bombings in New York, plus the sight of forty thousand protesters marching down Pennsylvania Avenue toward the White House, brought the president's men to a resolution: as soon as possible, Nixon would have to meet with the intelligence community chiefs. Better intelligence on the protesters and their foreign sponsors had to be produced. Rumors of possible kidnappings of high government officials and their families added a sense of urgency to the staff decision. In the aftermath of the April 1970 "incursion" by the U.S. military into Cambodia, the subsequent explosion of domestic outcries over this perceived expansion of the war in Indochina, and the tragic shootings of demonstrating students at Kent State and Jackson State universities, the meeting seemed all the more necessary.

In response to this violence, Washington officials held a series of seven important meetings on intelligence during June 1970. These meetings began on June 5 in the Oval Office with a conference between the president and the intelligence directors, during which Nixon requested the preparation of an intelligence report, the third one the White House had asked for on this topic in the past year, only this time with the added clout of a direct presidential order. The president wanted the report to outline what could be done to quell the violence. Twenty days later, the series of meetings would end in Hoover's office, where the intelligence directors gathered to sign the report for the president.

Huston was responsible for arranging the initial conference between the president and the intelligence leaders, having briefed Nixon in advance by means of a paper that he had prepared based on conversations with Sullivan. Sullivan's role was to tell Huston what changes in CI methods were desirable; Huston was expected to turn the desirable into the possible with the leverage he enjoyed within the White House as point man for domestic intelligence issues. If Huston could get the president to support Sullivan's views (shared by the other CI specialists), then presumably these views would become policy.

The session with President Nixon lasted less than an hour and consisted mainly of his remarks (based on briefing notes prepared by Huston) about domestic violence and his complaints about the quality of intelligence on student radicals and their foreign connections. At the end of his comments, Nixon leaned over to his long-time friend, J. Edgar Hoover, sitting to his right, and asked him to take charge of preparing a report on the security threat and intelligence shortcomings. The report was to include any tie-in between domestic groups and foreign influence, as well as a thorough discussion of restraints inhibiting better intelligence collection. The committee to draft the report would be known as the Interagency Committee on Intelligence (Ad Hoc).

Nixon ended with this observation to Hoover: "I understand you have a fine man there in Bill Sullivan. Isn't that right?" Hoover agreed. "Well, then," continued the president, "I think he ought to be chairman of the staff work-group." Sullivan and Huston were off and running.[9]

The president wanted the report completed within a month, so the Interagency Committee began its work the very next day, convening for reasons of convenience (large conference rooms and ample parking space) at CIA Headquarters in Langley, Virginia. As the staff began its work, their reactions were mixed. Some expressed delight at the turn of events. For years a group of discreetly unhappy CI specialists within the FBI had longed for the reinstatement of collection procedures terminated by Hoover. As one participant at the Langley meetings recalls:

> Hoover put us out of business in 1966 and 1967 when he placed sharp restrictions on intelligence collection. I was a Soviet specialist and I wanted better coverage of the Soviets. I felt—and still feel—that we need technical coverage on every Soviet in the country. I didn't give a damn about the Black Panthers myself, but I did about the Russians. I saw these meetings as a perfect opportunity to get back the methods we needed—and so did Sullie.[10]

The CIA, NSA, and most of the FBI representatives seemed to share this enthusiasm, with varying degrees of skepticism that the plans would ever get past Hoover. The military intelligence representatives were an exception; they wanted nothing to do with Huston's (perhaps more accurately, Sullivan's) plan. The Army was already facing the prospect of impending Senate hearings on its improper surveillance of civilians during the 1960s and vowed, according to its representative at the meetings, "to keep the hell out."[11]

Huston had his own agenda—to stop the antiwar radicals and to advance his career prospects at the White House—but he also saw himself acting in the capacity of a sympathetic White House staffer passing on to the president what the intelligence professionals wanted. "And I agreed with them," he remembers. "I say 'agreed.' After you work with somebody and you are convinced that what they want to do is right, you agree with them." He had no doubt that the CIA, the NSA, and others were in the venture for reasons other than strictly to improve domestic intelligence. Not that he always understood their objectives—especially the arcane intelligence collection needs of the NSA, with all of its fancy, high-tech electronic paraphernalia. "For all I know that thing [a policy directive on signal intelligence that the NSA was trying to sell Huston] could have authorized people to have free lunch in the White House Mess," Huston admits. "In other words, Admiral Gayler said, 'This what needs to be done,' and that's what I did."[12] Huston could help them; they could help Huston—a perfect symbiotic relationship.

(According to congressional investigations, a similar relationship evidently existed in 1985–86 between Lieutenant Colonel North and CIA Director William J. Casey during the Iran-contra operations. North sought funding for the contras. He had gotten to know these counterrevolutionaries [whom President Reagan once referred to as the "equivalent of America's founding fathers"] and passionately believed in their cause as "freedom fighters" [also the president's description]. North decided that support for the contras could be achieved only by skirting the Congress and its Boland amendment restrictions. In William Casey he found a willing ally. As North told investigators on the Inouye-Hamilton committees in July 1987, the CIA Director longed for a concealed "off-the-shelf, self-sustaining, stand-alone" capability to fund intelligence operations without having to go before a recalcitrant Congress for appropriations. North could help Casey establish a secret treasury for the CIA, and the CIA could help with the Iran-contra operations.)

At last, on June 25, 1970, the forty-three-page "Special Report" was ready to be signed by the intelligence directors, who met in Hoover's office in downtown Washington. Hoover called the meeting to order and began his normal routine for the review of official reports, which involved going through the document page by page and, for every page, asking each person around the table for comments. Each time Hoover came to Tom Huston, he would use a different name: "Any questions, Mr. Hoffmann? Any questions, Mr. Hutchinson?" and so on, getting the name wrong six or seven different ways. This was the director's way of showing his dislike for the rookie White House intellectual. Sullivan recalls that it was difficult to keep from laughing. He bit his lip and tried staring at the floor. Helms leaned back on his chair and winked at Huston behind Hoover's back. With some carping from Gayler and Bennett about a few passages (which greatly annoyed Hoover, who was unaccustomed to interruptions—though he accepted these minor changes grudgingly), the FBI director completed the reading and each of the four intelligence chiefs signed the document. Sullivan and Huston had achieved their goal—or so it seemed.

On June 26, an intelligence courier delivered the "Special Report on Intelligence" (hereafter the "Special Report") to the White House. For each of the major intelligence collection methods, the document laid out for the president the options of continuing the present restrictions, asking for more information, or accepting relaxations on the restrictions.

Should the president decide to lift the current restrictions, Huston advised a face-to-face "stroking session" with Hoover in which the president would explain his decision and indicate that "he is counting on Edgar's cooperation. . . ."[13] In this way, Huston continued, "We can get what we want without putting Edgar's nose out of joint." Though the director was "bullheaded as hell" and "getting old and worried about his legend," Huston predicted that he would "not hesitate to accede to any decision the President makes." He attached to this optimistic appraisal his own specific recommendations on the options the president should select in order to lift existing restraints on intelligence collection.

The Huston Plan Recommendations

Huston's recommendations, written under the heading "Operational Restraints on Intelligence Collection," made up the heart of his plan.[14] He offered advice on each operational section of the "Special Report," and then each of his recommendations was buttressed by a rationale taking up one or more paragraphs. His plan, as presented to the president under the classification "Top Secret," is outlined below, without the accompanying rationales (which concluded chiefly that the current state of intelligence collection against domestic radicals was inadequate and that all the methods that Hoover had abandoned had once been used with great productivity):

Communications Intelligence. Recommendation:[15]

Present interpretation should be broadened to permit and [sic] program for coverage by NSA of the communications of U.S. citizens using international facilities.

Electronic Surveillances and Penetrations. Recommendation:

Present procedures should be changed to permit intensification of coverage of individuals and groups in the United States who pose a major threat to the internal security.

ALSO, present procedures should be changed to permit intensification of coverage of foreign nations. . . . [the rest of this paragraph remains classified].

Mail Coverage. Recommendation:

Restrictions on legal coverage [that is, examining the writing and postmarks on the outside of envelopes] should be removed.

ALSO, present restrictions on covert coverage [that is, reading the message inside] should be relaxed on selected targets of priority foreign intelligence and internal security interest.

Surreptitious Entry [that is, the break-in option]. Recommendation:

Present restrictions should be modified to permit procurement of vitally needed foreign [a still classified section] material.

ALSO, present restrictions should be modified to permit selective use of this technique against other urgent and high priority internal security targets.

Development of Campus Sources. Recommendation:

Present restrictions should be relaxed to permit expanded coverage of violence-prone campus and student-related groups.

ALSO, CIA coverage of American students (and others) traveling or living abroad should be increased.

Use of Military Undercover Agents. Recommendation:

Present restrictions should be retained.

Beyond the lowering of specific operational restraints, Huston made two further recommendations:

Manpower and Budget. Recommendation:

Each agency should submit a detailed estimate as to projected manpower needs and other costs in the event the various investigative restraints herein are lifted.

Measures to Improve Domestic Intelligence Operations. Recommendation:

A permanent committee consisting of the FBI, CIA, NSA, DIA, and the military counterintelligence agencies should be appointed to provide evaluations of domestic intelligence, prepare periodic domestic intelligence estimates, and carry out the other objectives specified in the report.

As historian Theodore H. White has observed, the methods proposed by Tom Charles Huston reached "all the way to every mailbox, every college campus, every telephone, every home."[16] In his memorandum to the president on these methods, Huston raised—and quickly dismissed—questions about the legality of two collection techniques in particular: covert mail cover and surreptitious entry. "Covert [mail] coverage is illegal, and there are serious risks involved," he wrote. "However, the advantages to be derived from its use outweigh the risks."[17] As for surreptitious entry, Huston advised: "Use of this technique is clearly illegal: it amounts to burglary. It is also highly risky and could result in great embarrassment if exposed. However, it is also the most fruitful tool and can produce the type of intelligence which cannot be obtained in any other fashion."[18]

The young White House aide was, in a word, asking the president to sanction lawlessness by the U.S. intelligence agencies. This cavalier attitude toward existing statutes was not his alone; it was shared, evidently, by officials within the intelligence community. The recommendations made to the president, states Huston, "reflected what I understood to be the consensus of the working group [of interagency CI specialists]."[19]

Huston then waited expectantly for the president's decision. It came via H. R. "Bob" Haldeman, the White House staff director, on July 14. The president had approved each of the recommendations![20] Huston was elated. Only one problem remained: President Nixon informed Haldeman that he was too busy to meet again with Hoover and the other intelligence directors on this subject (as Huston had recommended). He preferred instead "that the thing simply be put into motion on the basis of this approval." Huston felt a certain uneasiness. He especially wanted the president to bring Hoover into the Oval Office in order to inform him directly of the decision, because, Huston later testified, "it seemed to me it would be easier maybe to get him to accept it."[21] Huston proceeded, nevertheless, to draw up an official memorandum bearing the news, and then he sent it out to the intelligence chiefs from the Situation Room, the White House Communications Center. The Huston Plan was now presidential policy.

When Hoover learned of the presidential approval, he "went through the ceiling."[22] He stormed down the hallway to Attorney General John Mitchell's office. "What is going on? Who does Huston think he is? Did the President really authorize this? Did you authorize this?" The questions popped out one after the other.

Mitchell was dumbfounded. It was the first time he had even heard of the existence of the Interagency Committee, let alone its "Special Report" or Huston's extreme recommendations. His immediate reaction was to agree with Hoover: the illegalities spelled out in the memorandum surely could not be presidential policy. Mitchell advised the FBI director to "sit tight" until President Nixon returned from San Clemente, his vacation home in southern California. The attorney general would then discuss the entire affair with the president.[23]

Richard Helms of the CIA went to see the attorney general about the President's memorandum on July 27. Helms was surprised to discover that Mitchell had only just learned of the Huston Plan a couple of days earlier from Hoover. "We had put our backs into this exercise," he told Mitchell, "because we had thought [the attorney general] knew all about it and was behind it."[24] As he had advised Hoover, Mitchell also told Helms to sit tight.

The President Takes a Second Look

Lyndon Johnson once complained that he often felt like a man with one hoe and three rattlesnakes. In the weeks that followed Huston may have felt the same way—without the benefit of a hoe. When President Nixon returned from the "Western White House" on July 27, one of his first conversations was with the attorney general. The message Mitchell delivered was that "the proposals contained in the [Huston] Plan, in toto, were inimical to the best interest of the country and certainly should not be something that the President of the United States should be approving."[25]

As Nixon later recalled, "Mr. Mitchell informed me that Mr. Hoover, Director of the FBI and Chairman of the Interagency Committee on Intelligence, disagreed with my approval of the Committee's special report. . . . Mr. Mitchell explained to me that Mr. Hoover believed that although each of the intelligence gathering methods outlined in the Committee's recommendations had been utilized by one or more previous administrations, their sensitivity would likely generate media criticism if they were employed."[26]

Mitchell also indicated, according to Nixon, that in his opinion "the risk of disclosure of the possible illegal actions, such as unauthorized entry into foreign embassies to install a microphone transmitter, was greater than the possible benefit to be derived."[27] Based on his conversation with Mitchell, the president immediately revoked his earlier approval. The Huston Plan had lived a short life—at least that is the way it appeared on July 27, 1970.

Exit Huston

Warned by Sullivan about the meeting between Hoover and Mitchell, and of Mitchell's visit to the president, Huston was expecting the call from Haldeman that came later in the day on July 27. The attorney general had come to the White House to talk about Huston's decision memorandum, Haldeman said. In response, the president had decided to revoke the memorandum so that he, Haldeman, Mitchell, and Hoover could "reconsider" the recommendations.

Upset, angered, and embarrassed about having to recall his decision memorandum, Huston walked quickly to the Situation Room. He handed the chief of the "Sit Room" the presidential order to rescind the memo of July 23. Huston was intense and agitated, recalls the "Sit Room" chief, and mentioned something about Hoover having "pulled the rug out" from under him.[28] Before the end of the day, each of the intelligence directors had been reached and asked to return the decision memorandum.

Though Tom Huston had suffered a major setback, he was unwilling to give up after all his hard work. On August 3, he went to Haldeman's office again and tried to persuade him that the president had to override Hoover's objections. He urged a meeting between Haldeman, Mitchell, and Hoover. Two days later, in a further effort to bring about this meeting, Huston put his views down on paper for Haldeman.

The memorandum, written under the title "Domestic Intelligence," ran five pages and was sharply critical of J. Edgar Hoover.[29] Huston reminded Haldeman that all the intelligence agencies and all of Hoover's own staff on the Interagency Committee supported the options recommended by Huston and approved by the president. Only Hoover had dissented. "At some point, Hoover has to be told who is President," wrote Huston. "He has become totally unreasonable and his conduct is detrimental to our domestic intelligence operations. . . . If he gets his way it is going to look like he is more powerful than the President." He warned further that "all of us are going to look damn silly in the eyes of Helms, Gayler, Bennett, and the military chiefs if Hoover can unilaterally reverse a presidential decision based on a report that many people worked their asses off to prepare and which, on its merits, was a first-rate, objective job." Huston added that he was "fighting mad," for "what Hoover is doing here is putting himself above the President."

Two more days elapsed, and on August 7, 1970, Huston sent a second, terser note to Haldeman.[30] The FBI director had left for the West Coast on vacation just as the new school year was about to begin; across the country, advised the note, student violence loomed as a real possibility. Huston again urged Haldeman to act: "I recommend that you meet with the Attorney General and secure his support for the President's decision, that the director be informed that the decision will stand, and that all intelligence agencies are to proceed to implement them at once." But by this time, recalls Huston, "I was, for all intents and purposes, writing memos to myself."[31] Haldeman remained unresponsive. Hoover had won.

Hoover's victory had many fathers. Huston believes that having the support of the attorney general was a large plus.[32] The president had a high regard for John Mitchell, an old political ally. When both Mitchell and Hoover concurred in their objections to the spy plan, Nixon no doubt saw little point in continuing the effort. Looking back, Sullivan—who deserves billing as at least a coauthor of the plan—saw other influences that worked in Hoover's favor. Sullivan believed that the president buckled under the pressure of the FBI director partly because Nixon and Hoover went back a long way, considered themselves old friends, and still socialized together frequently; and partly because the president owed his reputation from the 1950s as a staunch anticommunist to Hoover.

"Of course," added Sullivan, "Hoover had his files, too"[33]—an allusion to the private records the FBI director kept on Washington politicians, Nixon presumably among them. The director held another ace in the hole: he could always have leaked the Huston recommendations, bringing the whole enterprise to a sudden halt.

Huston recalls further that the opinions of Helms, Gayler, and Bennett carried far less weight at the White House than Hoover's. Neither the president nor Haldeman was well acquainted with Gayler or Bennett, and Helms was a virtual stranger at the White House—in part, Huston suggests, because of "the problems that he had with Mr. [Henry] Kissinger [the President's assistant for national security affairs] on foreign intelligence estimates." Finally, Huston concludes, "neither the President nor Mr. Haldeman had, in my judgment, any sensitivity to the operational aspects of intelligence collection."[34]

Subsequent memoranda written by Huston to Haldeman went unanswered throughout the month of August. Shortly after writing his August 7 memo, Haldeman informed Huston that John Dean—another young White House attorney (and former assistant to John Mitchell) soon to be caught up in the Watergate scandal—was taking over Huston's responsibilities at the White House for domestic intelligence. As Dean recalls, "Huston was livid."[35] His position at the White House having grown increasingly untenable, Huston resigned on June 13, 1971 and returned to Indiana to practice law. He continued to serve as a consultant to the White House, completing a study he had begun on the Vietnam War negotiations, and on October 7, 1972—in an appointment gilded with irony—he became a member of a Census Bureau Advisory Committee on privacy and confidentiality.

Hidden Dimensions of the Huston Plan

Reflecting upon the summer of 1970, Tom Huston remembers the atmosphere of duplicity as the most astonishing aspect of the meetings at CIA Headquarters during the writing of the "Special Report." On June 5, the president had sat across the table from the intelligence directors and asked them for a comprehensive report on collection methods used against domestic radicals. The resulting "Special Report," Huston argues, proved to be an exercise in deception by these officials. He and the president were deceived, because—without their knowledge—several of the operations that the agencies were asking permission to carry out in the future were already underway!

"I didn't know about the CIA mail openings, I didn't know about the COINTELPRO Program [an FBI internal security operation]," Huston states. "These people were conducting all of these things on their own that the President of the United States didn't know about. . . . In retrospect, we look like damned fools."[36] Former president Nixon has said he had no knowledge that a CIA mail-opening program was already in existence before June 1970 (though he was aware that the intelligence agencies read the outside writing and postmarks on envelopes for some individuals they found suspicious).[37]

Huston believes that part of the problem was bureaucratic game playing: "[T]he Bureau had its own game going over there. They didn't want us to know; they didn't want the [Justice] Department to know; they didn't want the CIA to know." And, across the Potomac, "the CIA had its own game going. They didn't want the Bureau to know."[38]

Agencies concealed programs from one another partially out of "interagency jealousies and rivalries," says Huston.[39] They wanted to conceal the fact that they were working on each other's "turf." For instance: "Mr. Hoover would have had an absolute stroke if he had known that the CIA had an Operation CHAOS going on."[40] Huston advances another possible motivation for concealment:

> I think the second thing is that if you have got a program going and you are perfectly happy with its results, why take the risks that it might be turned off if the President of the United States decides he does not want to do it; because they had no way of knowing in advance what decision the President might make. The President may say, "Hell no, I don't want you guys opening any mail." Then if they had admitted it, they would have had to close the thing down.[41]

The end result was that President Nixon remained unaware of the ongoing and, in some cases, illegal operations conducted by "his" intelligence agencies. If the sworn testimony of Vice Adm. John M. Poindexter before the Inouye-Hamilton committees in 1987 is correct, the NSC staff and the CIA also kept President Reagan in the dark about the diversion of funds to the contras. The two scandals—not to mention the Watergate burglary that ultimately destroyed the Nixon administration—reveal a troubling looseness in the chain of command at the very center of the government, stemming from insufficient attentiveness by the presidents, a reckless disregard for the law by key White House aides, and a willingness of officials in the intelligence agencies to take advantage of both for their own objectives.

The language in the "Special Report" concerning the CIA covert mail project (Operation HT Lingual) is a clear example of the concealment of an illegal intelligence-collection operation from the president. The section of the report dealing with mail plainly stated that "covert coverage has been discontinued." In truth, however, the CIA program established to read the international mail of selected foreigners and American citizens—including senators Frank Church, Hubert Humphrey, and Richard Nixon (when he was in the Congress), as well as scientist Linus Pauling and writer John Steinbeck, among a long list of others—continued to operate at the time of the Interagency Committee meetings. Former CIA director Helms has suggested, with strained credibility, that Huston may not have been told about HT Lingual because he was the White House contact man for "domestic intelligence. We thought we were in the foreign intelligence field."[42] Whatever the explanation, the Interagency Committee gave the president a misleading document.

James Angleton, who was in charge of the HT Lingual operation from 1955 until its termination in 1973, had other explanations for the misleading language in the "Special Report." "It is still my impression," he testified before Congress in 1975, ". . . that this activity that is referred to as having been discontinued refers to the Bureau's activities in this field . . . it is certainly my impression that this was the gap which the Bureau was seeking to cure."[43] The language of the report itself, however, does not reflect such a distinction, nor did Huston or Sullivan recall that this distinction was made at the time.

During hearings conducted by the Church committee in 1975, Angleton said that the concealment from the president was unintentional:

> MR. ANGLETON: Mr. Chairman, I don't think anyone would have hesitated to inform the President if he had at any moment asked for a review of intelligence operations.

SENATOR CHURCH: That is what he did do. That is the very thing he asked Huston to do. That is the very reason that these agencies got together to make recommendations to him, and when they made their recommendations, they misrepresented the facts.

MR. ANGLETON: I was referring, sir, to a much more restricted forum.

SENATOR CHURCH: I am referring to the mail, and what I have said is solidly based upon the evidence. The President wanted to be informed. He wanted recommendations. He wanted to decide what should be done, and he was misinformed.

Not only was he misinformed, but when he reconsidered authorizing the opening of the mail five days later and revoked it, the CIA did not pay the slightest bit of attention to him, did it, the Commander-in-Chief, as you say?

MR. ANGLETON: I have no satisfactory answer for that.

SENATOR CHURCH: You have no satisfactory answer?

MR. ANGLETON: No, I do not.

SENATOR CHURCH: I do not think there is a satisfactory answer, because having revoked the authority the CIA went ahead with the program. So that the Commander-in-Chief is not the Commander-in-Chief at all. He is just a problem. You do not want to inform him in the first place because he might say no. That is the truth of it. And when he did say no, you disregarded it. And then you called him the Commander-in-Chief.[44]

Further questioning Tom Huston on the subject of illegal mail openings, Sen. Church summarized the deception that lay at the heart of the Huston Plan episode:

SENATOR CHURCH: So we have a case where the President is asked to authorize mail openings, even though they are illegal. And quite apart from whether he should have done it, and quite apart from whether or not the advice of the Attorney General should have been asked, he acceded to that request, thinking that he was authorizing these openings—not knowing that his authority was an idle gesture, since these practices had been going on for a long time prior to the request for his authority. And after he revoked that authority, the practices continued, even though he had revoked it.

That is the state of the record, based on your testimony?

MR. HUSTON: Yes, I think it is.[45]

In retrospect, Huston reasons that if he and others in the White House had known that these intelligence options were already underway and that they had failed to produce results significant enough to curb domestic unrest, "it conceivably would have changed our entire attitude toward the confidence we were willing to place in the hands of the intelligence community in dealing with this problem."[46] He notes the obvious irony: intelligence in a democracy is supposed to provide policymakers with accurate information upon which to make decisions, but in June 1970 the top policymaker in the government was kept in the dark about certain sources of information that were already available.

In both the Iran-contra and the Huston Plan scandals, the nation's chief executive may as well have been a mannikin in a storefront display of the Oval Office. The president was an irrelevancy. In the Iran-contra scheme, the national security adviser cut out President Reagan, according to Poindexter's congressional testimony, in order to protect him from blame if the operation became public—a revival of the doctrine of "plausible deniability" discredited by the Church committee in 1975. During the Huston Plan episode, intelligence officials cut out President Nixon because they did not wish to risk the possibility that he might disapprove of illegal operations already underway. Whatever the rationale, the most important elected representative of the American people had lost control of the hidden—and therefore potentially the most dangerous—side of his government. In both cases, the abuse of power—the concern that most animated the nation's founders as they drafted a Constitution replete with safeguards—had raised its ugly head in a fundamental challenge to democratic procedures.

Lawlessness

Several of the techniques discussed in the "Special Report"—including covert mail cover and surreptitious entry—were, as Huston acknowledges, "clearly illegal."[47] And the legitimacy of other methods, such as placing on the NSA's "watch-list" the names of those Americans whose international communications were to be monitored, was highly dubious.[48] Yet former President Nixon does not recall "any discussion concerning the possible illegality of any of the intelligence gathering techniques described in the report during my meeting with the [Interagency] Committee [on June 5, 1970]."[49]

During congressional hearings by the Church committee on the Huston Plan, Sen. (and later Vice President) Walter Mondale (D, Minnesota) asked Huston whether any of the staff members who helped prepare the "Special Report" had objected "during the course of making up these options to these recommendations which involved illegal acts":

> MR. HUSTON: At the working group level, I do not recall any objection.
>
> SENATOR MONDALE: Do you recall any of them ever saying we cannot do this because it is illegal?
>
> MR. HUSTON: No.
>
> SENATOR MONDALE: Can you recall any discussion whatsoever concerning the illegality of these recommendations?
>
> MR. HUSTON: No.
>
> SENATOR MONDALE: Does that strike you as peculiar that top public officers in the most high-level and sensitive positions of government would discuss recommending to the President actions which are clearly illegal and possibly unconstitutional without ever asking themselves whether that was a proper thing for them to be doing?
>
> MR. HUSTON: Yes, I think it is, except for the fact that I think that for many of those people we were talking about something that they had been aware of, had been undertaking for a long period of time.
>
> SENATOR MONDALE: Is that an adequate justification?
>
> MR. HUSTON: Sir, I am not trying to justify. I am just trying to tell you what my impression is of what happened at that time.
>
> SENATOR MONDALE: Because if criminals could be excused on the grounds that

someone had done it before, there would not be much of a population in any of the prisons today, would there?

MR. HUSTON: No.[50]

Even though several important legal questions were (or should have been) involved in preparing the "Special Report" for the President, the Interagency Committee never sought legal counsel. (Huston and Angleton were the only two participants with law degrees; neither had practiced law.) The CIA general counsel was excluded because, Angleton has testified, "the custom and usage was not to deal with General Counsel, as a rule, until there were some troubles. He was not a part of the process of project approval."[51]

Avoidance of legal and constitutional questions seems to have been common throughout the intelligence community. As William C. Sullivan of the FBI has testified:

> During the ten years that I was on the U.S. Intelligence Board—A Board that receives the cream of the intelligence for this country from all over the world and inside the United States—never once did I hear anybody, including myself, raise the question: "Is this course of action which we have agreed upon lawful? Is it legal? Is it ethical or moral?" We never gave any thought to this realm of reasoning, because we were just naturally pragmatists. The one thing we were concerned about was this: will this course of action work, will it get us what we want, will we reach the objective that we desire to reach?[52]

Sullivan attributed much of this attitude toward the law to the molding influence of World War II on young FBI agents who had since risen to high positions. In a statement for the Church committee, Sullivan observed that during the 1960s there was "a war psychology. Legality was not questioned. Lawfulness was not a question; it was not an issue."[53] The following exchange took place during the deposition:

> SENATOR MONDALE: That carried on, unfortunately, after the war.
>
> MR. SULLIVAN: Senator, you are right. We could not seem to free ourselves either at the top or bottom, free ourselves from that psychology with which we had been imbued as young men, in particular, most all young men when we went into the Bureau.
>
> Along came the Cold War. We pursued the same course in the Korean War, and the Cold War continued, then the Vietnam War. We never freed ourselves from the psychology that we were indoctrinated with, right after Pearl Harbor, you see. I think this accounts for the fact that nobody seemed to be concerned about raising the question: "Is this lawful, is this legal, is this ethical?" It was just like a soldier in the battlefield. When he shot down an enemy, he did not ask himself: Is this legal, or lawful, is it ethical: It is what he was expected to do as a soldier.
>
> We did what we were expected to do. It became a part of our thinking, a part of our personality.[54]

As the behavior of Lieutenant Colonel North, Admiral Poindexter, and CIA Director Casey in 1985–86 illustrates, this "ends-justifies-the means" mentality lives on. For them, a hidden World War III with the Soviet Union—the cold war—required full use of the CIA, PM operations, and whatever else was necessary to defeat communism wherever it arose, irrespective of annoying statutes. This attitude seems to be less an ingrained part of the culture within the intelligence agencies—after all, most of these officials obey the nation's laws—or a perspective shared solely by World War II veterans (North and Huston were children at the time) than a result of viewing U.S.-Soviet relations in strict zero-sum terms—a perspective characteristic of the right wing in the United States. Rightists have a tendency to conclude that extreme measures, like the Huston Plan and the Iran-contra operations, are necessary for victory in the struggle against the USSR and its Marxist allies.

Exclusion

During the Huston Plan episode, neither Attorney General Mitchell nor anyone in his office was invited to the drafting sessions at CIA Headquarters or consulted during the proceedings. During his committee's hearings on the Huston Plan, Sen. Church asked Huston about the failure to consult with the Attorney General:

> SENATOR CHURCH: And it never occurred to you, as the President's representative, in making recommendations to him that violated the law, that you or the White House should confer with the Attorney General before making those recommendations?
>
> MR. HUSTON: No, it didn't. It should have, but it didn't.[55]

One reason for the absence of Attorney General Mitchell in the planning, Huston explained, was that this was an intelligence matter to be handled by the intelligence agency directors.[56] Huston later claimed, however, that he naturally thought that Hoover would check with Mitchell or his deputy before signing the "Special Report," just as General Bennett had cleared it with his superior, Deputy Secretary of Defense David Packard, and informed the secretary of defense, Melvin Laird.[57] Another reason for Mitchell's exclusion may have been the institutional animosity that existed between the professional intelligence establishment and the Office of the Attorney General. The former was interested primarily in the collection of intelligence and the protection of sources; the latter suffered, in Huston's view, from "prosecutor's mentality"—an interest in the collection of evidence for use in securing prosecution.

According to Huston, two approaches exist to handle the problem of violence-prone demonstrators. "One is the intelligence-collection approach where you try to keep tabs on what is going on and stop it before it happens," he has testified. "The other approach, which is perhaps the only tolerable one in a free society from a perfectly legitimate point of view, is you have to pay the price of letting a thing happen, and then follow the law and hope you can apprehend the person responsible and prosecute him according to the law."[58] In 1970, strong tensions existed between these two approaches—and they continue today.

Five years after his spy plan recommendations, Huston had come to the conclusion in testimony before the Church committee that, in the end, the preservation of a democratic society depended upon reliance on the law. For him, the sanctions of criminal law had replaced his earlier faith in the unrestricted collection of intelligence as the more appropriate response to the threat of violence in American society. The risk inherent in the latter approach was too great. In Huston's words:

> The risk was that you would get people who would be susceptible to political considerations as opposed to national security considerations, or would construe political considerations to be national security considerations, to move from the kid with the bomb to the kid with a picket sign, and from the kid with the picket sign to the kid with the bumpersticker of the opposing candidate. And you just keep going down the line.[59]

Credit Card Revolutionaries

Just as hidden from the president and Tom Huston as the CIA mail program, though more because of their own selective perception than because of duplicity on behalf of the intelligence community, was the reality of the antiwar movement that helped spur the writing of the "Special Report" in the first place. The threat-assessment section of the report was similar to the earlier assessments prepared for the White House in April and June 1969. Though more thorough, it also failed to produce much concrete evidence of foreign influence over domestic unrest. During the Church committee hearings, C. D. Brennan, a FBI witness, said the Bureau was never able to find evidence indicating that the antiwar protesters in the United States were financed by external sources. "I felt that the extremist groups and the others who were involved in anti-war activities and the like at that time were of the middle- and upper-level income," he testified, "and we characterized them generally as credit-card revolutionaries."[60]

Despite the lack of any substantial evidence of foreign involvement, the White House under both presidents Johnson and Nixon had persistently tasked the Bureau to unearth intelligence on foreign funding.[61] As in the earlier reports, however, the assessment section of the "Special Report" pointed to the danger of foreign connections that might develop in the future. Consensus here was high. Like Huston in the White House, the CI officers writing the report teetered on a slippery slope when they began to speak of the need to expand intelligence collection more because of potential rather than actual findings.

Here, then, were the main forces, hidden for the most part, that shaped the writing of the "Special Report" and the recommendations that Huston extracted from it. Those who sought to obtain official presidential authority to broaden methods of intelligence collection had failed, but they remained committed to their methods anyway. As with the contra diversion in 1986, national security officials were not to be dissuaded by the simple absence of presidential or statutory authority.

Aftermath: The End—or the Beginning?

In the wake of the Huston Plan, a secret meeting took place involving Hoover, Helms, Gayler, and Mitchell. On March 25, 1971, an FBI-CI officer wrote a memorandum for Hoover's information regarding a request from Attorney General Mitchell that asked the director to meet with him, Helms, and Gayler on March 31. The officer did not know the agenda for the meeting, but speculated that it would cover the subject of foreign intelligence as it related to domestic subversives.[62] The NSA, noted the memo, was already sending intelligence to the CIA and the FBI "on an extremely confidential basis" on the international communications of American citizens, but only as a byproduct of NSA's normal communications-monitoring responsibilities; this information was not further developed or pursued in any systematic way. The memo suggested that Helms and Gayler might have an interest in increasing intelligence gathering of this type. The Bureau's CI specialist reminded his boss that the principal source of FBI data on subversive activities was electronic surveillance and live informants; to supplement these methods, Hoover was advised to "take advantage of any resources of NSA and CIA which can be tapped for the purpose of contributing to the solution of the problem." The memo read like a fragment of conversation from drafting sessions on the "Special Report" held at Langley the previous June.

The proposed meeting took place on March 29 in Mitchell's office. Subsequently, Hoover prepared a memorandum for the files indicating that Helms had been primarily responsible for the gathering. The purpose of the meeting was to discuss "a broadening of operations, particularly of the very confidential type in covering intelligence both domestic and foreign." Gayler was "most desirous" of having the Bureau reinstate certain collection programs, and Helms spoke of "further coverage of mail."[63]

According to his file memo, Hoover rebuffed these approaches. He informed Helms and Gayler that he "was not at all enthusiastic about such an extension of operations insofar as the FBI was concerned in view of the hazards involved." Mitchell then intervened and asked Helms and Gayler to prepare "an in-depth examination" of exactly what collection methods they desired. After reading their report, Mitchell would then reconvene the group and, continues Hoover's memo, "make the decision as to what could or could not be done." Helms agreed and said that he would have the report prepared "very promptly."

The meeting represented a reenactment of the battle over the Huston Plan—this time with the inclusion of the major missing participant, the attorney general. The results were similar to the earlier outcome: a victory for Hoover. Yet clearly the war was not over. While neither Helms nor Gayler nor Mitchell recall this meeting, or the recommendations of the Helms-Gayler report, and while it is unclear whether such a report was ever actually prepared, one thing is certain: efforts to implement provisions of the Huston Plan persisted. The unlawful CIA mail opening continued; the list of American citizens on the NSA computerized "watch-list" expanded during the years from 1970–73; the age limit on FBI campus informants was lowered from 21 to 18; and the Bureau stepped up its investigations in the internal security field.[64] The intensified intelligence activities of the FBI included surveillance of "every Black Student Union and similar group, *regardless of their past or present involvement in disorders.*"[65] This project required the opening of 4,000 new cases. Also, the FBI placed members of the Students for a Democratic Society (SDS) under investigation, accounting for an additional 6,500 new cases.

The central FBI witness during Church committee hearings on the Huston Plan did not believe that the FBI ever told the president about this increased surveillance.[66] Nor, according to other witnesses, did the other intelligence agencies inform the White House about their expanded intelligence collection. With reference to the CIA mail program, former attorney general Mitchell has suggested that "the old-school-tie boys, who had been doing it for twenty years, just decided they were going to continue to do it."[67]

Looking back on the Huston Plan, President Nixon said in an official statement released in 1973: "Because the approval was withdrawn before it had been implemented, the net result was that the plan for expanded intelligence activities never went into effect."[68] It was not that simple, however. As the CIA's James Angleton has testified, "The Huston Plan, in effect as far as we [the CIA] are concerned, was dead in five days and therefore all of the other matters of enlarging procurement within the intelligence community were the same concerns that existed prior to the Huston Plan, and subsequent to the Huston Plan. The Huston Plan had no impact whatsoever on the priorities within the intelligence community."[69] Angleton continued, "People are reading a lot into the Huston Plan, and, at the same time, are unaware that on several levels in the community identical bilateral discussions were going on."[70] The former CIA-CI chief observed that since the creation of the CIA, "there had been constant discussion of operations and improvement of collection, so there is nothing unusual. . . . There were a number of ongoing bilateral discussions every day with other elements within the intelligence community which may or may not have duplicated the broad, general plan that Huston brought about."[71]

Angleton's disclaimers aside, that the president approved the Huston Plan—if only briefly—remains deeply troubling in itself, for some of its provisions clearly required violations of the law. That some of the intelligence agencies could conceal the existence of these operations from the White House, then continue—even expand—them after the president had revoked his authority is cause for profound concern. Without due regard for the sanctity of law—the formal expression of the people's will—democracy in the United States will be in extreme jeopardy.

The Huston Plan and the Iran-contra affair serve to remind us anew that individuals without respect for the law and for America's constitutional arrangements can come to roost at the heart of the government. Similar scandals can occur again—perhaps with even more disastrous results. No pat formula exists to guard against the abuse of power. The chief remedy lies in a more vigilant public that will choose its presidents with greater care, demand that their administrations understand the value of the law, insist that the White House and its agencies be watched closely by Congress and the press, and rise up in anger—in the voting booth and with expressions of indignation—whenever the social contract is violated.

EIGHT

The CIA in the Groves of Academe

The Katzenbach Report

The diction comes straight out of the Operations Directorate: "We exploit our territory [the campuses] as effectively as possible."[1] "He has excellent penetration into the student body."[2] "We have 200 assets in approximately 100 academic institutions."[3] "Exploit," "penetration," "assets"—terms that normally are associated with the dark arts of espionage have slipped easily into the vernacular of CIA officials responsible for intelligence activities on American college and university campuses. To what extent the Agency has adopted as readily the actual techniques of espionage on the nation's campuses has been an issue of lively debate since the mid-1960s. In 1966, *Ramparts* magazine exposed a secret CIA project at Michigan State University, funded at a level of $25 million, to train South Vietnamese police officials; the following year *Ramparts* further revealed that the CIA was the source of support for the National Student Association.[4]

These projects were hardly the beginning of CIA-academic relations. The CIA's ties to the academic world extended back to the day the Agency opened its doors in 1947. The new organization drew heavily upon the universities for brain power, and much of its bureaucratic cohesion stemmed from a network of "old boy" school ties—largely from the Ivy League. Harvard University historian William L. Langer, who had served as director of research for the OSS during World War II, set up the Agency's top panel of analytic experts (the Board of Estimates) and staffed it in part with former university professors. Harvard Law School professor Robert Amory was soon named head of the Directorate for Intelligence, a post he filled for almost a decade. Many other noted academicians joined the CIA or served as consultants.[5]

The new Agency quickly established student recruitment procedures, relying not only upon informal ties with friendly professors (some of whom had served in the OSS) but also working through formal and overt avenues on the campuses. Within a short time, CIA personnel officers moved out across the country. Their reach went far beyond the Ivy League corridor in the Northeast; as early as 1950, for example, a CIA recruitment office was established in Los Angeles to fish the teeming student reservoirs at UCLA, USC, and a dozen other colleges and universities in the Southern California basin.

During the 1950s and 1960s, the association between the CIA and academe became interwoven in ever more intricate patterns. As the Church committee reported in 1976, academicians (including administrators, faculty members, and students) carried out an assortment of intelligence-related activities. Among other things, they authored books and articles based on research financed by the CIA; spotted and assessed individuals for Agency use; served as "access agents" to make introductions between the CIA and potential agents or employees (foreign and American); and provided information to the Agency, both with and without prior instructions.[6]

The CIA's Bay of Pigs fiasco in 1961 stirred up early misgivings on many campuses about the Agency's judgment and competence, and this uneasiness grew with the CIA's deepening involvement in the Vietnamese civil war. Social scientists, especially political scientists with expertise on the developing nations of the world, proved to be the most vocal critics of the Agency during this early period and later (though a few worked secretly to assist the CIA in its Third World operations). Then, in 1967, the *Ramparts* expose sent a still darker cloud across this horizon. The cries of protest from the public and the Congress over CIA involvement at home with the National Student Association persuaded President Lyndon Johnson to establish a presidential commission with instructions to review Agency links with educational and private voluntary organizations in the United States. The Katzenbach commission, chaired by Undersecretary of State Nicholas Katzenbach, studied the problem for forty-three days and issued this recommendation: "It should be the policy of the United States Government that no federal agency shall provide any covert financial assistance or support, direct or indirect, to any of the nation's educational or private voluntary organizations."[7]

The chief response of the CIA was to shy away from some propaganda activities involving academicians. Following the issuance of the Katzenbach Report, the Agency quickly suspended support and publication in the United States of books and articles written by faculty members. By and large, though, the National Student Association scandal and the Katzenbach guidelines seem to have had little effect on CIA-academic relations. Professors continued to be hired for propaganda activities—though now their materials were targeted for foreign distribution only; recruitment contacts, both overt and covert, were continued; and graduate students and faculty gathered intelligence for the CIA at home and abroad as before. The number of campus people working with the CIA diminished, but still figured in the hundreds at over one hundred American colleges, universities, and affiliated research entities.

In 1986, newspaper reports disclosed that the CIA had clearly resumed its covert support for the research and writing of publications by scholars in the United States, despite the Katzenbach Report recommendation in 1967. The CIA provided Professor Nadav Safran, director of Harvard's Center for Middle Eastern Studies, $107,430 to write a book on Saudi Arabia (subsequently published by Harvard University Press). Professors Richard K. Betts and Samuel P. Huntington, also of Harvard, published an article in *International Security* (1986) based on research funded by the CIA without acknowledging in the journal the source of the research support—an omission required by the CIA at the time as part of the contractual arrangement.[8]

None of these writers believed his scholarly objectivity was affected by an association with the CIA ("Commitment to truth does not depend on the source of the funding but on the integrity of the scholar," declared Professor Safran[9]), and none acknowledged any sense of impropriety in the arrangement. The disclosures, nonetheless, raised misgivings elsewhere. "They're [the CIA] not supposed to operate within the United States," said Rep. Don Edwards (D, California), chairman of the House Judiciary Subcommittee on Civil and Constitutional Rights, "and as far as I'm concerned, this is operating within the United States."[10] The *Boston Globe* concluded, "The scholar who works for a government intelligence agency ceases to be an independent spirit, a true scholar."[11] And, in the opinion of the former *Ramparts* editor whose revelations had triggered the Katzenbach Report a decade earlier, that report's "solemn declaration of principle can now be tossed onto the scrap heap of failed national purpose."[12]

A primary reason why the Katzenbach Report failed to have a significant limiting effect on CIA-campus ties had to do with its lack of legal stature. Because it was neither a law nor even an executive order, but simply an executive branch "report," the document possessed little binding status. Moreover, it referred to "educational . . . organizations," leaving wide open the opportunity for clandestine arrangements with *individual* academicians. So, while stimulating a mild flurry of retrenchment, the cloud of the Katzenbach Report cast but a minor shadow across the multiple entanglements that had grown up between the colleges and the world of intelligence. The shouts for reform soon subsided, leaving behind in their faint echo little substantive change in CIA-campus relations.

Contemporary Relationships

Since 1947, each of the four CIA directorates and several subsidiary units within them have been involved in a wide range of associations with colleges, universities, research groups, think tanks, technical schools, secretarial schools, and even high schools throughout the country and abroad. The relationships defy simple description, resembling in their richness and diversity the past alliances the CIA has had with the media (see chap. 9). While the latter have diminished dramatically, ties with academicians continue to flourish—despite occasional internal CIA regulations promulgated on this subject in a partial bow to outside pressures.

Each of three recent CIA Directors—George Bush (1975–76), Stansfield Turner (1977–81), and William J. Casey (1981–87)—has authorized public guidelines on CIA-academic relations. The most succinct statement came from Bush. In a reply to an inquiry from the American Association of University Professors, he stated simply that the only campus ties sought by the CIA were with "the voluntary and witting [that is, willing] cooperation of individuals who can help the foreign policy processes of the United States." The Agency hoped to avail itself of "the good counsel of the best scholars in our country."[13]

Admiral Turner, the most reform-minded of all the Agency's directors, issued a much more comprehensive set of rules to guide the CIA, though they fell short of standards hoped for by some universities. The internal restrictions established by Turner invoked several opportunities and limitations, and warrant presentation in full:

> CIA may enter into classified and unclassified contracts and other arrangements with the United States academic institutions of higher learning as long as senior management officials of the institution concerned are made aware of CIA's sponsorship. CIA may enter into personal services contracts and other continuing relationships with individual full-time staff and faculty members of such institutions but in each case will suggest that the individual advise an appropriate senior official thereof of his CIA affiliation, unless security considerations preclude such a disclosure or the individual objects to making any third party aware of his relationship with CIA. No operational use will be made either in the United States or abroad of staff and faculty members of the United States academic institutions on an unwitting basis. CIA employees will not represent themselves falsely as employees of United States academic institutions. CIA personnel wishing to teach or lecture at an academic institution as an outside activity must disclose their CIA affiliation to appropriate academic authorities; all such arrangements require approval in advance from the Director of Security. Pursuant to Federal law, CIA will neither solicit nor receive copies of identifiable school records relating to any student (regardless of citizenship) attending a United States academic institution without the express authorization of the student, or if the student is below the age of 18, his parents.[14]

In a more streamlined version of his Agency policies toward academe, Turner advised the president of Harvard University:

> All of our contracts with academic institutions are entered into with the knowledge of appropriate senior management officials of the institution concerned.
>
> All recruiting for CIA staff employment on campus is overt.
>
> It is against our policy to obtain the unwitting services of American staff and faculty members of U.S. academic institutions.[15]

Then, in February 1986, Casey's CIA released a fresh set of rules in response to the Safran controversy. These rules were made public in a speech delivered at Harvard University by DDI Robert M. Gates. According to him, the CIA would continue to pursue a variety of relations with academe—though with some modification of past practices.[16] First, consultation with faculty on world issues would remain the most common form of CIA-academic contact. Second, the Agency would continue to hold conferences, pay others to hold them, and send representatives to conferences sponsored by others. Third, the CIA would carry on its funding of basic research by outsiders in the universities and think tanks, but this would remain "a very minor element" in CIA-academic relations. Fourth, the CIA would continue its scholars-in-residence program under which faculty members could conduct research (classified or unclassified) at the CIA. Finally, the Agency intended to debrief willing scholars upon their return from trips abroad (CIA practice since 1947).

The outline of CIA-academic ties presented by Gates amounted to nothing new. His accompanying remarks on rules, however, did. According to Gates, six rules would stand preeminent in this latest definition of CIA ties to academe:

• *Publication Review.* Scholars who had access to classified materials would now be required to have only that portion of a manuscript based on these materials reviewed by the CIA prior to publication, rather than the entire manuscript as before. This review, as before, was supposedly strictly for the purpose of assuring that no classified information was inadvertently revealed by the author—not to censure his or her beliefs.

• *Conferences.* Professor Safran's great sin, according to some critics, was financing a conference at Harvard University on the Middle East partially with CIA money without informing participants about this source of funding. When the hidden CIA sponsorship came to light, several participants refused to attend. Eventually, Harvard censored Professor Safran for his failure to report the CIA funding, as required by the university's guidelines established in 1977 (discussed later in the chapter). As a result of this flap, the Directorate for Intelligence pledged to tell individuals who organized conferences with CIA money that they "should" inform participants of the Agency's sponsorship in advance.

• *Classified Research.* As always, the CIA would continue to allow contracts for classified research, though, said Gates, the Intelligence Directorate had no such contracts in February 1986. If a university had explicit rules against classified research, however, the Intelligence Directorate would comply with these rules. Gates remained silent on the question of whether other directorates, like the directorates for Science and Technology or Operations, would bow to university rules of this kind.

• *Unclassified Research.* Gates emphasized that the CIA does not "commission or contract for books or articles." Rather it pays for research to be used by the Agency. In his vaguest prescription, he said that the Intelligence Directorate would try to accommodate the desire of scholars to publish this work after its use by the CIA. In some (unspecified) cases, the DDI said that prepublication review might be unnecessary for unclassified research—though why a review of such research might be required at all was left unexplained.

• *Acknowledged Sponsorship.* Gates offered several reasons why the acknowledgment of CIA funding for a particular research project could be detrimental to the national interest: "the possibility of difficulty with a foreign government by virtue of acknowledged C.I.A. interest in its internal affairs; the possibility that acknowledged C.I.A. interest in a specific subject—such as the financial stability of a particular country—could affect the situation itself; and finally, concern that readers might assume the scholar's conclusions were, in fact, C.I.A.'s."

The Agency nonetheless decided to allow acknowledgment in subsequent publications, unless the scholar requested privacy or—in one of those phrases that makes you want to hold on to your wallet—"we determine that formal, public association of C.I.A. with a specific topic or subject would prove damaging to the United States." The *New York Times* feared that escape hatches like this one "puncture the credibility of the C.I.A.'s avowals of candor. Given its reflexive passion for secrecy, the agency can be counted on to scent damage in the most innocuous information."[17]

- *Primacy of University Rules.* Finally, Gates said that the Intelligence Directorate would expect scholars to abide by the rules of their own institutions on whether they had to report their association with the CIA to their campus officials. In his view, it was not the responsibility of the CIA but rather of the academic institutions "to set such rules and to enforce them, and the responsibility of the scholar to comply."

While the Gates pronouncements clarified some ambiguities in the Turner guidelines, they failed to resolve the controversy surrounding CIA-academic relations. In the first place, Gates was clearly speaking only about the CIA's Directorate for Intelligence. Yet, a long list of Agency offices are involved beyond this directorate in pursuit of CIA campus objectives. Some employ elaborate covers for their activities, including proprietaries and "notionals" (that is, private commercial entities that exist only on paper); the purpose of this deception is to have several layers ("insulation," in CIA lingo) between the Agency case officer and his operation, should it run awry. Such tradecraft makes good sense from the perspective of cover, but the resultant kaleidoscope of false images can as readily fool legitimate overseers at home as they might adversaries abroad. The review that follows of key CIA entities with ties to academe indicates how limited Gates's focus at Harvard had been.

Directorate for Operations

For critics of CIA-academic relations, the most worrisome connection involves the CIA Directorate for Operations, for here is the "dirty-tricks" department of the Agency responsible for covert action abroad—everything from secret propaganda operations to assassination plots (in the past) and secret arms sales. What in the world, they ask, is this organization doing on the nation's campuses? The answer lies in the missions of two units within the directorate: the National Collection Division (NCD) and the Foreign Resources Division (FDR).

NATIONAL COLLECTION DIVISION "The key to effective intelligence collection," observed the CIA's director of the Domestic Contact Division or DCD (now called the National Collection Division or NCD), "is to ask the right person the right question."[18] The secretary of state, an assistant secretary, the CIA director, intelligence analysts, and sometimes the president himself will pose the "right questions," that is, establish the collection priorities. The right answers may lie abroad, to be plucked out of the air by the NSA or stolen from inner government councils by assets in the field. Or the person with the right answers may reside here within the United States.

To contact Americans who may be able to assist the CIA in its intelligence collection, the CIA established the DCD first within the Intelligence Directorate, and since 1973, it has been relocated within the Operations Directorate. Today its successor, the NCD, has field offices in some thirty-eight cities around the country and exists in an overt fashion, at least in the sense of appearing in the telephone book and having an office in the local federal building. What it does not do overtly is advertise its contacts on campus.

The bonds between the NCD and academe take on various colorations. They can be as limited as asking a professor home from a trip abroad to drop by the local NCD office to share his impressions (a "debriefing"), which the Intelligence Directorate does, too, on occasion. Or they may involve a formal, paid association in which the professor counts for the CIA the number of deep-water ships he sees in Luanda while doing research on African migratory butterflies. The work done by academicians for the CIA, then, may be voluntary or paid, and may include occasional or systematic collection tasks, recruitment tasks, or even covert action and counterintelligence.

A professor may be asked by the Agency through the NCD to attend an international conference of scholars (say, physicists). At the conference and under CIA instructions, he will seek out a foreign scientist (say, a South African) and at some point in their conversation weave in as subtly as possible a set of questions ("By the way, how well is the Becker jet-nozzle, uranium enrichment process really working out?"). If the scientist also happened to be a state minister or a well-placed government aide, the American professor could easily enter the realm of covert action or counterintelligence. He may pass a CIA propaganda theme along to the minister, thereby serving as an agent of influence (and engaging in covert action). He may also be used, though this would require infinitely greater skill and preparation, to plant the false notion that he himself might like to work secretly for the minister's government to spy on the United States. He could thus become a double agent who, in this guise, might be able to glean modus operandi and other information about the foreign government, including what its own agents may be after (in this instance he would be engaging in counterintelligence). These are only a few examples of the numerous ways in which the Agency might wish to employ an academician overseas in CA or CI operations. Most often, though, the usefulness of campus assets resides in the area of intelligence collection.

The DCD became an object of controversy in the past for collecting information (read spying) on American and foreign students enrolled at U.S. colleges and universities.[19] Presumably, the DCD and the NCD stopped targeting American students, but the Church committee reported in 1976 that the DCD had continued to report "leads regarding foreign nationals who could prove useful abroad or U.S. firms whose offices abroad could help the CIA."[20]

FOREIGN RESOURCES DIVISION No doubt the most secretive of the CIA entities that venture on campus is the Foreign Resources Division. According to published reports, the CIA established the FRD in 1963 under the chilling name Domestic Operations Division (DOD) and with the purpose of conducting secret operations against foreign personnel (including foreign students) within the United States. While the NCD gathers foreign intelligence at home openly (if discreetly), the FRD is strictly an enterprise for the clandestine collection of foreign intelligence. After the flap over domestic spying by the CIA (Operation CHAOS[21]), the name "Domestic Operations" carried too many negative connotations by half, and "Foreign Resources" was selected as a substitute. The mission remained constant, however: to spot and recruit foreigners within the United States for intelligence service against their own or other countries.[22]

Directorate for Intelligence

The Directorate for Intelligence has, in some ways, the most understandable need for a close relationship with America's universities. To interpret incoming intelligence as effectively and efficiently as possible, analysts in the Intelligence Directorate have a persistent need to stay abreast of current substantive and methodological advances in the wide range of fields reflected in university curricula. How does the African expert at University X interpret events in Angola, based on her recent visit there? What themes did the chairman of the agronomy department at University Y denote at the Third World Conference on World Food Production? What does the mathematics department at University Z think of the new variant of the Richardsonian arms-control model developed by the CIA Office of Strategic Research?

This list is endless. To meet some of its demands for information and interpretation, the CIA over the years has thrown out thousands of lifelines from Langley to the nation's campuses. Just as CIA officers have a lot in common with journalists (see chap. 9), so do they with professors, especially, shared professional analytical interests. The tangle of connections, both formal and informal, have ranged from bringing in full-time scholars-in-residence (four a year are common in the Intelligence Directorate) to occasional telephone conversations on this point or that. Most of the subsidiary units within the Intelligence Directorate have outside consultants at the universities and research institutes, numbering from twenty-five to fifty at any given time. About every three months, the directorate also brings in a visiting lecturer from a university campus or think tank.

As its chief noted in 1986, the Intelligence Directorate has engaged in "an intensified effort to reach out to the academic community, think tank of every strip, and the business community for information, analysis, and advice. . . ."[23] Among other things, this Directorate and others in the CIA have sponsored three hundred conferences since 1982 (compared to three or four each year before), and have sent, overtly, more than fifteen hundred analysts to conferences sponsored by others. Throughout the 1960s and 1970s (less so in this decade), the Directorate also released a significant number of unclassified economic analyses, lists of leaders in foreign countries, Foreign Broadcast Information Service (FBIS) "Trends" analyses (declassified and made available to the academic community within six months), and other information—especially on Communist nations—useful to university researchers. Unfortunately, the Reagan administration cut back sharply on the amount of CIA analysis made available to the public, including (since 1981) the termination of access to the FBIS "Trends" document—all for no apparent reason.

Directorate for Administration

A third directorate involved extensively in campus activities is the one for administration. Four offices in this directorate have a campus mission: the Office of Logistics, the Office of Training, the Office of Personnel, and the Office of Security.

LOGISTICS The Office of Logistics touches academe as a research contractor. It serves as a conduit through which the Intelligence Directorate requests ("tasks") academicians to generate, analyze, or evaluate data. Projects farmed out in this way may include assignments to develop personality profiles on known terrorists, to plot the likely trajectories of missiles fired from various locations, to draw more detailed maps of the Persian Gulf nations, and countless other possibilities. The CIA has many competent analysts within its own Intelligence Directorate, of course; but the world is large and complex, and often the Agency will have to go outside its walls to tap campus expertise.

TRAINING The Office of Training (OTR), as the name implies, is responsible for the technical, managerial, and intellectual development of CIA employees. Newcomers are trained in introductory courses, midcareers officers (GS 12-13) in a five-week Intermediary Course, and senior officers in a two-and-a-half-week Advanced Course (GS 14) and a nine-week Senior Seminar (GS 15 and above). These courses are taught at a location in rural Virginia (known as "the Farm" or "Isolation"), as well as at a site near Washington, D.C. This instruction is supplemented with lectures by well-known intellectuals presented at Langley Headquarters six or seven times a year (at around $500 a speech), among them professors (and former ambassadors) John Kenneth Galbraith and Edwin Reischauer of Harvard University.

At lower levels of pay, the CIA invites lesser lights to present lectures at their intermediate and senior training sessions, usually at the Senior Seminars where OTR tries to imbue top management with a broader and more sophisticated worldview. For the Senior Seminars (one week at the Farm, eight near Washington), OTR will fly in several professors to supplement the program of its own staff. These visiting lecturers are paid around $200 (plus expenses) for their half-day appearance and are asked to speak to their speciality within one of these general areas: domestic affairs (occasionally members of Congress and senior staff are invited to explain the mysteries of legislative oversight), foreign affairs, and science and technology. Brief background investigations, called NDIs ("no derogatory information" checks), are conducted before invitations are extended to these lecturers; but no full-field investigations are required—except for those invited to the Farm—since these lectures are unclassified (unless a lecturer who already has a top-secret clearance wishes to present classified information in his remarks). In one recent twelve-month period, OTR employed the services of forty-seven individual professors, representing thirty-two colleges and universities, as guests speakers in various courses.[24] Occasionally, OTR will bring in consultants for other purposes, such as reviewing curriculum or evaluating its language training.

The OTR has intersected the academic world in three other ways. During Admiral Turner's watch, it provided briefings for touring college students who wished to visit the CIA (in coordination with the Office of Public Affairs, which also gave briefings). It makes the arrangements for CIA employees to attend college part-time (1,500 CIA employees did so in one recent year) or full time (60 in the same year). And it develops contacts on the campuses to spot good linguists who might be recruited into the Agency, and to keep its list of visiting lecturers replenished. The strongest emphasis in the OTR campus-recruitment efforts has been to maintain close ties with college language departments, which are encouraged to refer promising graduate students to the CIA.

PERSONNEL The Office of Personnel has eight field offices in the country, all in the telephone books of their respective cities. Its primary mission is to locate and recruit qualified candidates for CIA staff employment. Each year the CIA has approximately a thousand job openings, 20 percent of which are clerical. The number of qualified applicants far outnumbers the available openings. Contrary to conventional wisdom, the loud criticism of the Agency beginning in late 1974 has actually increased the number and quality of perspective CIA employees, according to the director of personnel.[25]

In a recent year, the Agency received 37,000 letters of inquiry for jobs. The Office of Personnel conducted 12,500 interviews from this group. Winnowing the field further, 5,700 from among those interviewed were asked to provide a complete file on themselves, and the CIA then investigated their backgrounds. Finally, 429 professional staff and 605 clerical personnel were hired. In its search for good candidates, the Office of Personnel sends its staff across the country (40–50 percent of this staff is constantly on the road). Approximately 80 percent of its network of "spotters" are on the college campuses (often in the placement office, sometimes in the dean's office, and inevitably in the classroom), though some work with technical schools, high schools, and secretarial schools.

The CIA has what it calls an "aggressive" affirmative action program. During the Carter administration, it set two new records: of the 429 professionals hired, 80 (14 percent) were black or Chicano and 85 (20 percent) were women. While few blacks, Chicanoes, or women can be found at high management levels in the Agency, officials in the Office of Personnel point out this is only a matter of time as the new minority recruits climb upward.

The nation has 1,928 four-year colleges and universities and 1,147 two-year colleges, or a total of 3,075. From among these, the CIA tries to make some kind of contact with 750, steps up its mail and telephone links with 550 of these, and actually visits about 315. The Agency is particularly interested in those institutions with advanced programs in language, science, and foreign policy.

The contact points between the Office of Personnel and the campus appear to be as many as the CIA can manage. Entirely overtly, the Personnel Office representative will visit the campus placement office, perhaps to set up a CIA recruiting table or simply to drop off recruitment materials for posting on bulletin boards. The CIA recruiter may ask the permission of the campus placement director to contact academic departments and individual faculty. Sometimes the reception is enthusiastic. In a letter to the CIA, one placement counselor at a university wrote effusively:

> Thank you so much for your recent letter indicating current trends of interest for recruiting for the Central Intelligence Agency.
>
> This information will be most helpful to us in counseling with students interested in a CIA career. We are notifying appropriate faculty members in the areas of interest which you indicated so that they can also be prepared to visit with students expressing an interest in this type of work.[26]

A professor may gladly volunteer to assist the CIA on campus or at Headquarters. University of Wisconsin professor of law, Gordon D. Baldwin, has stressed, "Foreign intelligence gathering is vital to our common good," adding that if the CIA had been the recipient of more advice from academicians, "we might all have profited."[27] Political scientist Myron Rush of Cornell University evidently concurs. In 1976, he openly accepted a scholar-in-residence position at the CIA, on leave from Cornell, reportedly to write an analysis of Soviet succession possibilities.[28] Political scientist Raymond E. Wolfinger, University of California, Berkeley, reasons further that consulting relations with the CIA are nobody's business but the individual professor's.[29] Typical responses drawn from a survey of faculty at the University of Georgia are the remarks of two professors: "Maybe if the C.I.A. secured the service of a few university types, it wouldn't make such a bungling mess of its operations," and "This is up to each professor to decide for himself—should not be subject to a rule."[30]

In contrast, the campus environment can be hostile toward any academic connection with the CIA. Professor Rush's CIA appointment stirred up a whirlwind of protest among Cornell graduate students. In a resolution, they charged that faculty involvement with the Agency "undermines the trust necessary for the survival of the academic community and basic academic freedoms" and promotes a "chilling effect" on the free expression of opinion in the classroom.[31] At Brooklyn College, a faculty member was denied tenure and promotion in the 1970s, evidently because of brief contacts with the CIA. Before traveling to Europe, the professor had telephoned the Agency for advice related to his research project. The CIA declined to offer advice, but suggested he call again to share his impressions after the trip. This subsequent "debriefing" consisted of a fifteen-minute telephone call upon his return, which the professor's brother-in-law reported to the faculty at Brooklyn College.[32] And during the Vietnam War era, some CIA recruiters even encountered acts of violence directed against them on the campuses (see the next section).

Whether the environment is friendly or hostile, the CIA is unwilling to rely on overt contacts alone. The Personnel Office representative will have his own covert ties with selected academicians on many campuses or will secretly confer with those "on loan" to him from the NCD. Though the Personnel Office claims it would prefer to make all of its contacts overt, it points out that many professors will not accept an open relationship for fear of reprisal from their colleagues or students. The Office of Personnel emphasizes also that—at least since the Turner directive cited earlier—before an academic contact sends forward the name of a student as a possible recruit, the student must give his or her prior consent.

The Personnel Office reaches out to academe in two other ways: through advertising and intern programs. Slick, full-page ads are placed in college placement publications and campus newspapers; glossy brochures are mailed out with pleasant photographs of the CIA, looking every bit like another campus itself. The intern programs, based on a competitive SAT-like exam and a minimum 2.5 grade-point average, are enticing, too. The 1987 Summer Intern Program offered $2000, an expenses-paid stint at the CIA for eight weeks, and tuition assistance for the last year of college for those successful candidates willing to commit themselves to an eighteen-month hitch with the Agency after graduation. For many students, this is a highly lucrative inducement. With their consent, successful applicants undergo full security-clearance procedures. On the average, 22 percent of past summer interns have gone on to seek permanent employment at the Agency.

The Office of Personnel also conducts a Student Trainee Program, permitting qualified students to work at the CIA for a short period of time (say, one quarter to two years), return to their academic training, then back to the CIA, oscillating between the two to gain practical experience and earn enough money to pay for the completion of the academic degree. Approximately 50 percent of the student trainees stay with the Agency. Finally, the CIA—like most agencies in the federal government—has developed special programs to hire blind people and others with physical disabilities.

SECURITY The Office of Security, responsible for the protection of CIA personnel and facilities, entered the campus picture when CIA recruiters were threatened by protesting students during the Vietnam War era. At the request of the director of personnel, the DCI approved a policy (Project RESISTANCE) in 1967 ordering the Office of Security to protect the well-being of CIA recruiters. According to an internal CIA memorandum prepared for the deputy director of security, the necessity for security "was based upon the ever-increasing harrassment [sic] of Agency recruiters by various dissident groups on college and university campuses."[33] The primary impetus for the stepped up security was an episode at Columbia University, where a CIA recruiter was held hostage for several hours by student radicals.[34] The examples of student harassment feared, and often faced, by Agency recruiters have included the following:

- Passing out handbills saying "You may have already won: call———" [the telephone number of the CIA recruiter]
- Having campus reporters sign up for employment interviews
- Having dissidents sign up for employment interviews
- Picketing

- Making obscene and threatening phone calls
- Staging sit-ins or "be-ins"
- Obstructing halls and doorways
- Engaging in confrontation
- Holding CIA recruiters prisoners
- Blowing up the recruiting office (as occurred at the University of Michigan)[35]

The CIA documents cited above, released by the Agency under a Freedom of Information Act request, indicate that the Office of Security pursued its mission with zeal. From 1967 to 1973, the office developed files on various universities and colleges; secretly infiltrated meetings of protesters; recruited new informants; made advance visits to campuses "to determine attitude of dissidents"; monitored and reported on black student activities; and established a close working relationship with the local police—all the operations that Tom Charles Huston had in mind when he recommended CIA campus spying in his sweeping intelligence plan.

Directorate for Science and Technology

The DS & T has links to the academic world primarily from two of its divisions: the Office of Research and Development (ORD) and the Office of Development and Engineering (ODE). Of the two, ORD has more of an academic orientation than ODE (which has closer ties to technical developments in the private enterprise sector), but both have three types of contacts with college and university personnel: consulting arrangements, research contracts, and informal communications over the telephone (without payment).

The DS & T's contacts with academe differ from those of other CIA entities chiefly in the highly technical quality of the information sought. Recent ORD contracts, for example, involved payment to professors at a western and a midwestern university, respectively, for research papers on climate control and computer modeling. Whether or not such research contacts are disclosed depends, according to a recent deputy director of the DS & T, upon the attitudes of the university officials—particularly the contact himself.[36] Officials in the directorate say they are prepared to go along with disclosure if the academicians involved give their consent. That these contacts are often unwilling to disclose their cooperation with the CIA is said to be a function both of their fear of harassment by students and colleagues and a concern that their own research opportunities may be impaired (perhaps, for instance, Country X might refuse access to an acknowledged CIA contact, however innocent the research relationship might be).

The DS & T also operates a Joint Publication Research Service (JPRS), which provides translations of foreign technical articles. Fourteen percent of the JPRS staff are full-time faculty members at various universities.

The grayest area in this web of relationships between the CIA's technical directorate and academe occurs with the use of research institutes. The CIA has often gone beyond the universities to think tanks for research purposes but often ends up dealing with the same minds, since many university faculty are also members of these research facilities. Sometimes the CIA will have a research relationship with a professor who, to avoid sharing his profits with the university (say, for a lucrative hardware breakthrough), will establish a research institute and will work with the Agency through this entity rather than under the auspices of the university. That this approach may be used also to avoid CIA campus guidelines is resolutely denied by one recent DDS & T.[37]

Finally, the DDS & T (as is the case with all CIA directorates) enters the campus to conduct its own recruiting, over and above that done by the Office of Personnel. While the Office of Personnel uses a shot-gun approach to recruitment, each directorate adopts a "rifle" approach, zeroing in on specific types of candidates for which it may have a precise need.

Office of Public Affairs

The last CIA entity involved in campus affairs is the Office of Public Affairs (OPA). A recent OPA director has emphasized that its ties with academicians are open and minimal.[38] The OPA responds to campus requests for CIA speakers, provides unclassified briefings to outsiders on how the CIA operates, answers mail on CIA affairs, handles visiting groups, houses the Publications Review Board, and performs a host of public relations tasks.

In a letter to university professors who teach courses fully or partially on intelligence policy (over fifty courses with this content in the United States were listed in a survey published by the National Intelligence Study Center in Washington, D.C.[39]), OPA's coordinator for academic affairs extended

> an invitation to open up a dialogue on the teaching of intelligence and a sincere offer to help with your courses if you so desire. I can arrange for Agency speakers to visit your classes, or for visits here to CIA, or I can help with bibliographic and reference material. I also may be able to answer on short notice specific questions on the history of intelligence or other subjects. I enclose a brochure on maps and other reference material published in the last few years by the CIA.[40]

For some scholars, particularly those writing on international affairs and specifically on the KGB and other foreign intelligence services, CIA assistance, one supposes, might be a powerful attraction.

When all of these entities and activities are added up, a complex picture emerges. The campuses become hundreds of islands joined to the CIA by a multitude of bridges, overarching, sometimes intersecting, and often shrouded in a mist of secrecy, lost from sight. These bridges and their purposes are summarized in table 8.1.

Scholars as Spies

Two issues have generated especially heated debate over proper CIA-academic relations: covert recruitment on campus and academic involvement in covert op-

TABLE 8.1. Summary of Primary CIA–Academic Ties

CIA Subdivisions	*Primary Purpose of Association*
Directorate of Operations	
National Collection Division (NCD)	Collection
Foreign Resources Division (FRD)	Recruitment (foreign agents)
Directorate of Intelligence	Research consulting
Directorate of Administration	
Office of Logistics (LOG)	Research contracting
Office of Training (OTR)	Career instruction
Office of Personnel (OP)	Recruitment (staff)
Office of Security (OS)	Protection of campus recruiters
Directorate of Science and Technology	
Office of Development and Engineering (ODE)	Research contracting
Office of Research and Development (ORD)	Research contracting
Office of Public Affairs (OPA)	Public Relations

erations abroad. Several important and related issues are subsumed under these two.

Covert Recruitment

Harvard University and the few other institutions that have addressed the subject of CIA-academic relations do not object to Agency recruitment of American students, as long as the recruitment is carried out overtly. What is objectionable to many critics is the secret use of academicians to recruit individuals on campus, as possible CIA officers or staff employees, if American, or as infrastructure assets abroad, if foreign (the "targets" of greatest interest to the CIA covert recruitment program).

In the course of these secret efforts to spot, assess, and recruit on behalf of the CIA, academicians may involve themselves in a number of activities that critics have found objectionable. Taking advantage of the position of trust accorded faculty, for example, a professor may gather information for the CIA about a student's ideological views, his attitudes toward American foreign-policy objectives, his financial situation, and the like. Seminar discussions, office counseling, social gatherings, term-paper grading, and other contacts between faculty and their students provide several opportunities for agent spotting.

Critics have opposed the idea of covert recruitment on several grounds:[41]

- The practice, if permitted to continue, will weaken the free exchange of views, spoil the spirit of trust and candor on campus, and generally undermine the open atmosphere essential to the primary purpose of the university in a democracy, namely, the discovery and dissemination of truth.
- Covert recruitment represents an inappropriate intrusion of the government into the affairs of the university, particularly in light of the fact that some

universities (Harvard among them) have promulgated specific guidelines against this practice that the CIA seems prepared to ignore for the sake of "national security."

- Spying on campus, abetted by academicians, will result in a loss of public esteem for the university.
- The university should not participate in the recruitment of foreign students for operations that may result in the violation of laws within their own countries.
- The university should not be responsible for triggering a secret background investigation of individuals within the academic community.
- Foreign students studying in the United States should be treated to the same rights and respects extended to U.S. students.
- Academicians have a professional obligation to uphold the integrity of the university, which transcends the claim both of individual academic freedom to work secretly with the CIA if one so chooses and the broader assertion that the national good (as enhanced by intelligence activities) overrides the good of the university; critics argue that the true purpose of the university—the search for truth in an open environment—serves the national good to a greater extent than its involvement in secret intelligence liaisons.
- Covert relationships between academicians and the CIA are, by their nature, hidden; therefore, the university has no way of knowing to what extent the integrity of the American academic community is being compromised.
- Regulations controlling recruitment on campus by intelligence agencies must be more rigorous than for other organizations, since the former may use students for unusually hazardous assignments or for operations that may be illegal under the laws of another nation; therefore, requiring the CIA to report all its recruitment contacts is neither unreasonable nor unfair.

For all of these reasons, critics of CIA covert recruitment would prefer that the Agency—like other organizations—work openly through the campus placement counselors. Specifically, Harvard University has put into effect, as of 1977, the following guidelines on campus recruitment:

> Any member of the Harvard community who has an ongoing relationship with the CIA as a recruiter should report that fact in writing to the Dean of the appropriate Faculty, who should inform the President of the University and the appropriate placement offices within the University. A recruiter should not give the CIA the name of another member of the Harvard community without the prior consent of that individual. Members of the Harvard community whose advice is sought on a one-time or occasional basis should consider carefully whether under these circumstances it is appropriate to give the CIA a name without the prior consent of the individual.[42]

The CIA offers several refutations to those who oppose the policy of covert recruitment. The most important may be summarized briefly:

- The CIA should receive equal treatment on the campus with other extramural organizations.

- The interests of national security require the covert recruitment of foreign students (who would shy away from any form of overt recruitment).
- Academic freedom and self-governance permit an individual faculty member to make his or her own choices about covertly assisting the CIA.
- No conflict of interest exists between helping the CIA and the country, on the one hand, and performing academic duties, on the other hand.
- While the CIA would be willing in many cases to make its contacts on the campus overt, the academicians often refuse through fear of negative career repercussions.
- The CIA has a statutory obligation to protect its sources and methods.

As director of the CIA, Admiral Turner publicly dismissed Harvard's guidelines with the simple declaration: "Harvard does not have any legal authority over us."[43]

At the heart of the matter (certainly from the Agency's point of view) is the question of foreign student recruitment on American campuses. The CIA believes strongly in the importance of recruiting foreign students within the borders of the United States. To accomplish this task effectively the Agency also believes that covert academic contacts are necessary to focus the recruitment effort. Openly setting up a recruitment table at the campus placement center for foreign students will not work, since foreign students cannot let it be known they are working for the CIA against their own countries.

Once the CIA establishes a covert academic contact, the secret world of intelligence intrudes upon the open world of the university. The relationship may lead to operations quite alien to the traditions of free inquiry sacred to institutions of higher learning. An Iranian student in a political science class may well wonder on whose desk his term paper may ultimately land. The result of such concerns, argue critics, is a chilling effect on the freedom of expression in the university. As the CIA tries to develop more information on a foreign student, the degree of intrusion increases. A bona fide scholar from another land becomes the object of intense surveillance, his privacy cast to the wind.

In some cases, the foreign student may not be a bona fide scholar; he may be an intelligence agent. The CIA may be tempted to have one of its academic assets keep an eye on this suspected foreign agent. This approach, however, would be inappropriate, critics contend; suspected intelligence agents are, properly, the province of FBI counterintelligence, or subjects for criminal investigation by the Bureau under the espionage laws.

A central question, then, is whether the CIA really needs to recruit foreign students on the campuses. In a nutshell, critics say no, because the process inevitably draws in American academicians and thereby undermines their integrity and the reputation of the university. Moreover, America's universities should not submit foreign scholars to such intrusions by the U.S. government into their private and academic lives. Proponents say yes, because the United States needs spies abroad and foreign students may be highly placed officials in their governments one day. Further, the American campus is an ideal location to develop foreign assets, since the students are often lonely, low on cash, and critical of their governments. And, for the recruiter, the environment is far less dangerous than spy-infested Bonn or Baghdad. Finally, if a foreign student is not a serious scholar at all, but rather an agent sent by his government to spy on the United States, he or she is hardly worthy of the special legal protections afforded American citizens.

Even if covert recruitment of foreign students were to be prohibited, the problem of covert recruitment of American students for CIA staff employment would remain. Professors and other academicians continue to provide the Agency with the names of likely recruits, sometimes setting in motion security checks without the prior consent of the student. Some placement counselors apparently have even been paid for assisting the CIA in its campus recruitment efforts, an arrangement which does not seem to exist with other extramural organizations.

The harshest critics of CIA-academic relations argue against any ties whatsoever, not even a sideward glance. At the other extreme are those who have no reservations at all about any of the associations between the two worlds that have come to light over the years. And in the middle are a range of reforms offered by those who see some need for modifications in the current practices. As presented below in descending order of severity, these reforms move from fairly substantial changes down to only modest adjustments in the way things are done now:

- Recruit only CIA officers and staff (as opposed to recruiting foreign assets) and allow recruitment to be conducted only at Agency Headquarters. From Headquarters, placement materials could be sent out to the campuses, job applications reviewed, and the selection process carried out. The only recruitment tasks conducted outside Headquarters would be annual, regional interviews of top candidates (as determined by a review of the application files), held off campus in government buildings. No CIA recruiter would need to set foot on campus. Foreign student recruitment would be off-limits, at least on campus (or, some would insist, altogether). Faculty members could become involved in recruitment only as volunteers, who openly inform the campus placement office, their dean, and their departmental head that they are available to advise students on Agency careers. Moreover, faculty would never forward names to the CIA without permission from the student.
- Permit CIA recruiters on campus to recruit American citizens, but only in an overt way. All academicians would have to report any recruitment cooperation with the CIA to their dean and departmental heads. Covert recruitment of foreign students would remain restricted to off-campus locations, or might be off-limits altogether.
- Make all recruitment of American students overt, but the CIA would be permitted to recruit foreign students covertly. The only limitation on this covert recruitment would be that the CIA recruiters would have to conduct their efforts directly, without the use of academicians.

Covert Operations

On various occasions in the past, as documented by the Church committee, the CIA has called upon members of the academy to perform covert operations. According to a deputy director for administration (DDA), the cooperation of academicians has been "vital to the intelligence collection mission of the CIA."[44] These operations have included—in addition to intelligence collection—support operations, covert action (especially the writing of propaganda), and counterintelligence.[45] Critics have questioned whether academicians ought to be involved in any of these assignments. Most of the same arguments against covert recruitment are applied to covert operations as well: participation will undermine the integrity of the university, lead to the loss of public esteem, contaminate the open atmosphere of the university, and the rest.

Students usually see covert recruitment as the most important issue, while faculty generally worry more about covert operations. Students have a right to know who their teachers are and whether they have an affiliation with the government. Faculty members, in contrast, are often more concerned about whether their research pursuits might be jeopardized if scholars are suspected abroad as being spies; they fear that evidence of academic collusion overseas with the CIA in covert operations will slam down the gates of access for American researchers. One reason some academicians are said to be opposed to public identification of their assistance to the CIA in recruitment or other tasks stems directly from a fear of retaliation against them in foreign countries. They may be refused entrance to a country for research or lecturing, or, if allowed in, may be dealt with in unpleasant ways, from noncooperation in scholarly pursuits to even violent attacks.

As with journalists, academicians are less worried about losing their own objectivity through an association with the CIA than they are about possibly losing their credibility, both with their readership and professional colleagues at home and in dangerous regions abroad where their very survival may depend upon the legitimacy of their scholarly credentials. Just one professor shown to be an espionage agent calls into question—however unfairly—the integrity of every other member of the academic community. In a world of value trade-offs, argue the critics, protection of the academic community from suspicion of secret government collusion is of greater value to the nation than the limited contributions scholars may make to covert operations abroad. In a democracy, the preservation of the free and independent university is of incalculable importance. In this spirit, Harvard University's guidelines on the subject read:

> Members of the Harvard community should not undertake intelligence operations for the CIA. They should not participate in propaganda activities if the activities involve lending their name and positions to gain public acceptance for materials they know to be misleading or untrue. Before undertaking any other propaganda activities, an individual should consider whether the task is consistent with his scholarly and professional obligations.[46]

Academic involvement in propaganda operations has been particularly troublesome to many critics. Several of the leading lights in U.S. propaganda efforts during World War II were prominent American scholars (among them Harold Lasswell, the political scientist). Yet in such operations the campuses and the CIA are obviously at odds, especially outside the conditions of wartime. The raison d'être of the campus is to seek and tell the truth; often propaganda has precisely the opposite goal—to lie and mislead. ("The truth is so precious it must be protected with a bodyguard of lies," Churchill once said in defense of propaganda.[47]) For an academician to be involved in lying raises questions about his commitment to the basic purpose of the university. Critics question, too, whether scholars ought to associate themselves with projects that may involve blow back, whereby, the reader will recall from chapter 4, false or distorted information placed abroad circulates back to the United States. Blow back is especially troublesome for democracies, which rely upon an informed citizenry, because CIA propaganda drifts homeward to fool Americans as well as the adversary. The task of the scholar, in a word, is to inform society, not to deceive it.

As chapter 2 pointed out, most CIA propaganda is not intended to deceive but rather to disseminate broadly the American point of view on various issues. It may be that scholars can legitimately assist in this process, making a product meant to be true even more accurate through the use of objective research skills. Still, critics who are willing to condone academic participation in truthful propaganda efforts normally demand that scholars at least make known their association with the intelligence community. (The subject of academicians' researching and writing for the CIA raises other problems, too, chiefly regarding the ethics of government financing of books and articles.[48])

Stated succinctly, reform proposals with respect to covert operations usually include the following, in descending order of severity:

- banning all academic participation in covert operations
- allowing participation in truthful propaganda operations, but only if the connection to the CIA is publicly acknowledged
- limiting participation only to truthful propaganda activities
- allowing prebriefings and debriefings for scholars traveling abroad, but only occasionally and without systematic attempts to task (that is, explicitly assign intelligence collection requirements)

While covert recruiting and covert operations have been the two topics around which most of the controversy over CIA-academic ties has centered, several related issues provoke debate. Among the most disputatious has been the proper relationship between the Agency and the universities for research contracting and consulting.

Research Relationships

The Church committee wrote in its final report that the CIA "must have unfettered access to the best advice and judgment our universities can produce. . . ."[49] While this observation provides common ground for widespread agreement, not everyone concurs on what the rules should be (if any) when it comes to research connections between academe and the secret service. Harvard University, for example, prohibits its faculty from participating in classified research for the government—a rule that spread to many campuses in the 1960s when students protested vigorously against secret government contracts for, among other projects, the development of napalm and other war materiels used in Vietnam. Other campuses (the University of Georgia and Georgia Tech, for two) have no restrictions against classified research by their faculty. Some request only that a faculty member report consulting arrangements with the CIA in the annual "activities" summation for the department head.

The rules governing such matters (when they exist at all, which is rare) tend to be as diverse from campus to campus as the colors of football jerseys. Yet, as with the question of cooperating with recruitment efforts, critics are usually prepared to condone a research relationship between academicians and the CIA as long as it is overt. Harvard has adopted the following two guidelines for institutional and individual consulting arrangements with the Agency:

> Harvard may enter into research contracts with the CIA provided that such contracts conform with Harvard's normal rules governing contracting with outside sponsors and that the existence of a contract is made public by University officials.
>
> Individual members of the Harvard community may enter into direct or indirect consulting arrangements for the CIA to provide research and analytical services. The individual should report in writing the existence of such an arrangement to the Dean of his or her Faculty, who should then inform the President of the University.[50]

Beyond the proposal that research and consulting arrangements be made a matter of public record (and that means more than just being tucked away somewhere in dusty department files), other improvements in the research area advocated by reformers include the following:

- Along with the name and department of the researcher or consultant, the public record should state the subject of the project (along with a brief explanation of the objectives and methodology), its cost, and its duration. (The *Congressional Record* or some other accessible document could list this kind of information on all government research projects contracted with extramural organizations.)
- Certain kinds of research ought to be expressly prohibited, such as (to use known examples) the production of shellfish toxin, dart guns, and MK/ULTRA-type drug experimentations.[51] Precisely what research should be prohibited would have to be determined on a case-by-case basis. The intelligence committees in Congress should be kept appraised, in a timely fashion, of all significant CIA research projects (intra- and extramural). Moreover, all human subjects involved in a research project should be aware of their involvement and the concomitant risks, and should sign a consent agreement—standard procedures in academe that have sometimes been ignored by the CIA's in-house researchers.[52] A member of the academy remains so until he resigns, retires, or is dismissed from his or her college. This means that CIA-academic guidelines should apply when an academician is away from the campus on holiday, at a convention, on extended research leave or sabbatical, or perhaps wearing another hat as director of a private research institute.

"Unwitting" Use of Academics

According to CIA regulations, the "basic rule" guiding the Operations Directorate in its use of foreign agents is that "any consenting adult" may be used. Yet as the Church committee discovered, the CIA often had contracts with universities that were handled through a foundation front, so that the university "consented" to work with the foundation without knowing (or being "witting") that the CIA was involved at all. Critics uniformly agree that this unwitting use of individuals or institutions is improper and should be prohibited.[53] Harvard's guideline on this subject reads: "No member of the Harvard community should assist the CIA in obtaining the unwitting services of another member of the Harvard Community. The CIA should not employ members of the Harvard community in an unwitting manner."[54]

The directive issued on CIA-academic relations by Admiral Turner stated that "no operational use will be made either in the United States or abroad of staff and faculty members of United States academic institutions on an unwitting basis." The surface intent of this statement is unimpeachable, but a fine reading suggests nuances that are troubling to critics and raises questions that still need to be addressed. What, for example, is meant by "operational use"? Does this include recruitment on campus? Moreover, the phrase "staff and faculty members" leaves wide open (perhaps unintentionally) the use of students. Finally, nothing is said about the unwitting use of foreign students studying within the United States. These omissions could be corrected with the following clarifications:

- The CIA should not employ, or otherwise use in an unwitting manner, any one affiliated with an American academic community.
- The CIA should not use intermediary organizations in dealing with academicians without making it expressly known that the organizations are affiliated with the Agency. (The 1986 Casey rule on revealing CIA funding of conferences represents a step in this direction.)

Legislation Versus Professional Ethics

Another issue that has attracted considerable attention is whether CIA-academic relations ought to be guided by statute, professional codes developed by academicians themselves, or CIA regulations. Most critics within academe seem to agree that the best solution is a combination of all three (in contrast, journalists seem to prefer reliance on their own codes; see chap. 9). President Derek C. Bok of Harvard University, for instance, has stated that the source of guidance should depend upon the issue. When it comes to covert recruitment, Bok believes it is "the prerogative of the university" to decide.[55] Once the university sets its standards, however, Bok would insist that the CIA honor them (which he correctly states the Agency refused to do in 1978, chiefly because the CIA had decided—in Bok's words—"that individual faculty and staff members, as citizens of the

United States, should be free to serve the CIA and their country as they see fit"[56]).

With covert operations abroad, though, Bok sees an obligation for Congress—not the individual campuses—to set the rules, "because there the interests of all universities are inextricably bound up with one another."[57] A witness from the University of California, Dr. Richard M. Abrams, agreed with Bok during Senate hearings held in 1978. According to Abrams, it was necessary but not sufficient for universities to set their standards: "What we need now to complement this would be proscriptions against government agencies inducing this kind of behavior among member of the academic community."[58]

At least in 1978, the CIA apparently saw a need for legislative guidance, too. Dale Peterson, an Agency official, told the *New York Times* that he believed the disagreement between the CIA and Harvard had reached a point where "it is up to Congress to arbitrate it."[59] No congressional arbitration was ever forthcoming. The simplest legislative remedy would be for the intelligence committees to tell the CIA that, henceforth, its internal regulations should prohibit the use of academicians in covert operations, except in times of national emergency. The committees would need to establish a timely reporting requirement for any waivers and then check periodically to make sure the prohibition remained in effect.

Appropriate Senior Management

One of the more elusive phrases in CIA statements on academic relations has to do with whom the Agency consults when it wishes to establish a university contact. As the CIA put it during Admiral Turner's tenure: "All of our contracts with academic *institutions* are entered into with the knowledge of *appropriate senior management officials* of the institution concerned."[60] Commenting on the phrase "appropriate senior management officials," Sen. Daniel Patrick Moynihan observed, only partially tongue-in-cheek: "I taught at Harvard for a long, long while, and I was not cognizant of any person on that campus who I would refer to as an appropriate senior official."[61]

More worrisome still, the senior official himself could simply be a CIA asset, as appears to have been the case at the University of California. At Berkeley, Administrative Vice President Earl Clinton Bolton served as an "appropriate senior official" while carrying out covert contract work for the CIA (which consisted largely of advice on how to develop relationships on the campuses more effectively without having them revealed). With this and other examples in mind, the American Association of University Professors has concluded that "a practice which is wrong is not made right by informing 'appropriate senior officials.' "[62]

The difficulty with most university reporting requirements—such as notifying a senior official of a contract—is that the report becomes "public" only in a technical sense. Is it part of the "public" record to place information on a faculty member's CIA contract into a generally inaccessible file in the department head's office? Is it "public" to whisper into Vice President Bolton's ear that the CIA has *n* number of "penetrations" into the Berkeley campus? Closer to the spirit of openness is Harvard's effort to publish in its campus-wide *Gazette* all of the University's contractual relationships—though not individual faculty contracts (which are also deftly avoided in the CIA policy statement cited earlier).

Foreign Students

Throughout CIA statements on academe (as with its statements on the media) a basic premise seems to be that when it comes to foreigners, practically anything goes. Much of the Agency's interest in the American academy springs from its desire for special access to foreign students. The whole purpose of the Foreign Resources Division, recall, is to turn foreign scholars into spies for the United States. In the words of one critic, "the practice undermines the necessary trust between students and scholars. Foreign nationals in our institutions of higher education are just as entitled to that assurance as United States nationals."[63]

Another aspect of U.S. government relations with foreign students has to do with what obligation, if any, the intelligence community has in their protection (and the protection of American students for that matter) from operations carried out by foreign intelligence services. This is a CI responsibility largely in the hands of the FBI; but persistent stories that Iranian and Taiwanese students (among others) have been harassed by their governments' intelligence agents in the United States (as well as rumors suggesting the CIA has aided and abetted these operations) raise serious questions about what this country is doing to preserve the freedom of its campuses from intrusion by foreign government agents.[64]

Conclusion

Here, then, are the main issues at the center of the debate over CIA-academic ties. The treatment here has not been exhaustive. The imposition of secrecy orders on academic research is, for instance, another subject that has stimulated controversy.[65] The analysis does serve to indicate, however, how serious the risks can be for democracy when scholars work with—even become—spies.

One can agree with the critics who find the official Turner and Casey policy statements inadequate. They fall short and ought to be augmented on campuses by the Harvard University guidelines. Academicians and the CIA ought to remove themselves from covert recruitment roles. Further, it seems sensible for academicians to participate in CIA publication activities only by contributing objective research findings and—a strengthening of the Harvard guidelines—only as publicly acknowledged participants. Secret research is anathema to the traditions of the university in peacetime and should be eschewed by scholars. As for pre- and debriefings with the CIA, these sessions may be useful to the university researcher, as long as they remain open, scholarly discussions with intelligence analysts; but scholars must be wary of attempts at "tasking" by the Agency, which would move the researcher into the sphere of operations. All covert operations involving academicians should be prohibited by the campuses and by the CIA, and the intelligence committees should double-check periodically to ensure that the prohibition is being honored.

Admiral Turner's argument that any individual academician ought to be free to assist the CIA out of a sense of patriotism carries force. After all, few are against either academic freedom or patriotism. In the CIA-academic relationship, however, the possibility exists that the behavior of an individual member of the academy might discredit the entire community of scholars if, for example, a professor were apprehended overseas on an espionage assignment. Rules must be set to protect the reputation, and the opportunity for safe travel abroad, of the thousands of American scholars who are precisely what they proclaim themselves to be: free and independent minds in search of truth, without covert ties to this or any other government.

NINE

The CIA and the Media

Symbiosis

Deep within the Soviet Union, some five hundred miles from Mongolia, lies the town of Krasnoyarsk. Few locations on the face of the earth are so remote and obscure; yet in 1983 Krasnoyarsk became a name of importance in the annals of arms control, for here the Soviets began construction of a controversial "phased-array" radar station. The true purpose of this facility, hotly debated within arms control circles, held potentially sweeping implications for the future of arms talks and relations between the superpowers.

If the site had been strictly for the tracking of objects in outer space, as the Soviets maintained, the radar would have passed with little notice; but the possibility existed that the facility had broader objectives related to Soviet military capabilities. In the view of some experts, including top officials in the Reagan administration, this possibility was more accurately a probability. If they were correct, the Krasnoyarsk radar represented, at a minimum, a clear violation of the 1972 Anti-Ballistic Missile (ABM) Treaty and its protocol; at a maximum, it could be the first visible sign of a Soviet commitment to (in the words of a Department of State document) "an ABM defense of its national territory."[1]

Spotted by U.S. reconnaissance satellites in 1983,[2] Krasnoyarsk has revealed many of its secrets to America's watching mechanical eyes in space. In addition to knowing the frequency, direction, angle, and external structure the United States has also learned that it is not insulated with concrete ("hardened") or defended by interceptor missiles. Ambiguities remain nevertheless, providing leeway for argument about its ultimate intended purpose: space tracking, early warning, or—the troubling possibility—ABM battle management. Fueling uncertainty is the fact that phased-array radars are multipurpose so that they can be used for each of these tasks. As a classified report on Soviet noncompliance prepared by the Reagan administration puts it: while Krasnoyarsk and five additional phased-array radars in the USSR "probably have the size and the power to perform a battle management support role, we cannot ascertain whether they have the necessary data-processing capacity, and uncertainties remain about their actual performance characteristics."[3]

Krasnoyarsk, then, represent, an intelligence problem, no doubt one of high priority. Yet how can one resolve questions that have eluded the unblinking lenses of America's fly-over cameras? Ideally, one would want a trained observer on the ground at Krasnoyarsk—indeed, in the government buildings. This is hardly a matter of renting an auto in Rawalpindi and motoring across the Russian tundra for a look. The observer needs a good excuse for being there. Among the few, perhaps the only, people who might manage a plausible excuse are members of the American media; the Krasnoyarsk radar is news.

Say that the Soviets were willing to grant permission to an American television network to film a story out of Krasnoyarsk; perhaps the Soviets would like to demonstrate that this site really is for space tracking. How tempting it would be for the CIA to "tutor" the network reporter and camera person in advance on what to look for in Krasnoyarsk. Conversely, how tempting for the network to seek tutoring, for who beyond a few technical intelligence experts would be able to distinguish a space tracker from an ABM guidance system—the very heart of the news report.

Krasnoyarsk, in a word, illustrate the symbiotic relationship that often exists between the reporter and the intelligence officer. When he was chairman of the Intelligence Oversight Subcommittee in the House of Representatives, Les Aspin (D, Wisconsin) commented at length in an exchange with DCI Stansfield Turner on the "complex interrelationship"[4] between the CIA and the press:

> . . . it is a peculiar relationship because in some sense they are in the same business. Both the Agency and the journalists are out looking for information and both of them have something that the other one wants. You would love to use those journalists for cover and that kind of thing. And . . . you think, "If they would only be patriotic and only do what they are supposed to." And the journalists think, "If those guys would just give me the information and let me get my Pulitzer Prize." You have got great information that they need. They have got a sort of access and kind of ability to influence events that you would like. . . .[5]

This mutual attraction sketches back over the years to the early beginnings of the CIA. When Allen Dulles became director in 1953, one of his first steps was to explore the possibilities for a close working partnership with the press.[6] From his vantage point, the news media could help the intelligence community in two important ways: intelligence collection and propaganda. The press could provide cover for intelligence officers and assets abroad, as well as gather intelligence directly, and foreign press could serve as conduits for the dissemination of propaganda. One of Dulles's successors has explained the appeal of journalists to intelligence agencies:

> . . . [the journalist] can get to people that officials cannot get to . . . he can circulate in areas without bringing with him the official presence of the

> United States. He can speak to people who would not speak to a representative of the American Embassy . . . you need someone outside [the Embassy].
>
> Now, if you look around as to who can do that best, the journalist can circulate perhaps the best. . . .[7]

So began an association between the arts of espionage and journalism that would grow increasingly interwoven. To critics of this alliance, the line between intelligence officials and news correspondents seemed in danger of vanishing altogether, undermining the cherished ideal of a free and independent press embodied in the first amendment to the Constitution. For advocates, the alliance was good for the CIA, good for the reporters (the proper tips could advance their careers), and therefore good for the country. Besides, cooperation with the government was, for the journalists like any other citizen, the patriotic thing to do. Wherever one stood normatively on this issue, one fact was incontrovertible: the CIA-media relationship had evolved by the late 1950s into a complicated matrix of people, activities, and bonds of association.

Several major issues emerge from an analysis of CIA ties to the media. Should the CIA be allowed to use press credentials as a "cover" (rationale) for their presence abroad? Should CIA officials and the media swap information? Should media personnel gather information for the CIA in the course of their own work? Should the media cooperate with the CIA by providing additional support overseas, such as recruiting agents or serving as intermediaries between assets and case officers? Should the media, American and foreign, cooperate with CIA propaganda operations? Should the media cooperate with the CIA to suppress sensitive news stories that, in the Agency's opinion, might be detrimental to national security?

In each of these instances, the CIA and the media have answered yes at one time or another, but the noes have become more frequent in recent years. The reasons for this change of heart are examined in detail in this chapter. The central democratic issue at stake here is the reputation of the American free press. The media in this nation are widely respected and trusted throughout the world, especially compared to the government-controlled news agencies in the USSR. Yet if the U.S. media become servants of a government espionage agency—providing credentials, information, and other aid—can that valued reputation endure? Can the world continue to distinguish clearly the *New York Times* from *Pravda* as an organ of unfettered expression of fact derived independently of government influence? In the competition for friends and allies in the world, will the United States have lost one of its great attractions as a democracy? Indeed, could even American citizens trust their media as sources of reporting on government activities?

To questions like these raised by critics of too close a relationship between the CIA and the media, advocates offer this argument: in a world of bitter conflict with deadly rivals, should not the media cooperate like all good Americans to protect and advance the interests of the United States through assistance in times of need to government agencies, including the CIA? As this chapter attempts to show, black-and-white answers to these questions have proved difficult and unsatisfactory.

Reporters as Spies

"The only thing we do," said William Colby shortly before his retirement as CIA director in 1976, "is [hire] the odd fellow who makes a living in some foreign capital selling a little story here and there. We may also be able to use him."[8] One is left with the impression that the CIA merely recruits a few journalists now and then. While this may have been the case in January 1976, at the height of the Senate investigation of the CIA, the Agency over the years has had extensive ties with hundreds of American and foreign journalists. Carl Bernstein, famous for his investigative reporting on Watergate for the *Washington Post,* claims that "more than 400 American journalists" secretly carried out assignments for the CIA from 1952 to 1976.[9] The Church committee reported a figure of 50 American journalists for the same time period.[10] The CIA concedes a more modest figure of "some three dozen."[11]

The only way to settle definitely the true number would be for the CIA to reveal the names of these journalists, as demanded by some members of the press.[12] This the CIA has steadfastly, and quite appropriately, refused to do on the basis of the wording in the 1947 National Security Act that charges the DCI to protect "sources and methods."[13] Even with its subpoena powers, the Church committee yielded, in face of CIA intransigence, in its demands for access to complete information on journalists used by the Agency, settling instead for detailed files on a sample of twenty-five individuals (half or more of whom were foreigners employed by American news agencies) with names and other identifying data excised.[14]

The Bush Directive

The question of numbers draws one immediately into a definitional thicket about what is meant by an "association" or "relationship" with the intelligence community. A statement released by DCI George Bush in February 1976, supposedly to mollify press and public concern over CIA use of the media, indicates how even the apparently simple word "journalist" can lead to extensive definitional and even philosophical arguments. In a further tightening of restrictions on CIA media relations initially invoked by William Colby in 1973, Bush declared as one of his first public statements as DCI that henceforth the "CIA will not enter into any *paid or contractual* relationship with any full-time or part-time *news correspondent accredited* by an *U.S.* news service, newspaper, periodical, radio or television network or station."[15] This went considerably beyond Colby's willingness to bar from CIA recruitment only the staff of general-circulation, U.S. news organizations, but not the press associated with more obscure outlets.[16]

Placing aside for the moment the phrase "paid or contractual" (which left the door wide open for voluntary associations of all stripes), omissions and qualifications in the Bush directive seemed to raise as many troubling questions as it may have settled. First, it was clear from the word "accredited" that so-called stringers (occasional contributors) and freelance writers—that is, individuals who may work for the media without being on the permanent payroll or having an explicit contractual affiliation to a media organization—would continue to be fair game for Agency recruitment. This represented a rather sizable exception; such well-known writers as Tad Szulc and Seymour Hersh have been freelancers at one time or another in their careers. Further, any non-American journalists would remain potential targets of CIA blandishments, presumably including journalists in various allied nations with a free press.

The phrase "news correspondent" also seemed to allow broad discretion for the CIA to use individuals in the press who possessed full accreditation but happened not to be the scribblers one often think of as "correspondents," "reporters," or "journalists"; that is, news managers, technicians, cameramen, even accountants and the like escaped the Bush prescription. As Bush himself acknowledged, his directive left undisturbed more than half of the existing CIA ties with American media personnel.[17]

The directive did change the pattern of associations, however; following the Bush directive the trend, according to Agency officials, was toward the cultivation of nonaccredited writers within the more specialized media (trade journals and commercial newsletters, for instance). The CIA continued unabated its search for recruits within the foreign media, though the press working for English-speaking allies of the United States (Great Britain, et al.) were supposedly considered off-limits by tacit understanding among the intelligence services of the various nations.[18] The Church committee commented on this extensive CIA contact with foreign journalists, referring to

> a network of several hundred foreign individuals around the world who provide intelligence for the CIA and at times attempt to influence foreign opinion through the use of covert propaganda. These individuals provide the CIA with direct access to large numbers of foreign newspapers and periodicals, scores of press services and news agencies, radio and television stations, commercial book publishers, and other foreign media outlets.[19]

This network was part of what Frank G. Wisner, an early CIA propaganda specialist, liked to call his "mighty Wurlitzer," upon which he would pipe out around the world various anticommunist themes as part of the Truman administration's policy of containment.[20]

Also untouched by the Bush directive was the CIA's lively interest in the development of its own media proprietaries abroad. Past examples would include CIA funding for the West German magazine, *Der Monat,* the blue-blooded British periodical *Encounter,* and the Rome *Daily American.*[21]

Still dissatisfied with Bush's attempts to pare down the connections between the intelligence community and the press, individual journalists and media associations continued to clamor for tighter restrictions. The American Society of Newspaper Editors (ASNE), for example, called for a CIA "hands-off rule" for *all* American *and* foreign journalists. "The power of America's commitment to freedom resides in its example," stated the January 1977 ASNE *Bulletin.* "The CIA should exhibit the American commitment to free press abroad and at home alike."[22] Among others, the Fund for Investigative Journalism, the Overseas Press Club of

America, and the Washington Independent Writers also complained vigorously to Bush about his stringer and freelance exemptions.[23]

The Turner Directive

When Stansfield Turner became DCI in 1977, he felt obliged to expand the Bush restrictions further still—though substantially short of the outer limits sought by the press organizations referred to above. Issued on November 30, 1977, the Turner directive, which remained in effect without modification by his successor, William J. Casey,[24] represented a response, said the CIA, to "a number of additional and relevant points . . . raised" (presumably a reference to press criticism) since issuance of the Bush directive.[25] Acknowledging that the "special status afforded the press under the Constitution[26] necessitates a careful policy of self-restraint on the part of the Agency," the CIA pledged not to:

> a. enter into any relationships with full-time or part-time journalists (including so-called "stringers") accredited by a U.S. news service, newspaper, periodical, radio, or television network or station, for the purpose of conducting any intelligence activities. The term "accredited" means any full- or part-time employee of U.S. or foreign nationality who is formally authorized by contract or by the issuance of press credentials to represent himself or herself either in the U.S. or abroad as a correspondent for a U.S. news media organization or who is officially recognized by a foreign government to represent a U.S. news media organization;
>
> b. without the specific, express approval of senior management of the organization concerned, enter into any relationships with non-journalist staff employees of any U.S. news media organization for the purpose of conducting any intelligence activities;
>
> c. Use the name or facilities of any U.S. media organization to provide cover for any Agency employees or activities.[27]

An addendum to this policy statement emphasized that the latest CIA pronouncement on its relationship to the media would continue to allow even accredited journalists "the opportunity to furnish information which may be useful to his or her Government"—as long as the information was provided on a strictly voluntary (unpaid) basis.[28] In what critics would view as a suspicious escape hatch, the addendum ended with this proviso: "No exceptions to the policies and prohibitions stated above may be made except with the specific approval of the DCI." The worry arose immediately that any of the new limitations could vanish overnight with the DCI's nod, perhaps in a secret codicil never brought to the attention of the intelligence committees on Capitol Hill, let alone the press.

The provisions of the Turner directive tightened the restrictions on CIA-media relations in three ways. First, it defined more precisely what types of journalists would be off limits to CIA recruitment efforts, clarifying the Bush directive by explicitly adding stringers and foreign nationals *accredited* by a U.S. media organization. While Bush probably meant the inclusion of these journalists anyway, Turner removed any ambiguity on that score. The new directive, though, failed to respond to the question of freelancers. Clearly, it remained open season

TABLE 9.1. CIA Directives on Media Relations

Prohibition	CIA Director: Colby (1973)	Bush (1976)	Turner[a] (1977)
Relationship with Staff of General Circulation U.S. News Outlets	X	X	X
Paid Relationship with Part- or Full-Time Accredited U.S. News Correspondents		X	X
Any Relationship with Journalists (Including Stringers) Accredited to U.S. News Outlets			X
Relationship with Nonjournalist Media Personnel unless Approved by News Management			X
Use of U.S. Media as Cover			X

[a]The Turner directive remains in effect.

for them, as well as for foreign journalists without a formal contract or press credentials from a U.S. media outlet (the prerequisite for "accredited" status). Second, the Turner directive promised to use nonjournalistic staff members of U.S. media organizations (cameramen, et al.) only with "the specific, express approval of senior management in the organization concerned." Third, the CIA would no longer use the credentials of real U.S. media outlets for purposes of intelligence cover (see table 9.1).

To what extent these restrictions settled the debate on the proper scope of CIA-media relations depended upon one's point of view. Despite Turner's attempt to reduce dissatisfaction within the press corps over the Bush directive, the ranks of the critics remained deep. The CIA's lingering ties to freelancers, nonjournalist staff employees, and the foreign press in democratic regimes continued to be the main bones of contention.

The Problem of Freelancers

Joseph Fromm, deputy editor of *U.S. News and World Report* and an overseas correspondent for twenty-eight years, has pointed to the difficulty of defining the word "freelancer." Finding himself unable to distinguish clearly between "busy freelancers" who write frequently for major U.S. news outlets and the stringers and journalists covered by the Turner directive, he "would be inclined to apply the blanket prohibition to those who claim to be freelancers."[29] Former U.S. ambassador to Nigeria, William Trueheart, and others agreed with this position in testimony before the House Intelligence Oversight Subcommittee.[30] The definitional problem, though, continued to trouble at least one member of the subcommittee. "What is a freelancer?" asked Rep. Bob Wilson (R, California). "How big is a horse? You know I just don't think it is possible to include it and I am frankly glad it is not included."[31]

Those opposed to the use of freelance writers by the CIA offer seemingly uncomplicated solutions to the problem of defining the term. A freelancer is "a journalist who does not draw a regular wage," offers newspaper editor Gilbert Cranberg.[32] "The key is whether a person identifies himself as a freelance journalist," Morton Halperin suggests; he would extend coverage not only to freelancers but to editors as well.[33] Barbara Raskin, president of the Washington Independent Writers, asked the CIA simply to prohibit any of its employees from presenting themselves as "independent freelance writer[s]."[34] Robert Myers, publisher of the *New Republic* (and earlier a CIA officer for twelve years), recommended a prohibition against CIA recruitment of any U.S.—but not foreign—freelancers.[35] In a word, many observers urged the CIA to remove the freelance exception, while others found the word too slippery to define and regulate.

Nonjournalist Staff

Some critics also hoped that the CIA would move away from its insistence on access to nonjournalistic employees for intelligence operations—the technicians, cameramen, teletype operators, and the like. Tad Szulc sees this as a "dangerous gap or loophole."[36] Seasoned correspondent Daniel Schorr notes that "a typical television network bureau in a large foreign capital employs correspondents and editors," and, like Szulc and Mike Wallace of CBS, he favors their inclusion in the "hands-off" category.[37]

From the vantage point of management, this desire to widen the circle of media people placed off-limits to the CIA stems from concern about two problems: the access of U.S. reporters to foreign news sources and the safety of employees abroad. If members of the American press corps are suspected of being intelligence agents in disguise, foreign nations may be reluctant to allow their entry or may limit their travel within the country. (Plainly, some level of suspicion will always remain despite proclamations by the U.S. government to the contrary, but prominent media representatives argue that the existence of CIA directives prohibiting certain relations is better than nothing. The directives have simply failed, in their view, to go far enough.[38]) If a television technician were to "count airplanes," states Philip L. Geyelin of the *Washington Post* (who himself was a CIA officer for one year), and it were to become known, "it would surely affect access, and that is the thing that . . . I think a lot of us are worried about."[39] An Associated Press (AP) general manger speaks to the second question of safety: "[W]e have a very real problem . . . concern for the safety of 500 people on our staff who must work in areas where any suggestion of collusion, not only with the CIA but with the American government in general, would be quite dangerous, and has been quite dangerous."[40]

Foreign Journalists

The attention devoted to the proper CIA relationship with foreign journalists has been as heated as the debate over freelancers and nonjournalist staff. During Senate hearings on this subject, *Milwaukee Journal* editor Richard Leonard observed, "Failure [to prohibit paid relations with foreign journalists] would make a mockery of our efforts to steer developing nations away from a press with strong government influence. Failure to do so could lead to expulsion of United States correspondents from foreign nations. Most certainly, CIA activity involving foreign newsmen would have a chilling effect on news sources overseas."[41] The Board of Directors of the Society of Professional Journalists (Sigma Delta Chi) passed this resolution in 1978: "We now call on Congress and the President to state positively that the CIA will not employ foreign journalists."[42]

Two years earlier the ASNE had unanimously adopted a similar resolution.[43] "Until [the CIA] accepts the same policy with respect to paying off foreign journalists and fouling foreign news media," declared ASNE president Eugene Patterson in an impassioned statement before the Intelligence Oversight Subcommittee, "nobody in this world can credit the truthfulness of the American claim to stand for a free, untainted alternative to manipulated news."[44]

Again a definitional problem arose: what was meant by the foreign press? Surely, said former DCI Colby, "we do not need, for example, the self-inflicted wound of being barred from intelligence operations targeted against TASS."[45] Should a Soviet or PRC "journalist," who, chances are, is in the intelligence business, be barred from recruitment by the CIA? Few, if any, reformers would go that far. Daniel Schorr has expressed the commonly held view that "if the CIA could get a subeditor of *Pravda* as an agent, would I mind? Hell, no I would not mind. I would love it . . ."[46]

Many thoughtful journalists, though, have had serious reservations about extending CIA operations to those nations that have a free press (some 30 percent of the total, according to one estimate[47]). The CIA "should not subvert a free press . . . where a free press exists," argues Philip Geyelin of the *Washington Post*, seconded by Jack Nelson of the *Los Angeles Times*.[48] A consensus seems to exist among several critics that much depends upon the political system in which the foreign journalist works. Hands off the British and the New Zealanders; hands on the Soviet journalists. The line, however, is not always simple to draw. Schorr, among others, does not believe it is feasible to equate the U.S. press with the press in other countries ("the first amendment of our Constitution applies to citizens of this country"), in part because of the practical difficulty of rating from year to year which presses are really free.[49]

Another practical matter pushes the argument in the opposite direction, away from using most foreign journalists even in authoritarian regimes: the problem of cost-effectiveness. If the CIA "rents" a foreign journalist one month as a way to plant propaganda in his newspaper, only to have him rented by the KGB the next, what benefit does the United States receive? As former U.S. ambassador L. Dean Brown notes, "[F]avorable stories in those countries where the press is bought are so heavily discounted that I am not sure it is worth it. The only one benefiting from that is the publisher who is getting money from everybody to put the material in the paper."[50]

This propaganda side of the issue raises the whole question of why the CIA should be used at all for the covert dissemination of information when the United States has an agency in almost every country, the U.S. Information Agency

(USIA), with the avowed mission of spreading information about the American view on this issue or that. One experienced former CIA official sees the use of foreign journalists for propaganda as "a marginal thing."[51] (For other intelligence operations, however, he suggests that a relationship might prove useful, as in this agent-recruitment scenario: "If you had a French journalist somehow related to a Hungarian journalist with a brother-in-law in the KGB who was anxious to defect, it seems to me that this is an obvious way to use a journalist.")

The short answer from the CIA in favor of covert propaganda is that a story planted through a local reporter is apt to have more credibility with the indigenous population than an official USIA news release. Colby explains: "[T]here are some situations in which an individual in a foreign country is going to accept what USIA says as the official view of the American government about what it is doing, but if [he] reads what a local journalist-columnist writes as his opinion, then he is going to say, well, that man speaks or thinks as I do. He is a co-national of mine, and that is his view, and that seems to make sense."[52] Pointing to West Africa as an example, Ambassador Brown concedes that "in areas somewhat hostile or where press is otherwise controlled . . . it may be very difficult, very difficult no matter what overt methods you have, to get something through [explaining the American point of view]."[53]

So one is left with a mixed record. In some circumstances, the use of foreign journalists seems to have marginal utility at best, and, arguably, taints the purity of America's professed belief in freedom of the press—especially when the CIA is caught trying to recruit journalists in democratic regimes. In other circumstances, use of foreign journalists may be the only way to counter U.S. adversaries in what Colby refers to as the "international ideological contest going on."[54] The sensible position seems to be to examine propaganda proposals on a case-by-case basis, taking these considerations into account during the approval process. The use of CIA foreign-media operations, however, to manipulate electoral outcomes in democratic nations, as occurred in Chile from 1963 to 1973, does appear quite clearly to go beyond the pale of acceptability for those who endorse the idea of democracy.

This comparison of American and foreign journalists has at its core what Ambassador Trueheart refers to as "the basic double standard" flowing directly from the 1947 National Security Act. "It has been accepted for 30 years now that that Act authorizes the CIA to conduct espionage abroad," he states. "Espionage involves violating the laws of foreign countries. On the other hand, we do not authorize anyone or any agency to violate U.S. laws."[55] In light of this, the CIA has felt it proper (or at least necessary) to issue—however belatedly—restrictions regarding American journalists, but also, in keeping with the double standard underlying its founding statute, to continue the recruitment of foreign journalists (with the British apparently representing one of few exceptions).

The CIA clings, almost desperately, to its claimed prerogative to recruit foreign journalists, as well as its more controversial use of nonaccredited and nonjournalist American media personnel, for a central reason: concern over concealing its operations overseas. Colby has complained bitterly about the diminishing cover abroad.[56] As one junior intelligence officer has said, "If the restrictions continue, will we soon be allowed only to collect intelligence from pimps and prostitutes?"[57] In a kind of domino theory of falling cover and contacts abroad, the CIA fears that other groups will be placed off-limits: missionaries,[58] students,[59] businessmen, artists, travel agents—who knows where it will stop?

Often the buck is passed to the businessman. Keep the journalist and the scholar pure, and let the overseas offices of corporate America hide the Agency's spies and otherwise help with operations. Unsympathetic to the CIA's jeremiad, journalist-writer Ward Just recommends "lawyers who are traveling, businessmen [who] are traveling, bartenders, gamblers, you know, all the way down the pike. . . . The Agency acts as if somehow they are denied [journalistic cover], that the clandestine services are going to collapse, which is nonsense, absolutely bloody nonsense."[60] Fromm of *U.S. News and World Report* also observes that "other countries' intelligence organizations . . . use businessmen. . . ."[61]

These suggestions fail to dispel uneasiness about the CIA's problem of foreign cover for its operations. One former U.S. ambassador emphasized during legislative hearings: "I can just see someday a hearing going on here with the American Bankers Association, which says they [the CIA] really should not be allowed to have a relationship with bankers, American bankers, or any American firms abroad, because that, once again, is a divided loyalty on the part of that person."[62]

Two remedies have widespread support, though each has distinct risks. One is to require more U.S. government agencies to hide an intelligence officer or two within their overseas missions[63]—a suggestion these agencies are usually cool if not hostile toward, for it causes their personnel to fall under blanket suspicion by other nations. A second option is the "NOC" or nonofficial ("deep") cover—someone (perhaps a businessman) working outside the U.S. embassy compound holding down a regular job and running intelligence operations on the side. Recall from chapter 4 that, in the view of one experienced ambassador, the NOC is "the only effective thing"—though it is "dangerous" because the undercover officer lacks diplomatic immunity.[64]

In its search for the proper relationship with journalists (and others), the CIA has had to worry about more than what types of individuals might be off-limits. Precisely what intelligence activities are fair to ask citizens to carry out has been equally disputatious.

Foreign Intelligence Activities

Though intelligence officers and journalists are as a whole more interesting than the people one usually encounters on a train or at a cocktail party, they are attracted to one another for much more practical purposes than lively social discourse. Journalists seek information from the CIA; the CIA would like information from journalists, too—and far more. Table 9.2 presents in outline form the scope of activities that nurture the symbiotic relationship between the media and the CIA.[65]

TABLE 9.2. CIA–Media Intelligence Activities

Exchange of Information
- Swapping and confirming stories
- Prebriefing and tasking (selecting intelligence targets)
- Debriefing
- Gaining access to files/outtakes

Support
- Hosting parties
- Providing safehouses
- Acting as couriers

Agent Work
- Spotting
- Assessing
- Recruiting
- Handling

Propaganda

Exchange of Information

INFORMAL EXCHANGES At the lowest, most innocuous level are activities that virtually every intelligence officer and journalist I have spoken to, as well as almost every public pronouncement on the subject, find acceptable—indeed, inevitable—namely, confirming stories and swapping information.[66] "I do not believe the exchange of information in any way makes a journalist a Government agent," says Jack Nelson of the *Los Angeles Time.* "It is a time-honored way of journalists going about the business of finding out and reporting what is going on in Government."[67] Daniel Schorr puts it even more simply and generically: "I do not think that under the first amendment that you can interfere with the right of an American to talk to anybody he or she wants to."[68]

In Nelson's view, though, the relationship is proper only if it is voluntary and does not compromise the reporter's sources or freedom to pursue a story as he or she sees fit. The CIA, of course, might try to use the reporter and might be more friendly to reporters who are "cooperative." But this is a hazard journalists face in dealing with every government agency. "Why should the CIA be different from the State Department, the Department of Agriculture, or the White House, for that matter?" asks Nelson, who is resigned to this reality.[69] *Caveat emptor.*

PREBRIEFING AND TASKING A trickier situation arises with CIA briefings for journalists. One form (the "prebriefing") takes place before a journalist embarks upon a trip to write a story. He or she may drop by CIA Headquarters (or, less frequently, by a CIA station abroad) to speak with experts about the country or countries on the journalist's itinerary. The CIA may be only one of several agencies the journalist checks with before going abroad; the Department of State is usually high on the list. The purpose is simply to gain as much knowledge as possible beforehand, in order to make the trip as useful and efficient as possible. "Very helpful," "terrific value," say two prominent American journalists about CIA prebriefings.[70] The CIA has been pleased too. According to former CIA deputy director for intelligence (DDI) Ray Cline: "[B]y far the great preponderance of contacts between CIA and journalists have been the debriefings and prebriefings, and . . . they have been unpaid, voluntary, and mutually beneficial."[71]

Not everyone has been so sanguine. Prebriefings are "a witless practice," concludes Ward Just, freelancer writer and novelist.[72] For Morton Halperin, the primary danger with prebriefings relates to who is giving the brief. During the CIA's CA operations against Allende, prebriefings on Chile were often given by officers of the Covert Action Staff (CAS) rather than analysts from the Intelligence Directorate. The CAS's officials were inclined to use these opportunities, Halperin believes, to advance the Agency's "worldwide [propaganda] campaign to create an image of Allende as a man who would use violence and end democracy in Chile."[73] Halperin accepts the idea of prebriefings, but only if conducted by objective analysts—not psy-war specialists.

Prebriefings remain controversial less because of the fear that they might lead to propaganda than because of an uneasiness that they might slip all too easily into a tasking. The reporter comes to CIA Headquarters to inquire about the latest developments in Iran, where he will be traveling next week. A top CIA Iranian analyst provides an excellent briefing, replete with personality profiles on leading Iranian figures, data on oil production, a progress report on regional disputes, and more. Near the end of the briefing, the analyst pauses. "While you're over there, I wonder if you could keep on eye out for ———?"he asks. Or: "If you are able to get an interview with the Ayatollah Ruhollah Khomeini, could you ask him about ——— and let us know his answer when you return?"

What was once a prebriefing has strayed, in the minds of some critics, into a "laying on" (tasking) of intelligence requirements through one or more intelligence questions—however subtly interwoven into the prebriefing. The philosophy is: I've bent over backwards to help you with a damn good briefing, now how about your helping the CIA—and your country—by being a good fellow and asking a few key questions overseas, and keeping your eyes open (How frail did the Ayatollah look?). If these assignments are carried out, beyond what the correspondent might normally have been expected to learn anyway in the course of his or her job, the activity may have subtlely shifted from reporting to espionage.

For Nelson of the *Los Angeles Times,* the key to deciding how far to let this process go resides in the nature of the questions: "I see nothing wrong with [the CIA officer's] suggesting 100 questions. Now, I might see two out of the 100 that I would think were newsworthy, and I might ask those two; but, I would ask all 100 if I thought they were newsworth."[74]

According to most reporters who have addressed such matters publicly, crossing the forbidden line would be asking questions abroad for the CIA that were really not newsworthy and would never end up in the reporter's story. As Daniel Schorr has testified, when the CIA says " 'Please do not put it in any news story, because that would blow the whole thing,' that is where you would stop [cooperating with the Agency]."[75] Or, in Loory's words: "I think we [journalists] do our best when we report for our readers. If the CIA can find any value in my dispatches, terrific, I am all for it, but I do not think that I should disclose them the contents of my notebooks beyond what I report in my dispatches."[76]

DEBRIEFING Having embraced prebriefing as having "terrific value," Loory rejects out of hand the idea of returning to CIA Headquarters after a trip for "debriefing." This, in his view, would make one a government agent.[77] Halperin would simply insist that any debriefing be done by analysts, not operational people.[78] Some reporters apparently find the debriefing as valuable to them as prebriefing. A reporter often wants to debrief himself to an experienced intelligence officer, says former DDI Cline, "so as to get some idea of whether or not he has valuable information."[79]

OUTTAKES Of high interest to intelligence officers and others in government are visual data gathered by photo and electronic journalists abroad. Much of what they gather never sees the light of day; one out of a thousand photographs may make its way into the newspaper or magazine; a few feet out of several hundred feet of film or videotape may appear on the evening news. The rest ("outtakes") go to the news morgue, filed for reference but rarely used.

When a U.S. ambassador was killed in Cyprus, television cameras happened to have recorded the murder. The Department of State asked for access to the tape and was able to identify who had done the shooting. Similarly, Colby and other CIA officials argue that they ought to be allowed to "look at the full take on some particular set of scenes that were in an area of great importance to us. . . ."[80] Some journalists respond with an emphatic no.[81]

Support and Agent Work

Most reporters are comfortable with the swapping and confirming of stories, and with prebriefings, but consensus over the activities in table 9.2 disappears quickly thereafter and, indeed, most reporters who have addressed these issues openly reject all the remaining activities listed in the table—from tasking to agent work and propaganda. Only one other activity would probably garner widespread approval among journalists: extraordinary service to the country when its vital interest is jeopardized. An example would be when, in the past, reporters have acted as intermediaries between hostile parties in times of crisis—a valuable role played by NBC's John Scali during the Cuban missile crisis when official communications between the U.S. government and the Soviet Embassy had become unreliable.[82]

Intelligence support activities and agent work are anathema to most journalists. Other than perhaps hosting parties, these relationships would clearly fall within the realm of espionage. Nevertheless, some journalists have been prepared to cross this line. When David Phillips owned and operated a newspaper in Chile, he handled agents for the CIA (and later joined the Agency). He also recalls another American reporter actually keeping radio communications equipment hidden away for the CIA in case of emergency.[83] Such cases, though, appear to be rare.

Propaganda

The participation of journalists in CIA propaganda activities raises several controversial questions, some of which have been alluded to already. One conspicuous hazard is that the Agency's effort to recruit a foreign journalist will simply backfire. The next day's papers might carry a story by the reporter not about the theme the CIA wanted printed, but about the recruitment attempt, with a "CIA attempts to undermine free press" lead-in. The embarrassment to the Agency and the United States may not have been worth the effort—not to mention the public identification of a CIA officer or at least his intermediary (a "cut-out" in spy talk). Such complications lead some experienced journalists to recommend, in the words of Herman Nickel of *Fortune,* that if "the U.S. Government has a policy of trying to influence opinion in other countries, then let it be done by the USIA."[84]

This advice takes on added weight when one considers that CIA propaganda programs have had at best a mixed record of achievement. A former CIA operative with over a decade of experience in Asia recalls that Agency propaganda programs there were "not very valuable."[85] Indeed, some of the operations have been ridiculous, rivaling World War II attempts to drive Nazi infantry mad by dropping pornography into their foxholes.[86] One U.S. ambassador recollects a program to televise propaganda to the people of Cambodia. An expensive transmitter was established in Vietnam. Only much later was it ascertained that in all of Cambodia only three television sets worked, each one in the palace of Prince Shianouk. On another occasion, the CIA attempted to spread propaganda to mainland China from Taiwain. This time the Agency managed to figure out in advance that few radios existed in the PRC; so the CA experts tied transistor radios to balloons, sent them aloft from Taiwan, whence the prevailing winds were to whisk them across the Formosa Straits into the grasping hands of Communists starved for news from abroad. The winds reversed and carried the balloons back to Taiwan.

Still, the temptation remains to carry out covert propaganda. While this temptation is fueled no doubt by internal bureaucratic forces, especially a CAS that wants something to do, quite rational and persuasive arguments exist for the occasional use of covert propaganda in special circumstances. Ambassador L. Dean Brown, for the most part a critic of excessive CIA propaganda, presents a case. Suppose in Country X a series of articles appear in the local press accusing the United States of some unfair allegation, say painting a distorted portrait of race relations that suggests slavelike conditions for American blacks, replete with misleading photographs and the like. To set the record straight (without glossing over the unfortunate conditions under which many poor blacks, whites, and others do live in this country), one would like the opportunity to present a more complete portrait. In some countries, though, the chief of state may prohibit the open USIA distribution of propaganda. He may be unfriendly toward the United States; he may even be in on the KGB-sponsored propaganda in these hypothetical (but not uncommon) press articles in Country X. In such a case, Brown argues, rather than risk having the citizens of Country X become fully alienated toward America, countering KGB disinformation with truthful CIA press placements—if possible—might be of some use.[87]

Perhaps the most disquieting danger in the CIA use of the media lies in the phenomenon of blow back or replay (mentioned in Chaps. 4 and 8), that is, the return to the United States of Agency propaganda planted abroad—the brainwashing of the American people by one of their own secret intelligence agencies, to put it in harsh, Orwellian terms. Although Colby has argued that blow back happens "very rarely,"[88] and former DDI Cline could think of no instances "where the U.S. public might have been confused or disturbed by false or distorted information,"[89] others disagree. In its study of the Agency's CA program against Chile, the Church committee discovered an internal CIA Operations Directorate memorandum on blow back. Though the CAS had apparently planted anti-Allende propaganda with foreign journalists in Chile for them to send back to their own countries, the memo concluded with evident pride that "replay of Chile theme materials" surfaced in the *New York Times* and *Washington Post* (Church committee files). Ray Cline concedes that it "used to worry me a lot that false CIA propaganda about mainland China might fool China experts in the Department of State, skewing their analyses." He says that he and others tried informally to keep such stories from spreading throughout the U.S. government, but virtually no procedures existed to monitor and head off the replay.[90]

Especially disconcerting are alleged attempts by the CIA to plant propaganda directly into the American media. A resentful former CIA officer, Frank Snepp, claims—based on his eye-witness observations while serving in Saigon for several years—that the CIA and the U.S. ambassador there (Graham A. Martin) intentionally fed fabricated horror stories to the U.S. press about an impending Communist "bloodbath" against South Vietnamese civilians in an effort to "generate sympathy for the South Vietnamese cause." (As it turned out, fabrications—if indeed they did occur—were hardly necessary, for the evidence indicates that upwards of sixty-five thousand South Vietnamese civilians were murdered by the Communists. It could be, contrary to Snepp's view, that these American officials quite honestly anticipated the carnage that would come and openly expressed their concern.[91])

Snepp continues, "We would leak to [a small group of key U.S. journalists] on a selected basis, draw them into our trust, into our confidence, and then we could shape their reporting through further leaks, because they trusted us." One *New York Times* reporter, according to Snepp, "bit all this hook, line and sinker. . . ." Snepp also charged that he knew a "foreign journalist" (unnamed) on the CIA payroll whose pieces appeared in the *New York Times* and other American newspapers. In a recent interview, Snepp added that the CIA station in Saigon sometimes wrote entire articles on the Vietnam scene for the British periodical *The Economist*, as well as other English-language publications.[92]

No doubt the CIA will continue to argue a case for covert propaganda. The intelligence committees on Capitol Hill, though, would be wise to view with skepticism themes that could have an adverse effect on America's own citizens—especially the use of plainly false or misleading information. "I don't believe that putting misleading information out as news is ever justifiable . . ." said Secretary of State Henry Kissinger in 1976. "I would think that any information that is placed through any American governmental organization should be such that it could be published here without misleading the American public."[93] Here ought to be the First Rule of Covert Propaganda. The Second Rule ought to be: use it sparingly and only when the USIA is unable to accomplish the job alone.

Bonds of Association

For some observers, the nub of the relationship between the CIA and the media lies in the bonds that tie, particularly if money changes hands (see table 9.3). "Any paid relationship is especially odious," declared Nelson of the *Los Angeles Times* during Senate hearings in 1976.[94] Nicholas Daniloff, at the time with United Press International (UPI), agreed that a paid association "will inevitably create a powerful incentive in the newsman's life, and this incentive can serve as a lever of manipulation."[95] William Colby acknowledges this possibility. "It would affect [the journalist's] independence if he were sent [abroad] and paid to come back with the right information," Colby told the House Intelligence Oversight Subcommittee, "but it wouldn't affect his independence if he just comes home and lets the Government know."[96]

Reformers often point to the use of money as the only tangible aspect of the CIA-media relationship that Congress could control; tampering with voluntary associations strikes at cherished constitutional privileges.[97] One problem with this approach is where to draw the line: Should the CIA be able to pay for a journalist's lunch, cocktails, a flight to Antananarivo, a small stipend? Lunch, dinner, or drinks, all right, conclude most media advocates, but, in the words of one, "anything beyond that is absolutely out of the question."[98] A more serious problem with using money as a yardstick lies in the apparent fact it has hardly been the solitary, or even the most important, motivation in the long-standing flirtation between journalists and spies. As former DDI Cline recalls (in a view corroborated in my interviews): "[T]he sums of money which I am aware of ever having provided to U.S. newspapermen for expenses have been picayune and insignificant."[99]

In 1976, the Church committee reported: "About half of the some 50 CIA relationships with the U.S. media were paid relationships, ranging from salaried operatives working under journalistic cover, to U.S. journalists serving as 'independent contractors' for the CIA and being paid regularly for their services, to those who receive only occasional gifts and reimbursements from the CIA."[100] So money has played a role, but clearly the bonds of association have been based on far more than simple pecuniary gain. Strictly voluntary ties have been plentiful, sometimes occasioned by a sense of patriotism. The well-known columnist and Washington social lion, Joseph Alsop, routinely wrote articles (often with CIA information and nudging) designed to promote favorable political outcomes for the United States in Asia. In defense he declared, "The notion that a newspaper man doesn't have a duty to his country is perfect balls."[101]

Proscribing the recruitment of "paid, bribed, subverted members of the media

TABLE 9.3. CIA–Media Bonds of Association

A. Monetary bonds ("hard currency")
  - Regular salary, plus expenses
  - Small retainer
  - Piece-rates
  - Expenses only
  - Valuable gifts

B. Nonmonetary (social and psychological) bonds ("soft currency")
  - Entrapment (getting a journalist "on a string")
  - Secrecy agreements/employment contracts
  - Information
  - Social ties
  - Ideology (the "Communist threat" presents an extraordinary situation necessitating association)
  - Patriotism
  - Status ("hobnobbing" with secret agents)
  - Charisma of recruiting agent
  - Affinity through shared experiences (foreign travel, foreign friends)

to be agents for American intelligence," the ASNE states, too, that it opposes "any cessation of unpaid relationships between reporter and spook or between spook and reporter. It is conceivable in some extreme circumstance that a reporter might consider it his duty to step forward and give information to the CIA just as a reporter who has witnessed a murder would step forward without hesitation to be a witness in the trial of the murder."[102] Columnist George F. Will concludes that "journalists are, if not citizens first, at least citizens, also." He encourages reporters to reject the idea that voluntary contact with the CIA is "inherently, meaning 'in all situations,' impermissible."[103]

Beyond appeals to money and patriotism call other, perhaps even stronger, siren songs to journalists. Schorr notes that "there can be forms of compensation [from the CIA] more valuable than money, [such as] scoops and even occasionally helping to get a Pulitzer prize."[104] Less dramatic, but possibly more telling on a day-by-day basis, are certain human (social and psychological) attractions that draw the journalist toward intelligence officers: the vanity of being picked to help the government on a secret mission,[105] the desire to boast to friends about clandestine activities,[106] friendship,[107] and the plain fascination that the world of spies carries for journalists (and many others).[108] "Journalists are kind of spies manque, and kind of vice versa," concludes Ward Just.[109] In sum, the bonds of association between journalists and spies are varied and plentiful. Prohibiting CIA payment to reporters would clearly fail to touch a wide range of ties, many quite acceptable to both groups.

Nor are easy remedies available for the issues presented earlier in this chapter. The CIA's relations to freelancers, as well as the blurred line between briefings and tasking, continue to elude simple solutions that satisfy both the democratic requirements for a free press and protection of our society against foreign threats. Most public testimony on this issue has been wary of attempts to proscription through law; rather, intelligence and press spokesmen have agreed for the most part that journalists, their managers, and their associations will have to police themselves.[110] For its part, the CIA has promised to honor its current regulations on the media, though critics continue to worry about the last sentence in Turner's directive: "No exceptions may be made . . . except with the specific approval of the DCI." Turner, at least, seemed to indicate that he would never make an exception without informing the oversight committees in Congress.[111] His successors have avoided the topic altogether.

Prior Restraint

The preceding discussion has focused on the main issue raised by CIA-media relations in a democracy: the separation of media activities from intelligence operations, so necessary to preserve a free press as a central pillar of democratic rule. The relationship presents other quandaries, which may be treated only briefly here. One is the proper right of the CIA to conduct surveillance against reporters (and others) to see if they are security risks untrustworthy of contact with the Agency. Should the CIA be allowed to conduct background investigations of American journalists without their authorization? One would think not. Certainly one must question the CIA's saturation surveillance of columnist Jack Anderson, reportedly spied on by sixteen Agency security people each day throughout 1972 inside the United States.[112]

Efforts by the CIA to suppress the publication of books, articles, and news stories related to intelligence has been another contentious issue bringing the intelligence community into conflict with journalists and other writers. This is a vast and tangled subject (often labeled "prior restraint"), warranting more attention than can be allowed here; but an examination of relations between the CIA and the media can ill afford to neglect at least a glimpse at ways in which the CIA has tried to control the press through means other than leaks and propaganda.

Among the most well known CIA attempts to suppress a news story is the case of "Project Jennifer."[113] In 1974, the CIA conducted a top-secret and unique intelligence collection operation: the raising of a Soviet submarine from depths in the Pacific Ocean never before attempted by divers. The cover for the mission was the Summa Corporation, owned by billionaire recluse Howard Hughes. A Summa subsidiary, Glomar Marine, was—as a deep-seabed mining company—a natural for the project. The CIA further equipped the company's ship, the *Glomar Explorer,* with the latest diving technology and a highly trained crew (the operation reportedly cost $550 million), then set sail.

The *Explorer* crew managed to place thick cables around the submarine and began to raise it slowly. As the sub approached the surface, the cables slipped and cut into its hull; half the prize fell back to the ocean's bottom. The captured half, however, contained (among other things) two nuclear-tipped torpedoes: the mission had partially succeeded. Hope remained for a second trip back to retrieve the rest of the submarine.

Meanwhile the story of the trip began to seep out in Los Angeles, probably as the crew hit the local bars after their risky time at sea. The *Los Angeles Times* wrote a short piece on the mission. When DCI Colby saw the article, he was alarmed that a second trip would have to be abandoned if subsequent articles disclosed the Summa cover. He immediately telephoned the *Los Angeles Times* and other leading papers (some of which, until then, had discounted the reports) to stifle the story. Virtually all of the news managers accepted Colby's argument; in a rare occurrence, they set aside their normal devotion to the principle of printing whatever news came to their attention. Columnist Jack Anderson, though, was unwilling to play along. He published the story, sensing (perhaps rightly) that it would not stay bottled up for long anyway. The public learned about the mission; Anderson tallied another scoop; and the CIA had to cancel plans to capture whatever bounty remained in the rest of the submarine (further efforts to "mine" for the sub might have led to a confrontation with Soviet warships now alerted to the true nature of the *Glomar Explorer*).

Anderson has defended his frequent willingness to publish, despite executive strictures to the contrary, with these words: "What qualifies a lowly reporter to judge whether a bold military venture is bound to end in catastrophe and whether to publish the plan before it becomes a *fait accompli?* Certainly I am not competent to outguess the Joint Chiefs of Staff. But I am in close touch with military experts whom the Joint Chiefs themselves consult."[114] While this answer is perhaps more comforting than if Anderson or other reporters failed even to seek the judgment of specialists before publication, the entire philosophy of investigative reporting is bound to trouble those who keep the nation's secrets. Such ad hoc surreptitious consultations with "military experts" may lead to mistakes. It has been claimed, for instance, that an earlier Anderson revelation of "Operation Guppy"—American audio surveillance of Soviet discussions in their Moscow limousines—caused the termination of a useful collection operation (an allegation that Anderson rejects[115]).

On other occasions, revelations by Anderson and other reporters may have helped the country by alerting Congress and the people to misguided plans within the bureaucracy. Anderson is particularly proud of his reporting in December 1971 about the India-Pakistan conflict. He thought America was drifting dangerously close to a confrontation with Russia and China that could have led to World War III; he believes that he helped to focus public attention on the crisis, persuading President Nixon to back away from further involvement.[116]

Clearly "good" secrets exist that reporters should not reveal and government agencies have a legitimate right to contain, if possible; but "bad" secrets exist, too, that ought never to have been kept from the public in the first place. Under the heading "good" might come such secrets as the blueprints of weapons systems, the specifics of international negotiations (especially sensitive fall-back positions), communications from foreign governments, candor in private conversations among government officials, the details of troop and ship movements in time of war,[117] and intelligence sources and methods. On subjects like these, the government may be on solid ground in asking reporters to desist. To aid the CIA in its protection of sources and methods, the Congress went so far as to pass a law in 1982, entitled the Intelligence Identities Protection Act, which prohibits identification of U.S. undercover intelligence officers and agents.[118]

A second and less legitimate set of secrets, however, contains those that attempt to conceal decisions that are unconstitutional, unlawful, or wrongful. Examples here would include, Sen. Frank Church once argued, the covert action in Laos, the CIA-FBI mail-opening program, and the CIA assassination plots.[119]

A precise listing of good and bad secrets is unlikely to satisfy government officials, reporters, and civil libertarians alike. In the Pentagon Papers suit, the Supreme Court embraced the standard of "irreparable damage" to the nation as a guide to what stories should be contained by the government; but the Court provided limited insight into what this phrase might encompass in the future—indeed, who could?[120] The basic questions of who has the right to know and who has the right to reveal will never be settled in any definitive way, at least in democratic regimes. Just as Congress and the executive branch will always engage in a struggle with one another over the appropriate balance of powers between them, so, too, will reporters and intelligence agencies ever jostle over the proper disclosure of information. In the effort to balance three values—the right of the government to maintain some secrecy, the right of the public in a democracy to be informed, and freedom of the press—the establishment of a clear, immutable standards of precedence is all but impossible. These values are intertwined in complex ways, and judgments must be made on a case-by-case basis, with careful appraisal of circumstances at the time. None of the claims can be dismissed a priori.

So far the balancing of these claims has worked out reasonably well. Even an experienced observer such as former secretary of state Dean Rusk can recall no disclosure of national security information by a reporter that led to serious harm to the nation.[121] The security procedures of the government and the professional integrity of American journalists stand as powerful checks against irreparable damage through improper disclosure.

Conclusion

The upshot of the close attention paid to CIA-media relations in recent years has been for both the press and the intelligence community to be more sensitive to the democratic and ethical questions posed by their association. The debate has been, in the view of one participant, an important exercise in consciousness-raising.[122] As both sides seek new adjustments in their relationship, Joseph Fromm sums up as well as anyone four sensible principles for guiding CIA-media ties: "No pay, no tasking, no violation of confidence, and no conscious reporting of disinformation."[123]

The issue of journalism and secret intelligence agencies in American society ultimately reduces to a central point: reporters in this country have a unique tradition, a freedom to pursue their profession with fewer constraints than in any other nation. This tradition is highly valued by reporters and others. "For years I worked abroad and I had to work beside the *Tass* man or the *Pravda* man or the *Isvestia* man or whatever the Russian was," recalls Harrison Salisbury, the seasoned *New York Times* correspondent, "and I knew he was an agent for the KGB automatically, all the Russian correspondents were; this was my assumption. I felt a superiority and pushed this as much as I could and emphasized that American newspaper men were not agents of their government in any way. I thought it gave us a special place in the world. . . ."[124]

Indeed it does. The late Supreme Court justice William O. Douglas once observed that the United States was admired abroad not "so much for our B-52 bombers and for our atomic stockpile, but we're really admired for the First Amendment and the freedom of people to speak and believe and to write, have fair trials." Here was the "great magnet."[125] The United States must have the protection afforded by conventional weapons, a nuclear deterrent, and a strong intelligence service, but this nation must not lose sight of what it is trying to protect. High on this list will continue to be the vital difference between the American media and its pale counterfeit in the Soviet Union.

IV

INTELLIGENCE IN A DEMOCRATIC FRAMEWORK

TEN

Congress and the New Intelligence Oversight

Dawning

The Huston Plan, coupled with the additional shocks to the body politic resulting from the Watergate scandal and the failure in Vietnam, created a fresh skepticism about executive power in the United States—an attitude reminiscent of those held by America's early forebearers.[1] Hardly had the public caught its breath from the Watergate revelations when yet another scandal struck. As noted in chapter 1, *New York Times* reporter Seymour M. Hersh exposed in December 1974 the existence of Operation CHAOS, the CIA domestic surveillance program ordered by the Johnson and Nixon administrations and aimed at Vietnam War protesters.[2] In clear violation of the CIA's founding charter (the 1947 National Security Act), CHAOS further inflamed a sense of mistrust in the public mind. Hersh's charges of "massive" domestic spying by the CIA, accompanied by a series of other allegations—from questionable covert actions in Chile (the anti-Allende operations) to Agency involvement in domestic publications (the CIA-Praeger publishing connection)—triggered demands, as the reader will recall, for a thorough congressional investigation and prepared the way for quick passage of the Hughes-Ryan Act to tighten supervision over covert action. Passed on December 30, 1974, this law represented the first measure since the creation of the CIA in 1947 to place formal controls on the Agency.

In January 1975, President Gerald R. Ford created a commission chaired by Vice President Nelson Rockefeller to investigate the Hersh charges. Critics looked upon the commmission as merely an effort to defuse the controversy and dissuade the Congress from the pursuit of a more serious inquiry.[3] The Senate insisted on its own probe and, three weeks after the creation of the Rockefeller commission, established a special committee to investigate the entire intelligence community.

With Sen. Frank Church (D, Idaho) at the helm, this panel began an examination of the CIA that would result sixteen months later in the creation of a permanent committee to supervise the intelligence agencies. By the spring of 1975, the House had started up its own serious investigation, led by Rep. Otis Pike (D, New York). This chamber, too, would eventually establish (in June 1977) its own permanent committee to supervise the intelligence community.

With the Rockefeller commission (which proved to be a more serious enterprise than its early critics had anticipated) and the two legislative probes, the intelligence community suddenly found itself standing in the unaccustomed—and uncomfortable—glare of public scrutiny. The "Year of Intelligence" had brought the torch of democracy into its dark halls.[4] Whether this new light was helpful or harmful to the Republic was, and remains, a topic of often bitter dispute. Former president Richard M. Nixon accused Congress of nearly "throwing the baby out with the bath water."[5] Some former intelligence officials complained of the "straitjacket" that Church and Pike had tailored for the intelligence community.[6] In contrast, civil libertarians applauded the congressional efforts to restore constitutional restraints on America's secret agencies.[7] Two intelligence directors openly welcomed the legislative attempts to define more clearly, in William Colby's words, the boundaries "within which [the CIA] should, and should not, operate."[8] For Adm. Stansfield Turner, the involvement of Congress in intelligence policy would mean, among other things, that legislators would share the responsibilities for intelligence policy and "ensure against our becoming separated from the legal and ethical standards of our society."[9]

Whatever one's views on this turn of events in 1975–77, one fact was clear to all: the nation had entered a new era in the conduct of its intelligence policy. As outlined in chapter 1, the first era of modern American intelligence, the Era of Trust, had lasted from 1947 until the Hersh disclosures in 1974. During this period legislators were inclined to treat the intelligence agencies with benign neglect, punctuated with an occasional wrist-slapping when an operation ran awry and caused a brief public stir (as with the Bay of Pigs disaster and the flap over CIA funding of the National Student Association[10]). The Hersh exposes and the ensuing domestic "Intelligence Wars" between the executive and legislative branches ushered in a transitional period: the Era of Skepticism (1974–76), when disillusioned legislators—stunned by the scope of the CIA's operations at home—sloughed off the scales of complacency that had covered their eyes in favor of a hard, if belated, look at the intelligence agencies.

When the congressional investigations led to the establishment of permanent supervisory committees on Capitol Hill for the monitoring of intelligence policies, a third period began: the Era of Uneasy Partnership (1976–86), which would soon be shattered by the Iranian arms scandal. During this third era, Congress insisted upon an annual legislative budget review, periodic hearings and reports, and continuous interaction with the intelligence bureaucrats—a far cry from the infrequent contact between the two during the Era of Trust. The new relationship was one that the secret agencies often found difficult to accept.

The new intelligence partnership between Congress and the executive fell short of objectives hoped for by the Church committee. The Carter and the

Reagan administrations rejected the committee's call for a comprehensive intelligence charter to augment the 1947 National Security Act with detailed restrictions on intelligence activities—indeed, over two hundred pages of legal restraints. Both administrations—and even most legislators—considered the proposed charter too comprehensive and restrictive.

Executive orders, secret internal regulations, and an occasional statute would become the substitute for a Grand Charter. In 1980, the Congress passed into law the most important of these statutes: the Intelligence Accountability Act (or, less formally, the Intelligence Oversight Act), less than two pages long but packed with significant provisions—including the requirement of prior notice to Congress on all important intelligence operations. (See the text in the Appendix.) Here was a small watchdog, but it did have several sharp teeth.[11]

Supervision or monitoring of the executive branch by Congress usually goes by the awkward name "legislative oversight." The keys to effective oversight (or review) are access to information about executive branch operations, a set of standards for evaluating the operations, and motivated legislators to carry out these tasks and to insist on corrections when necessary. The 1980 Oversight Act addressed the first problem of information by requiring the executive branch to give prior notice to the Congress of all important intelligence activities. The second problem of standards has never been worked out by Congress in any detail. And the third problem of motivation defies a statutory remedy, dependent as it is upon the attitudes of individual legislators.

Even the attempt to gain access to information, as mandated by the Oversight Act, has proven to be contentious. The Carter administration immediately tried to back away from full compliance with the prior-notice provision for covert action. Key legislators stressed during floor debate on the 1980 bill that prior notice must be given to Congress; but Carter, when signing the law, observed that "there are circumstances in which sensitive information may have to be shared only with a limited number of executive branch officials, even though the Congressional oversight committees are authorized recipients of classified information."[12] Though the president promised such circumstances would be few in number and approached the issue in "a spirit of accommodation," this landmark law began its history on, at best, an ambiguous note—one symptomatic of the tension between the branches over foreign policy and national security affairs.

The challenge by the Carter administrative to legislative prerogatives over intelligence never resulted in acute institutional conflict, though, because the administration never pushed its case to an extreme. Intelligence liaison between the branches remained reasonably smooth, and seldom did the administration fail to report to Congress in advance of a covert action (discussed later in this chapter). More serious communications problems would occur during the Reagan years. For the most part, this breakdown in cordial relations with the Congress stemmed more from the cantankerous personality of Reagan's DCI (William J. Casey) than from the president's policies or his omnibus executive order on intelligence—with the large exception of the Iran-contra affair which, in one fell swoop, undermined trust built up over the years between the two branches on intelligence policy.

This scandal aside, Reagan's use of the CIA was basically similar to Carter's, only with a bigger budget and a greater fascination with large-scale PM operations. The Reagan executive order on intelligence was also similar to Carter's, except for a troubling section that seemed to allow CIA surveillance in the United States of individuals possibly in possession of information about key foreign events (like potential terrorist incidents). In response to congressional alarm over this ambiguous provision, the CIA promised to report to Congress whenever this authority was invoked.[13] Regardless of such assurances, this portion of the Reagan executive order on intelligence represented a disquieting counterreform—potentially the most dangerous since the new oversight began in 1975. Along this misty road could appear, clothed in flimsy legal robes and the veil of counterintelligence, another Huston Plan or Operation CHAOS.

Uneasy Partnership

A review of highlights in the Era of Uneasy Partnership provides a sense of how Congress and the executive attempted to adjust to the new experiment in intelligence partnership. One way to appraise the quality of legislative oversight for intelligence is to examine closely the behavior of the congressional committees charged with this supervisory challenge. What follows is a look at the House Permanent Select Committee on Intelligence (HPSCI, or "hip-see" in Hill venacular; also called the Boland committee, after its chairman, Edward P. Boland [D, Massachusetts]) during its first year of existence, 1977–78. The analysis will be based upon four considerations: the personalities of the committee members, their prior experience, the committee's organizational attributes (structure), and outside pressures buffeting the committee (its environment). The chapter subsequently offers a broader comparative portrait of both the House and the Senate intelligence committees from 1976 to 1988, along with some thoughts on the response of the CIA to this unparalleled legislative involvement in intelligence policy.

HPSCI was established in July 1977 with thirteen members (nine Democrats and four Republicans) and four subcommittees: Oversight, Legislation, Evaluation, and Program-and-Budget Authorization. The chairman sat on all the subcommittees, each of which also included in its membership two other Democrats and one Republican. The broad authority of HPSCI clearly provided sufficient work for every member. Among other things, the committee was supposed to develop new charter legislation for the intelligence agencies, review their existing and newly developed administrative guidelines, study their bids for annual funding, examine the managerial soundness of new structural arrangements instituted by DCIs, evaluate intelligence results and the integrity of the methods behind them, appraise hardware innovations to determine which machines collect the most intelligence for the least cost, and serve as an ear for whistleblowers.

The sheer number of intelligence entities in the federal government—some forty-five—is more than HPSCI can hope to watch systematically. The monitoring that was accomplished by the committee in its first year involved changing constellations of legislators and their staff, guided by different objectives, using

TABLE 10.1. Degree of Congressional Involvement in Oversight: The House Intelligence Committee, 1978[a]

Involvement	*Score Range*	*Members*
High	8–10	The Top Four (A, B, C, D)
Moderate	3–7	The Middle Six (K, L, M, N, O, P)
Low	0–2	The Bottom Three (X, Y, Z)

[a]The rankings in this table are derived from the author's observations, interviews, and documentary analysis. The extent of vigorous oversight was estimated according to the amount of time devoted by members to various monitoring tasks: coordinating staff projects; reading staff reports and agency communications; initiating briefings, hearings, and meetings with agency officials; attending briefings, hearings, and other committee meetings; brainstorming with staff and members on oversight matters; studying for hearings; visiting the agency and its field offices; and any other attention given to the administration of agency programs.

Various checks were used to test the accuracy of the author's judgments. In early December 1978, the members were initially assigned scores between 10 (highest involvement in the oversight duties discussed in this study) and 0 (lowest involvement) based on the author's observations over a year. Other observers close to the committee (whose anonymity was assured) were asked subsequently during the same month to evaluate the members according to time invested in oversight, using the same 10-through-0 scale. Each of the six individuals comprising this "panel of experts" arrived independently at precisely the same three clusters discerned by the author (though the rankings of members within each cluster varied among the panelists), adding to the reliability of the data presented here. A successful effort was made in this polling of other observers to include individuals with diverse partisan and ideological views.

Also, committee records were examined to determine the extent of oversight activity among individual members, including the transcripts of briefings and hearings (almost always classified) to see who attended meetings and asked questions; committee tabulations kept on when members visited the committee premises and for how long; and the correspondence file to see which members wrote letters to the agencies. In each instance, these indices supported the impressions of the author and the panel of experts.

different approaches, and achieving various degrees of thoroughness. Frequently, the constellations were small, comprised of a single congressman and one or two staffers gathered together to write a report, visit a field station, hear a briefing, or conduct a hearing; sometimes a full subcommittee might combine; sometimes staffers or members from more than one subcommittee would join forces for a short period on a discrete project; and, least often, the full committee would sit in judgment on an agency program or decision.

Amidst these complicated human interactions, one phenomenon stood out in stark relief: some members were much more diligent monitors than others. This point is illustrated in table 10.1, where HPSCI members are arrayed into three clusters according to how much time they invested in the monitoring of CIA programs in 1978 (the first full year of the committee's existence).

Personality

In the view of political scientist Morris Ogul, "The motivations of members are more central to oversight efforts than are structural factors."[14] Charles S. Bullock III has shown that, with respect to HPSCI assignments, "the overriding motive" among members is not reelection concerns but an "interest in committees offering opportunities to participate in shaping public policy on controversial topics."[15] Similarly, observation of and interviews with participants both suggest that nothing explains the separation of the Top Four in table 10.1 from their colleagues so well as sheer interest in the committee's subject matter. This interest revealed itself in a variety of ways: Congressman A wading through thick briefing books on the CIA, writing detailed comments in the margins; Congressman B spending long afternoons in the isolated committee rooms, tracing the Agency budget line by line; Congressman C, seemingly frozen to his chair, sitting through each hour of lengthy CIA briefings, patiently asking for clarifications; Congressman D, the committee gadfly, pouring out a stream of provocative questions during closed hearings, driving CIA officials to a point of exasperation; all four coming by the committee rooms regularly to discuss intelligence developments with the staff. These men were intrigued by the committee's subject matter and wished to learn more.

This personal interest not surprisingly led to activity. While other members showed an occasional spurt of enthusiasm for program monitoring (one member was particularly fascinated by CIA parapsychological research, for instance), the Top Four wore away at the job of intelligence oversight like water dripping in a hidden cave. This is not to say that the Top Four were interested only in monitoring the CIA, or that others shared none of this concern; rather, the variations in interest were a matter of degree—discernible differences in the center of gravity among the three groups of table 10.1.

Another personality influence has to do with the degree of sympathy or antipathy (or love or hate, to use the extremes) felt by the member toward an agency's policies. Ogul puts it succinctly: "Congressmen are seldom eager to monitor these executive activities of which they approve."[16] HPSCI, newly created, had little by way of a corporate memory for judging CIA programs and personnel. The reputation of the Agency had been blackened, of course, by the domestic spying, assassinations plots, and questionable drug experimentations revealed in press accounts and congressional investigations during 1974–76. This historical backdrop heightened the skepticism of several HPSCI members, but no one on the committee (and few in the entire House) was fundamentally hostile toward the intelligence community.

Instead, distrust of programs and personnel was again a matter of degree—and a matter of individual interest in specific problem areas. Congressman A could be bitingly critical about some aspect of covert action; Congressman D, livid at the ambiguity of some Agency guidelines; Congressman O, enraged at delays in access to CIA documents. For the most part, though, the infant committee had no record of bad experiences with the Agency. The HPSCI relationship during this early period was chiefly one of cautious acceptance by most members (and a warm embrace by a few). Had the two shared a longer history together, punctuated perhaps by a few misbegotten projects or uncovered mendacity, the question of affinity may have loomed larger.

Even the DCI, Adm. Stansfield Turner, was fresh to his job and had done nothing to turn the committee against him—though his early, seemingly harsh personnel decisions raised some committee eyebrows (with a callous form letter he informed over eight hundred intelligence officers that they should plan on early retirement). So did his candid refusal, in his first formal appearance before the HPSCI, to share all intelligence information with the committee, even if pressed (such as the specifics of certain sensitive intelligence-collection "sources and methods").

While the committee as a whole was without a track record, all but one of its members (a freshman) had established definite voting patterns on national security issues. The National Security Index, based on selected roll-call votes, showed Congressmen X, Y, and Z—the Bottom Three—with high indices (averaging 87), indicating strong affinity with legislation supportive of the national security agencies in the government.[17] Congressmen A, B, and D rated considerably lower (averaging 33)—though Congressman C upset the symmetry with a maximum NSI score of 100.

Observers of HPSCI also thought the degree of personal stamina among individual members was an important influence in oversight. Some public figures, as Barber has illustrated,[18] are human cyclones, attacking problems without rest; others are comparatively lackadaisical and accomplish little. Barber refers to this difference in personal energy investment as "active" versus "passive." Few congressmen matched B and C in their capacities to stalk doggedly the CIA's budget requests; and Congressman A, by all accounts, was a bona fide "workaholic." Other energetic men, though, channeled a smaller portion of their energies toward intelligence oversight. As a generalization one may fairly conclude that strong personal stamina is a necessary but not sufficient condition for oversight. Certainly, all the Top Four were on the active side, and Congressmen Y and Z, at least, were universally regarded as relatively passive personalities.

Experience

Some individuals are better equipped to engage in oversight as a result of special training. Certain kinds of prior employment enhance the effectiveness of congressional overseers, and if the member has an intimate understanding of an agency's program through prior contact, he starts with an oversight advantage. Nonetheless, prior exposure to intelligence issues had a mixed effect on HPSCI members. Congressman A's stint as a military systems analyst, his experiences as a staff aide to an aggressive legislative overseer, his successful style as a gadfly on another House committee, and his participation in the 1975 House intelligence investigation all combined to whet his appetite—and sharpen his skills—for intelligence oversight on HPSCI. The experience of Congressman Z as a member of the House International Relations (now Foreign Affairs) Committee and Congressman X as a member of the Armed Services Committee had, however, a different result: a relative indifference to HPSCI oversight, stemming in part from a confidence developed over the years that they already understood and approved of CIA operations.

Congressmen B and C came particularly well prepared to HPSCI for intelligence oversight duties in the budget area. Both also served on the Subcommittee on Defense of the House Appropriations Committee—the unit that reviews CIA budget requests. The compatibility between their positions on HPSCI (where they served on the Budget Subcommittee) and on the Appropriations Subcommittee on Defense could not have been more perfect—two pieces of a puzzle fitting together. Unique in the House, they could follow the CIA budget through both the authorization and appropriations phases at the critical subcommittee levels.

While previous exposure spurred on some members and slowed others, no exposure at all produced mixed results, too. Like a boy on his first trip to a candy store, Congressman D was alternately dazzled and bewildered by all the new intelligence programs before him; their novelty provoked him into searching questions and extensive review of Agency operations. But the curiosity of other untutored members failed to be equally piqued; they preferred to concentrate on other legislative opportunities that they perceived to be more interesting or more rewarding to their careers.

Structure

Until recently, oversight tended to be discussed predominantly in terms of institutional arrangements and resource allocations, and despite the increased awareness of personality and experience as prime forces, structural variables remain abundant and significant.

DECENTRALIZATION The chairman permitted a certain amount of devolution of authority to the four subcommittee chairmen. Subcommittee staff were hired by the chairman and ranking minority members of each subcommittee; subcommittee jurisdictions, though loosely defined, were honored for the most part. Furthermore, subcommittee members had direct access to agency documents and personnel, and they had no difficulty obtaining travel support for field inquiries.[19]

But, officially, before the subcommittee chairman could proceed publicly, the committee rules required the consent of the full-committee chairman, one of the four members on each subcommittee, and a majority of the subcommittee—a three-fourths majority! This requirement was occasionally ignored as some subcommittee chairmen charted their own courses with little consultation. Congressman A, something of a loner and from previous encounters on another House committee scarcely on speaking terms with Congressman X, his GOP counterpart on the Oversight Subcommittee (another commentary on the importance of personality in committee behavior), was conspicuous in his willingness to test the outer limits of his first House subcommittee chairmanship. The full-committee chairman alternated between sharply pulling in the reins and begrudgingly loosening them a little. To the extent he allowed subcommittee autonomy, HPSCI oversight of the CIA prospered.

FORMAL POWERS Delegation of responsibility by the chairman to junior members serves little purpose unless the members have formal investigative powers at their disposal. The HPSCI subcommittee chairmen were able to conduct oversight hearings and receive reasonably good Agency response ultimately because the formal rules of the committee backed these activities with subpoena powers and the authority to slash CIA budgets. The committee never had to use its subpoena powers in 1978, and trimmed the Agency budget only marginally; but these powers remained important latent weapons without which oversight would have been so much shadowboxing.[20]

RESOURCE ALLOCATION The relationship of staffers to committee members was as complex as one might expect in a relatively unstructured organization like the Congress. In this fluid setting, the interests, abilities, and backgrounds of individual staffers, as well as their relationships with one another and with the committee members, became highly important.[21]

The staff of the Budget Subcommittee, for example, was a smoothly functioning unit of four professionals and a secretary, all of whom came from similar backgrounds: careers in the CIA or military intelligence. This background was shared, too, by the senior staffer on the Evaluation Subcommittee, the minority counsel, half the support staff, and the staff director of the full committee (who was primarily responsible for bringing onto the committee this large contingent of intelligence professionals). The glue of common experience bound together these elements of the HPSCI staff. Other coalitions, smaller in size, similarly grew up around ties that predated HPSCI employment, such as two individuals on different subcommittees who had worked together previously in the Senate.

The importance of these affinities resided in the effect they had on the behavior of committee members. As chairman of the Budget Subcommittee, Congressman B had at hand an experienced, compatible staff that served him well (though its strength was also its potential weakness: a common intelligence background with the hazard of lingering "company" loyalties—a danger the group seemed to resist successfully in its first year). In contrast, each of the other subcommittees had conflicts among some of the staff that sometimes required arbitration by committee members and drained staff energies in interpersonal negotiations.

Beyond these complexities of group dynamics, the staffing pattern for this medium-sized committee (twenty-three employees, ten of whom were clerical) illustrates the limits of resource allocation for oversight. First, partly by design and partly by natural evolution, several of the minority staff soon became overseers (not to say obstructors) of the majority staff, rather than devoting their time to the monitoring of agency programs. Second, the Budget Subcommittee was the only one with four professional staffers; the one or two staffers on the other subcommittees were insufficient to maintain genuine watchfulness. (On HPSCI, no one monitored the FBI with regularity.) Third, scant use was made of congressional support agencies, such as the Congressional Research Service (CRS) or the General Accounting Office (GAO)—partly because of security complications but mostly because the HPSCI leadership remained content with its own small staff. Fourth, a few staff positions were distributed as personal or political rewards without regard for qualifications.

PARTISANSHIP With Democrats in control of the White House and Congress, one might expect the GOP members of congressional committees to be tougher overseers than the majority party.[22] Such was not the case for HPSCI monitoring of the CIA. Three Democrats (or 33 percent of all the Democrats on the committee) and one Republican (25 percent of all the Republicans) were among the Top Four overseers. The Bottom Three comprised one Democrat (11 percent) and two Republicans (50 percent). Moreover, the Democrats tended to ask the most critical and exhaustive questions.[23]

PREROGATIVES Jealousy can be as powerful between institutions as between rival lovers. On HPSCI attention to oversight was immediately enhanced whenever a member had the impression that the CIA was flouting the prerogatives of Congress to obtain information from the Agency or to examine any topic it wished. One normally mild-mannered, moderately active member, for example, became a stalking bear when, occasionally, information sought by his staff assistants was denied. Once, with disdain, the committee halted an ambiguous briefing by the CIA on a sensitive program and forced the briefers to return later equipped with more thorough documentation to explain the matter in greater detail. In short, hell has no fury like a committee scorned by an executive agency. However, such instances were rare for HPSCI in 1978.[24]

ASSIGNMENTS The number of additional House duties held by HPSCI members outside the committee is important for understanding, in part, the less than vigorous oversight efforts of some. One only moderately active member was high in seniority on the important Appropriations Committee (and chairman of one of its subcommittees); Congressman Z was chairman of a large and busy committee; and Congressman X was a ranking Republican on a key committee. Several among the Middle Six also held high positions on other committees. Each of these responsibilities was clearly of greater concern to them than intelligence. In contrast, none of the Top Four held any vital leadership posts on other committees; for two, HPSCI provided the first subcommittee chairmanships in their careers.[25]

In the case of two HPSCI members among the Top Four, though, other official responsibilities actually seemed to enhance their dedication to intelligence monitoring. Already members of the Appropriations Subcommittee on Defense, HPSCI simply provided them another boat and additional ammunition for duck hunting in the national security waters where they were already experienced overseers. Their special working relationship on HPSCI, honed by previous shoulder-to-shoulder participation on the Appropriations Committee, was reminiscent of the bipartisan cooperation once displayed by the chairman and ranking minority member of another money committee also known for its careful selection of "responsible" members: the Ways and Means Committee.[26] The shared background of the two HPSCI members nurtured a mutual respect and trust that paved the way for a cohesive approach to intelligence monitoring on the HPSCI Budget Subcommittee.[27]

JURISDICTION The great breadth of HPSCI jurisdiction makes it virtually impossible for members to monitor each agency under the committee's auspices. A few new hardware items and one or two particularly sensitive programs outside the

CIA drew close scrutiny; but otherwise, during the committee's first year, the CIA was the chief object of oversight. Every CIA covert action brought to the committee (as a result of the Hughes-Ryan Act) was reviewed in detail; CIA institutional subdivisions, programs, and the views of top personnel were examined closely, through multiple briefings, access to operating documents, and extensive questioning in closed hearings; programs in the field were observed at firsthand; and the like.

What HPSCI monitored most closely depended largely on what struck the fancy of its members. For Congressmen A and D, attention focused on the CIA as the kingpin of the intelligence community as well as the agency most blamed for abuses in the 1975 investigations. For Congressmen B and C, the primary fascination was with the intelligence budget process—a natural extension of their work on the Appropriations Committee.

OBSCURE TRADECRAFT "The more technical and complex the subject matter is perceived to be," writes Ogul, "the less likelihood of oversight."[28] One of the central reasons some intelligence agencies escape congressional monitoring is simply that their modus operandi baffles the generalist. Few members of Congress are technocrats, fewer still intelligence specialists. The esoteric computer capabilities of the NSA could consume months of a member's time for full comprehension; none is willing to make the investment. Easier and more interesting to grasp is the stuff of spy novels: the covert actions and the daring espionage missions—largely the preserve of the CIA.

Another topic lending itself more readily to discussion and analysis by politicians is the proper relationship between secret intelligence agencies and the rights of American citizens. The first major public hearings held by HPSCI were on the ethics of contacts between the CIA and the news media. Generally legislators concentrate on what they (and their constituents) find meaningful and understandable. Members of HPSCI once dwelled for two hours on Turner's personnel decisions (some of those fired were constituents of committee members), later spending a fraction of this time on a technically complex, expensive, and controversial hardware issue that in fundamental ways could affect the future security of the United States.[29]

OVERSIGHT SUBCOMMITTEE The creation of special oversight subcommittees also seems to have an influence on the incidence of oversight. "Once established," notes Aberbach, "many members of these subcommittees and their staffs would want to make something of this assignment. . . ."[30] The existence of a Subcommittee on Oversight within HPSCI gave this subject special salience and helped legitimize the monitoring inclinations of its chairman, one of the Top Four. He could draw upon this official cachet to pursue his interest with added authority. Moreover, the explicitness of the subcommittee name enhanced the sense of responsibility felt by the chairman and his staff for shouldering the oversight chores of the committee. The Oversight Subcommittee held many more hearings in public related to oversight than did the other subcommittees (though the Budget and the Evaluation subcommittees held several closed hearings designed to monitor, respectively, the cost and the quality of intelligence). Finally, the existence of an Oversight Subcommittee provided a conspicuous in-box for the whistle-blower; with a formal apparatus available for handling charges of abuse, fewer allegations were probably discarded or ignored by the full committee than might otherwise have been the case.

OUTSIDE RESOURCES A few HPSCI members drew upon resources outside the committee to further their oversight objectives. Two of the Top Four had access to staff and materials of the Appropriations Subcommittee on Defense. Occasionally, in preparation for hearings on "CIA and the Media," one of the Top Four sought suggestions from an experienced journalist on his office staff. Another among the Top Four had a personal staff aide sufficiently knowledgeable on some intelligence issues to prepare questions for him for use in committee hearings. But these were rare occurrences; for the most part, members relied on their own wits or turned to their colleagues and HPSCI staffers for assistance. One of the Top Four was like a complex nerve ganglion receiving and processing information from various regions of the body politic and seeking out views on intelligence from throughout the Washington community (though chiefly from those on the moderate-to-liberal side of the political spectrum). A former Hill staffer and executive branch official, this member had informed contacts throughout the government unavailable to many members; as a member of a defense-related House committee for several years, he had compiled a useful telephone list of experts to call about national security issues and often used these sources to help generate new ideas for oversight.

PROCEDURES One of the most significant structural variables affecting HPSCI—and every other congressional committee—are procedures based on statutes or chamber rules.[31] Statutory reporting requirements, for example, may require an agency to inform a committee before or soon after it makes a policy decision; this, in turn, may trigger a committee briefing, hearing, or even investigation. The appropriations cycle, with its specific deadlines, also forces committees to take steps according to a timetable. Clearly, intelligence reporting procedures and budgetary requirements stimulated oversight from HPSCI. The Budget Subcommittee was expected to report to the full committee and the latter to the House on its recommendations at a specific time in the fiscal year. (A classified version of the HPSCI report was kept on the committee premises where House members could read it, if they wished; few bothered, and fewer still understood the alphabet soup of arcane intelligence acronyms.) To have its report ready, the Budget Subcommittee staff thoroughly combed the intelligence community's fiscal performance and future plans for months before the deadline.

The primary additional procedural stimulus affecting HPSCI was the Hughes-Ryan Act reporting requirement. This statute made the NSC report "in a timely fashion" to Congress on all CIA covert actions approved by the president. Once reported, these operations almost invariably inspired closed hearings by the Oversight Subcommittee (with wide participation by most HPSCI members) to evaluate their merits.

ELECTORAL VULNERABILITY The standard definition of a "safe" congressional seat is a winning percentage of 55 percent or better in the last election. According to this standard, all HPSCI members were secure, but in reality virtually all members are edgy and feel vulnerable. Most are concerned with improving their lot in the home district.[32] This translates into spending more time there, as well as on projects with visibility at home (which HPSCI usually failed to offer).

Based on their last election returns, Congressmen D, X, and Y were the most vulnerable, and this condition may have helped turn the attention of the latter two away from intelligence oversight. Unlike the others, Congressman D may have calculated that there were useful headlines in intelligence; but he also may simply have been less worried than the others about recent election returns. His biyearly tallies had gone up and down like a roller coaster over the years (for example, 49 percent in 1970 and 71 percent in 1974, then down to 57 percent in 1976), while X and Y had experienced a marked erosion in their popularity since the 1960s. Congressman D aside, the other members of the Top Four had strong margins of victory (though not consistently the strongest on the committee).

Environment

CITIZENS Unlike most public organizations, the intelligence agencies operate in an environment relatively free of interest groups. The only three with any visibility to speak of on Capitol Hill in 1978 were the Center for National Security Studies (CNSS, directed by Morton H. Halperin and located in Washington, D.C.); the Coalition on Government Spying (subsequently called the Campaign for Political Rights, a subsidiary of the Youth Project, located in Washington, D.C.); and the Association of Retired Intelligence Officers (now called the Association of Former Intelligence Officers, or AFIO, located in Virginia near Washington, D.C.). Halperin's Center and its ally, the Campaign for Political Rights, were critical of many (though not all) aspects of the intelligence community, while AFIO was generally fully supportive. None had much of a systematic presence on Capitol Hill, but representatives of these groups were frequently invited as witness when HPSCI or the Senate Select Committee on Intelligence (SSCI) held public oversight hearings.

Of the three, the Center and the Campaign were the most aggressive, occasionally telephoning selected members and staff, preparing briefing material on intelligence matters, publishing newsletters and position papers, making a film on intelligence issues, and, above all, carrying to the public (especially students) its essentially liberal view on intelligence reform. The AFIO appears to be less of a lobbying group than an old-boy pep club that collects dues, publishes a newsletter, and hosts an annual convention; spokesmen for AFIO, however, have provided insightful testimony for committees working on intelligence reform.

During its first year of existence, HPSCI received more "noise" from individual citizens with grievances about the intelligence community than from organized pressure groups: intelligence officers claiming they had been fired unfairly; scientists pursuing patent disputes involving the NSA; an intelligence technician bringing charges of radical discrimination; reporters confidentially passing along names of other reporters supposedly once, or still, on the CIA payroll; campus newspaper editors wanting help with their Freedom of Information Act (FOIA) requests filed against the CIA; occasional "crazies" having visions of FBI agents hiding in their chimney flues; a former CIA officer offering a tip on the real identity of the "mole" (or enemy agent) burrowed deeply within the structure of the Agency; people sending letters from across the country inquiring or complaining about one intelligence agency or another.

Contacts from individual citizens stimulated the exercise of oversight on the committee at the staff level; members, however, seldom engaged themselves in these ombudsman chores unless one of their own constituents was directly involved. The major exception arose from the repeated complaints that Turner had improperly dismissed senior intelligence officials. The hue and cry grew sufficiently loud that the full committee questioned the DCI extensively at closed hearings on two occasions. A spate of elaborate CIA personnel charts and nimble management theorizing by the director laid the matter to rest, at least within HPSCI.

The most active lobbying of HPSCI, however, came not from individual citizens but from discrete groups within the intelligence community itself: agencies justifying budgetary objects, official legislative liaison personnel skillfully building rapport with members and staff (a skill refined in recruiting agents overseas) to grease the program skids, administrators jockeying for favor on the Hill—the entire range of "subgovernment" coalition building well-known to students of politics.

THE MEDIA As demonstrated by the *New York Times* story of December 1974 on CIA domestic spying, media allegations of wrongdoing can throw the Congress into a frenzy of oversight. Though allegations in 1978 were few and modest in comparison to the *Times* bombshell, they always stimulated a reaction of varying intensity inside the congressional intelligence committees. The SSCI was invariably more aggressive in its response to press stories. With a staff over twice the size of HPSCI's and with members such as senators Walter D. Huddleston (D, Kentucky), Gary Hart (D, Colorado), and Charles McC. Mathias (R, Maryland)—all seasoned investigators from the abuse-oriented Church committee (as were most of the senior staffers)—the SSCI was prepared to probe thoroughly any serious allegations (in 1978, at any rate).

In contrast, HPSCI—while not ignoring press stories on intelligence—was more inclined to have a jaded outlook on intelligence exposes. Typically, HPSCI would ascertain (over a secure telephone) the Agency's response to a news item; more often than not, this ended the matter. Across the Hill, the standard operating procedure of the SSCI was to establish a task force of staff investigators, devote days or even weeks to an analysis of the allegation, and produce a formal report for its committee members.

One of the stories that HPSCI examined (if only briefly) dealt with an alleged CIA plot to assassinate the president of Jamaica. The Agency adamantly denied the charge and the case was dropped. As often voiced by the chairman, the general attitude of HPSCI (and quite likely of the country as a whole at the time) was that the investigations of 1975 had gone into past abuses at sufficient length; the time had come to put this earlier era behind us and look toward the future. In the chairman's words during a House debate, "This Committee is, frankly, less interested and less oriented toward dragging out past abuses than it is [toward] present operations and legislation."[33] This philosophy was a central reason behind the concentration on reforms in HPSCI "CIA-Media" hearings rather than on the ferreting out of the names of journalists who may have once worked for the CIA (a kind of witchhunt strongly but privately recommended to the Subcommittee on Oversight by a few well-known newsmen).

EVENTS Major events can be as important as investigative journalism for opening the floodgates of legislative oversight. The surprise bombing of Pearl Harbor by the Japanese led to the first major congressional inquiry into U.S intelligence in the modern era. In 1979, the unexpected overthrow of the Iranian Shah sent HPSCI and SSCI in search of explanations for this intelligence failure; both committees probed the case extensively and issued statements critical of the CIA.

But the event that had the greatest influence on HPSCI occurred before its birth and, over the short run at least, did more to dampen than enhance intelligence oversight. That event was the unfortunate history of the Pike committee, which conducted its 1975 probe of the intelligence community amidst leaks and constant controversy. Eventually most of the committee's findings (still in draft form) were leaked to a journalist and published in *The Village Voice,* despite a heavily lopsided vote in the full House against their release.[34] The result was a deep-seated distrust in the House toward new adventures in intelligence oversight; it had placed its faith in the Pike committee and that experience had gone awry. Once burned, twice shy, the House delayed the creation of HPSCI until July 1977—more than a year after the establishment of the SSCI. When HPSCI finally was created, the message from the House was clear: no more fiascos. The new committee would have to stay in line; the honor of the House was at stake.

The experiences of 1975, then, left a deep imprint on the new committee. Its primary objective seemed to be to convince the House that it was not Son of Pike; that things were under control. No leaks, no headline grabbing, no unnecessary battles with the executive branch. From the start, the chairman's dictum was oversight, yes; but in moderation and—as much as possible—out of the public limelight. The highest priority was for the committee to establish its credibility. As Boland put it before the House at the time of his appointment as chairman: "After this body's recent experience with previous select committees on intelligence, we must first prove to the Senate, to the President, and to ourselves that we can handle the job."[35]

What emerges from this review of circumstances affecting legislative oversight is a complicated picture: different forces had different effects on different members. In his study of legislative oversight, Vinyard wrote: "Since the Committee members are not equal in power nor equally interested or attentive to duties, oversight may be carried on in the name of the committee by a few."[36] Similarly, in the case of HPSCI, the Top Four carried a disportionate amount of this responsibility.

A single glimpse into one committee during one year, of course, is inadequate to test meaningfully any of the observations presented here; the best such a limited methodology can achieve is the generation of hypotheses for more extensive investigation. To this modest end, a summary impression may be offered. Vigorous oversight on HPSCI was associated particularly with the substantive interest of a member (or members) in intelligence issues, disapproval of an agency's programs, and surplus stamina to devote to program monitoring (all functions of personality); past experiences stimulating an interest in intelligence; and concomitant House assignments related to intelligence matters (a function of structure). Low levels of involvement in oversight were most apt to prevail among members who generally approved of an agency's programs and who had pressing obligations and interests on other committees or subcommittees.

The Periodicity of Oversight

One of the many arguments over intelligence oversight is whether Congress should retain separate House and Senate committees or compress the two into a single Joint Intelligence Committee in the fashion of the old Joint Atomic Energy Committee. A joint committee is strongly favored by the CIA, on the ostensible grounds that two committees require more Agency time in briefings and that one committee is apt to be more leak-proof than two. A more compelling reason from the vantage point of the CIA, however, may be that one committee would be easier to co-opt.

Co-optation, that is, the succumbing of legislators and their staffs to the blandishments of executive officials to the point where little meaningful review of executive programs takes place, is a common phenomenon in the government. The co-optation becomes all the easier when the bureaucrats have to confront only one legislative committee. Pulling the wool over the eyes of two is still possible, of course, but harder.[37] An examination of legislative oversight since 1976 suggests that two committees may very well provide a more reliable form of oversight, because the diligence of any one panel—the House or the Senate—has fluctuated over time.

In 1978, for instance, the House Intelligence Committee focused on covert action abroad and the relationship between the CIA and domestic groups (like the media) at home. The Senate, while not ignoring these subjects, concentrated on how well the Agency performs its primary mission of collection and analysis. And when one committee fumbled its duties, as the Senate did when it failed to find out about the CIA operation to mine piers in Nicaraguan harbors, the other committee could pick up the ball—as did the House, in this instance, with its more thorough probing of the mining episode. Thanks to the existence of two intelligence committees, when one has been lax, the other has been available to carry on real oversight—just as the second engine on a twin-engine airplane can take over if the first fails.

At the beginning of the Era of Uneasy Partnership, the Senate Intelligence Committee was far more vigorous in its exercise of oversight than its House counterpart. The Senate Committee drafted the key legislation during this period (the Electronic Surveillance Act of 1978[38] and the 1980 Intelligence Oversight Act) and had a larger, more experienced, and aggressive staff. The House Committee was by no means dormant, however; among other activities, it held thorough hearings on CIA-media relations (in open session) and CIA covert action (in closed session), and issued an occasional critical report on the CIA.

As Chairman Boland grew more confident in his knowledge of intelligence, he became more assertive in his role as the House's chief overseer for this policy. When the Reagan administration turned to covert action as a means to carry out its attack against the Sandanista regime in Nicaragua, Boland reached the acme of his public criticism of Agency policy with his amendments to curb this PM operation. At about this same time, the Senate Committee became comparatively passive as an overseer. Barry Goldwater (R, Arizona) had assumed the chairmanship and was known to favor less legislative oversight. Indeed, as a member of the Church committee, he opposed the establishment of the very committee that, ironically, he now chaired. As he remarked during the historic debate on the War Powers Resolution in 1973, Goldwater preferred the executive branch to run foreign policy; for him, a partnership between the branches was something to be avoided. This notion also applied to intelligence policy: Congress apparently was supposed to delegate responsibility to the president—a return to the Era of Trust. "In my opinion, the Boland amendment is unconstitutional," Goldwater concluded in 1983. "It's another example of Congress trying to take away the constitutional power of the President to be Commander in Chief and to formulate foreign policy."[39] This attitude suited like-minded DCI William J. Casey, the *deus ex machina* selected by the Reagan administration to head a CIA still in trauma from the intelligence wars of 1975.

In 1984, Goldwater changed his tune somewhat after discovering that Casey had been less than forthright in his briefing to the Senate Committee on the CIA operation to mine Nicaraguan harbors, or, more accurately, piers within the harbors—a slippery distinction used by the CIA, the reader will recall from chapter 6, to deny, at first, the mining operation when asked by legislative overseers about *harbor* mining. For the remaining months of his chairmanship, Goldwater displayed a more skeptical attitude toward the CIA, reinforced by the persistent unwillingness of Casey—for whom oversight was gall and woodworm—to accept the idea of a genuine partnership with Congress. "The business of Congress is to stay out of my business!" grumbled a defiant Casey to a group of visiting scholars (myself included) at CIA Headquarters in June 1984. At roughly this same time, he was entering into an agreement with the SSCI to improve the CIA's oversight cooperation!

Goldwater's successor, David F. Durenberger (R, Minnesota) proved to be more serious about the task of oversight than his early critics had foreseen. Beginning cautiously as a new and inexperienced chairman, Durenberger within a year assumed a tough stance and publicly criticized Casey for his failure (among other things) to give the CIA a "sense of direction." Angered, Casey accused Durenberger of conducting oversight in an "off the cuff" manner that had led to "repeated compromise of sensitive intelligence sources and methods." This charge spurred the ranking Democrat on the committee, Patrick J. Leahy (Vermont), to rally behind Durenberger, scorning the CIA for its "yearning to go back to the good old days" and for wanting "to destroy the two [intelligence] committees."[40]

CIA-legislative relations thus reached, in November 1985, a new low point in the Era of Uneasy Partnership; but from the point of view of oversight, it was clear that the Senate committee, following a period of dormancy under Goldwater, was for the time being alive and well. Durenberger and Leahy, its leaders, exhibited no interest in a return to the Era of Trust. After the flap over the CIA's failure to report adequately on the Nicaragua mining operation, the Senate committee in an exchange of letters with the Agency insisted on tougher reporting requirements: all changes in policy had to be reported, reports had to be presented in greater detail, and the CIA had to give the Senate committee an annual accounting of all covert actions currently underway.

On June 6, 1984, DCI Casey signed an agreement with the committee to this effect. Further informal clarifications were reached between Casey and the SCCI throughout 1984, and in the following spring the Senate committee added two more requirements. Its members wanted detailed briefings, in advance, on counterterrorist operations, like the U.S. attack on Libyan forces in the Gulf of Sidra that had recently taken place without the committee's knowledge, and more detailed briefings on covert actions that held a high risk of public exposure. Then, in June 1986, Casey signed another agreement promising to inform the committee about third-party involvement in covert actions (that is, U.S. agencies other than the CIA, or foreign countries).[41]

The Senate committee seemed to have reached full stride in its determination to fulfill its oversight responsibilities (see fig. 10.1). It faltered somewhat, though, in 1986 as Chairman Durenberger went through a stage of personal emotional turmoil (what some close observers referred to as a "mid-life crisis") that disrupted his public life as well.

In the meantime, on the House side, the see-saw was down, as Lee Hamilton (D, Indiana), plagued by sometimes bitter internal partisan squabbling on his panel and himself going through a period of learning in his first committee chairmanship, eased into the role of Boland's successor (1985). As he became more accustomed to his new position, the level of intelligence oversight in the House of Representatives began to increase. As Leslie Cockburn notes, though, even when Hamilton had reached his stride, he (and his committee colleagues) displayed insufficient attention to the rumors pointing to an Iran-contra connection (or, at least, an NSC staff-contra connection), accepting administration disclaimers too much at face value.[42]

Then, in November 1986, the bottom fell out of CIA-legislative relations. The Reagan administration's covert sale of arms to Iran in exchange for assistance in the release of U.S. hostages held by terrorists in the Middle East leaked to the press and quickly achieved the proportions of a scandal. Beyond the obvious issue of questionable judgment regarding the swap of arms for hostages, which seemed to invite more hostage-taking any time a terrorist organization needed more arms, stood the possible violation of several statutes. Antiterrorist laws prevented arms sales to terrorists, and the 1980 Intelligence Oversight Act clearly envisioned reporting to Congress on covert action—which was never done in this instance.[43]

The finding for the arms sale also violated the expectation of the Hughes-Ryan law that the president would approve all covert actions in writing, not orally. Congressional investigators discovered in 1987 that the finding had been written only long after the operation had been initiated—an improper (and novel) retroactive finding. Further, it soon became clear in hearings held by the Inouye-Hamilton committees that the CIA—or at least DCI Casey and some of his officers—were well aware of these "irregularities" but failed to report them to Congress, as required by the 1980 Oversight Act. Finally, and most devastating to the oversight process, investigators found that Lt. Col. Oliver L. North of the NSC staff had diverted funds from the arms sale through a Swiss bank to the contra army in Nicaragua, an obvious violation of the Boland amendment of 1985, which set strict prohibitions on U.S. assistance to the contras. "If a White House can decide that a law passed by Congress is inconvenient, and simply set out to circumvent it," wrote a thoughtful commentator, "then our constitutional system is finished."[44]

The intelligence "partnership" between the branches, such as it was, had now been subverted by this unreported covert action. In the place of partnership came a fourth (and current) phase: an Era of Distrust (1987–). The Inouye-Hamilton committees sorted through the rubble of the Iran-contra scandal—all too gingerly in the view of critics[45]—to see how much damage had occurred. Also, at this time, through normal membership rotation, the permanent intelligence committees came under new leadership: Chairman Louis Stokes (D, Missouri) in the House and Chairman David Boren (D, Oklahoma) in the Senate. While relations between the CIA and the Congress had not returned all the way to square one (routine staff-level oversight continued throughout the contretempts—and at a higher level of thoroughness than experienced during the Era of Trust), the events of 1987 clearly produced a long slide backwards as the new oversight leaders on Capitol Hill tried to work out what their relationship with the CIA would be in the wake of the scandal. One of the first moves by Rep. Stokes was to introduce a bill on February 4, 1987, designed to guard against future Iran-contra abuses. His proposed legislation (House Resolution 1013) required explicitly what had been only implicit in the Hughes-Ryan and 1980 Intelligence Oversight acts: *written* presidential approval of important covert actions.

The Stokes bill also sought a change in what had become the most controversial aspect of intelligence oversight: when to report to Congress. The expectation on Capitol Hill had passed from the occasional report on covert action during the 1947–74 era, to reports "in a timely fashion" under Hughes-Ryan from 1974 to 1980, to prior notice under the 1980 Oversight Act. Now Stokes was prepared to give the executive branch forty-eight hours in which to report on covert action (the same standard employed in the War Powers Resolution for the introduction of U.S. troops into areas of hostility)—but only in cases of "extraordinary circumstances." Under normal circumstances, prior notice remained the expectation.

Critics of the Stokes bill attacked it from different sides. For the CIA, the proposal was too restrictive; binding the president with a specific time limit for reporting would discourage the robust use of the secret service. For others, just the opposite: The forty-eight-hour leeway represented a retreat from prior notice and, therefore, a diminution of meaningful oversight. These core arguments would be amplified by both sides as other reform proposals joined the legislative docket with H.R. 1013 in reaction to the Iran-contra affair. In the wake of the Stokes proposal, the most prominent reform initiatives came from Senators Boren and William S. Cohen (R, Maine), the leaders of the Senate Intelligence Committee; from Sen. Arlen Specter (R, Pennsylvania), a member of the Intelligence Committee; and, in a revised and expanded version of H.R. 1013, from Representatives Stokes, Boland, and Matthew F. McHugh (D, New York), members of the House Permanent Select Committee on Intelligence (HPSCI).

The Boren-Cohen bill (Senate bill No. 1721, introduced on September 25, 1987) and a similar Specter bill, which it eventually subsumed (S. 1818, introduced on October 27, 1987), represented major efforts to improve the 1980 Oversight Act in light of lessons derived from the Iran-contra scandal.[46] Like the Stokes bill, the Boren-Cohen proposal embraced the idea of written findings and a forty-eight-hour reporting clock in extraordinary circumstances.

Again, with these bills, the forty-eight-hour provision led to fireworks among critics. The Reagan administration preferred its own remedy, established in the Iran-contra aftermath. In 1987, the White House issued a new National Security Decision Directive on Special Activities (NSDD 286) which also prohibited oral and retroactive findings; but it allowed indefinite delays in reporting to Congress at the discretion of the NSC, following periodic review (every ten days after the presidential authorization of a finding).[47] The Congress and the White House were clearly at loggerheads: one for prior notice—or, at most, a two-day delay—and the other for indefinite delay. That the Senate felt strongly about its position was reflected in the voting for passage of the Boren-Cohen bill on March 15, 1988: 71 to 19 in favor, more than enough to override a potential presidential veto.

The debate heated up as HPSCI took up the Stokes-Boland-McHugh initiative (H.R. 3822, which had absorbed the key features of Stokes's earlier proposal as well as the Boren-Cohen bill in the Senate). After extensive hearings, agreement between the committee and the executive branch seemed to have been reached on all the key proposals except one: the forty-eight-hour bugbear. Like the Senate, the HPSCI insisted on prior notice, with a two-day grace period in emergencies; the Reagan administration, the CIA, and some others wanted greater presidential discretion over reporting.

In testimony before the HPSCI, Clark Clifford explained why prior notice, or at the latest two-day notice, was an important principle:

> One of the principal shortcomings of the Iran-contra affair was the failure of the President to notify the intelligence committees of the government's activities. The oversight process could have served a significant, salutary purpose: giving the President the benefit of the wisdom of those who are not beholden to him, but beholden like him directly to the people, and prepared to speak frankly

to him based on their wide, varied experience. Had the President taken advantage of notifying Congress, he and the country might well have avoided tremendous embarrassment and loss of credibility.

Clifford concluded that to release the president from his obligation to report because of "exceptional circumstances," as determined by his own discretion or the NSC's (which he selects), "would make any notification requirement meaningless."[48]

The administration stood steadfastly against a fixed period of time for CA reporting as an infringement on the president's constitutional prerogatives. In the words of a Justice Department letter to the HPSCI, H.R. 3822 "would unconstitutionally intrude on the President's authority to conduct the foreign relations of the United States."[49] The White House fielded powerful allies for its point of view, including an alliance of Zbigniew Brzezinski, George A. Carver, Jr., William Colby, Richard Helms, Henry A. Kissinger, and Brent Scowcroft—all of whom cosigned a letter opposing H.R. 3822 because of its "rigid, inflexible notification provisions."[50] In separate testimony, Stansfield Turner also registered opposition to the forty-eight-hour clock. "When . . . risk to human life is diminished sufficiently," he advised, "is when it is timely to notify the Congress. . . ."[51]

Republicans within the HPSCI itself rebelled against the fixed reporting time, calling the bill "impractical, unworkable and dangerous."[52] They preferred the approach advocated by another anti–H.R. 3822 witness, Lloyd Cutler, former counsel to President Jimmy Carter. In his view, "you have got to allow a certain amount of initiative to the President, although you require him to account to you afterwards."[53]

The prime example presented by those opposing H.R. 3822 was President Carter's experience during the Iranian hostage-taking of 1979–80. While the Carter administration failed to report in advance on covert actions in only three instances, those special cases illustrated for opponents precisely why a fixed-clock would be a mistake. Two delays were associated with findings in support of the ill-fated mission to rescue American hostages from the U.S. embassy in Tehran, an operation which collapsed in the Iranian desert, victim of various snafus. Though this particular effort failed, operations of this kind nevertheless require secrecy to succeed, and Carter refused to risk a leak about the scheme by notifying Congress in advance. The intelligence committees were told only six months after the two related findings were authorized.

The Carter administration was more successful in the third instance. In this case, the administration rescued six U.S. diplomats from Iranian abductors, thanks to help from the Canadian embassy in Tehran and an accompanying CIA covert action that provided fake documents and other assistance for their escape. The Canadian government, the reader will remember from chapter 6, offered to assist on one condition only: that the Carter administration would not report in advance to Congress and possibly jeopardize the Canadians involved. Carter complied, the operation succeeded, and, three and a half months after the finding (the time required for training and implementation of the covert action), the CIA briefed the intelligence committees.

Proponents of H.R. 3822 found the only other example of a failure to report in advance (and, indeed, evidently the solitary example since the passage of the 1980 Oversight Act) a more instructive and alarming case: the absence of any report whatsoever in the Iran-contra episode until the arms sale had leaked to the public through a Middle East periodical—fully ten months after the operation began. Here prior notice was withheld, pointed out the HPSCI, "not because of risk to life, but in order to hide a bizarre and troubling policy reversal."[54] This is what the Boren-Cohen and the Stokes-Boland-McHugh legislation hoped to prevent in the future.

As for the examples drawn from the Carter years, the supporters of H.R. 3822 emphasized that the "Gang of Eight" reporting opportunity had not yet been established by the 1980 Oversight Act; therefore, Carter faced the unhappy prospect—which clearly troubled Canadian officials, too—of reporting to eight congressional committees, a foolish requirement corrected by the 1980 law. Moreover, added supporters, prior notice—employed in the overwhelming number of cases (all but four) in the Ford, Carter, and Reagan administrations—had clearly worked without endangering sensitive operations. The intelligence committees could be relied upon to treat covert actions with utmost care.

To those who remained skeptical about the trustworthiness of legislators, Stokes's predecessor, Rep. Lee Hamilton, offered these words:

> What an astounding admission that argument is. It says, in effect, that our most distinguished Members, our leadership, cannot be trusted. . . . The Intelligence Committees regularly receive information concerning intelligence programs and covert actions that directly involve risk to human life. These secrets, and covert action secrets, are protected, and protected well by the Intelligence Committees.
>
> Dozens, even hundreds of people in the Executive Branch know of covert actions. A few more senior Members do not expand the risk.
>
> Moreover, if Congress grants that some things are too sensitive to tell Congress, is not the whole oversight process irreparably undermined? We have a government of co-equal branches. Secrets are not Executive Branch secrets; they are U.S. Government secrets. Each branch must be responsible for them. Each House of Congress must act to discipline any Member or staffer of the Intelligence Committee thought to have breached his duty to protect information. But the Congress cannot, without seriously eroding its powers, accept the notion that some secrets can be held only by the Executive Branch, and not by the leadership of the Congress. . . .[55]

The administration argued vigorously before the HPSCI and arrayed an impressive list of witnesses on its behalf; but, as in the Senate, its arguments failed to impress more than a small group of diehards who viewed the reforms as, in the words of their leader on the HPSCI, merely a "feeding frenzy" in the wake of the Iran-contra affair by the "sharks of congressional supremacy in the field of foreign affairs."[56] With all six GOP members in dissent, the Democratic majority on the HPSCI reported H.S. 3822 favorably on May 11, 1988.

The Intelligence Oversight Act of 1988, as the House bill and its Boren-Cohen counterpart in the Senate were less formally known, had now cleared major hurdles on the way toward passage. Yet, the bill failed to come to the House floor for a vote. As Speaker James Wright, Jr. (D, Texas) became embroiled in a controversy over whether he had revealed classified information in

FIGURE 10.1 Legislative Oversight of the CIA, 1947–87 (estimated)[a]

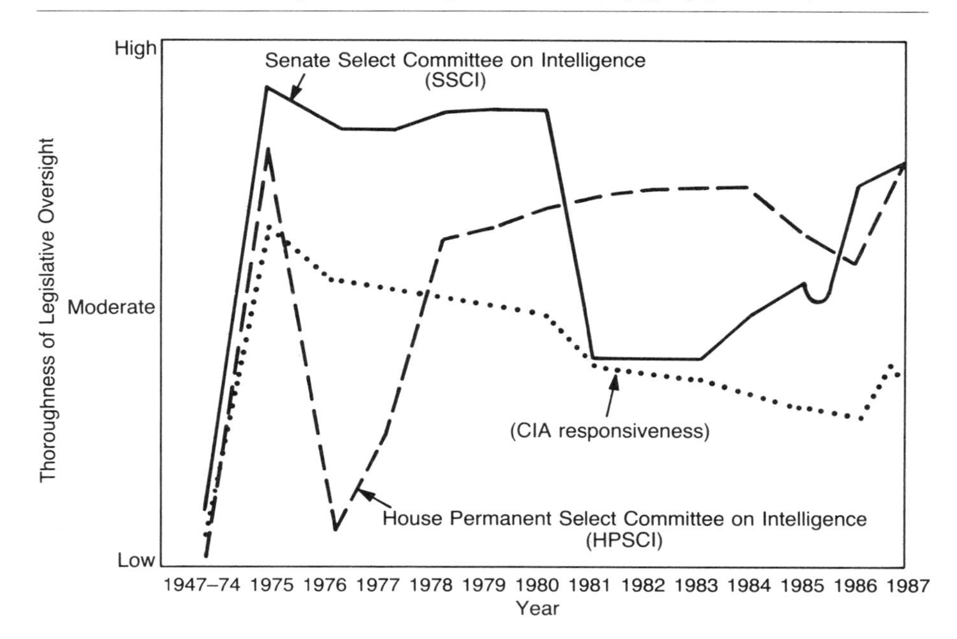

[a]Based on interviews with members and staff of the congressional intelligence committees, CIA officials, and informed observers; personal observation of the committees; and evaluations in the popular and scholarly literature.

remarks about a CIA operation in Nicaraqua, supporters of the bill postponed further efforts to seek its passage until the next session of Congress when the Intelligence Oversight Act of 1989 might have a better chance of success.

When the next session arrived, the Speaker withdrew the bill in an apparent gesture of good will toward the new Bush administration. In return, President Bush reportedly promised to consult with legislators in a "timely" fashion about covert actions. This represented an abandonment of prior notice in favor of the old Hughes-Ryan reporting requirement—a substantial victory for opponents of the proposed oversight bill.[57]

Regardless of "honeymoon" compromises, many legislators had lost faith in the institutional détente that had begun to crystallize before the Iran-contra revelations. That experiment had blown up in their faces.

The Response from Langley

Just as the Congress has fluctuated in the vigor of its intelligence oversight (see fig. 10.1), so, too, has the CIA responded unevenly in this past decade of experimental partnership. As the following summary indicates, the Agency's responsiveness has been influenced by the same four basic factors of personality, experience, structure, and environment.

Personality

The personalities of DCIs and how well they interact with key legislative overseers is of great importance to the intelligence partnership urged by reformers. From the year of its inception in 1947 until the Year of Intelligence in 1975, the Agency had a basically cordial and trusting relationship with the small handful of congressmen responsible for intelligence oversight. "Congress was informed in the way it wanted to be informed at the time," recalls former DCI William Colby, "in other words, only a few legislators."[58] Often during this period, this meant only Sen. Richard B. Russell and Rep. Carl Vinson, each a Georgia Democrat and chairman of his chamber's armed services committee. Among the few others brought into the oversight loop was Sen. Leverett Saltonstall (R, Massachusetts), who would now and then carry out his duties by way of gentlemanly chats with DCI Allen Dulles in his spacious living room over the weekend, or sometimes at Dulles's home. Both dwellings were periodically "swept" by the CIA security force to check for possible listening devices planted by the KGB or other hostile intelligence services.[59]

The relationship between Congress and the CIA was civil, casual, and permissive. Legislators seldom asked for sensitive information and the Agency seldom gave. The possibilities for friction were consequently minimized and "cooperation" was simple.[60]

With the advent of the investigations led by Church in the Senate and Pike in the House, Congress suddenly became assertive in the intelligence field and demanded access to information it had never before received. The top leaders of the Ford administration were reluctant to assist these inquiries. CIA Director William Colby, however, thought it necessary at the time to be relatively forthcoming; he feared that Congress might have responded to intransigence by dismantling the CIA.[61] The superficial comity of the earlier era became, under fire, deeper cooperation—particularly with the Church committee.

In the aftermath of the investigation, this cooperation remained genuine until the election of the Reagan administration—though the Pike committee confrontations left a strong residue of uneasiness at the Agency regarding the motivations of some congressional overseers and their ability to keep secrets. Attempts are sometimes made to return to the previous style of having the CIA director whisper only in a chairman's ear, but for the most part the chairmen of the House and Senate intelligence committees have resisted this approach and required full committee participation in sensitive briefings.

While eventually forthcoming in most instances (the Iranian affair stands as the most egregious exception), the Agency has been skillful at footdragging (of course, it hardly invented this classic bureaucratic response and has no monopoly on the market). Staff requests for documentation may take three to nine weeks to fulfill, even if they are not especially sensitive—unless a member personally involves himself directly and unequivocally. The CIA uses this time ostensibly to search for the documents; but it also gives the Agency leeway to ascertain the intentions of a committee and prepare defenses.

On some matters the CIA has politely refused to cooperate, such as in reporting to a committee its intelligence agreements with other nations. On other matters the Agency has temporized, evidentally hoping the committee would forget or lose interest in a subject; if the committee persisted, the Agency would finally respond. This occurred in 1978 with a HPSCI request for internal guidelines on covert action. In still other situations the CIA has sought refuge in a cloud of imprecise language, usually under the guise of requiring "necessary administrative flexibility." Covert-action findings, for instance, have often been unduly ambiguous, and reporting on the mining of Nicaraguan harbors was muffled as well as misleading. And then with the Iranian affair, the Agency pushed the bonds of cooperation past the breaking limits, eschewing partnership altogether.

But in the overwhelming majority of instances, presidents and the CIA have accepted the inevitability of more rigorous monitoring by the Congress. The Agency has fielded a talented and attentive corps of legislative liaison officers and generally seems to view the intelligence committees as a more or less sensible—and certainly inescapable—board of directors (a board, incidentally, whose institutional halves never convene as a whole and whose members rarely cross the Hill to speak to one another). As Admiral Turner put it in a message to his field offices around the world (partially cited earlier in this book): "Oversight can be a bureaucratic impediment and a risk to security. It can also be a tremendous strength and benefit to us. It shares our responsibilities. It ensures against our becoming separated from the legal and ethical standards of our society. It prevents disharmony between our foreign policy and intelligence efforts. It helps us build a solid foundation for the future of our intelligence operations."[62] This remarkable statement stood as a hallmark of the new attitude toward intelligence oversight—until DCI Casey came to office in 1981. One can scarcely imagine a similar message clacking out from CIA Headquarters during the tenure of Allen Dulles or Richard Helms. Differences in attitude toward oversight among Agency directors are bound to have some effect on how cooperative relations are with the Hill. Turner's views provided an important endorsement for the democratic monitoring of the CIA, while Casey's "bite-and gouge" approach made this task much more difficult for legislators.[63] The main limit on oversight, however, will remain less the personality of individual CIA officials (as important as Casey proved that to be) than the willingness of congressional overseers to study the issues, dissect the budgets, ask—persistently—the hard questions, and be willing to use sanctions against a recalcitrant CIA, from denial of funds to public criticism in the powerful forum of legislative hearings.

In this regard, a comparison of testimony by William Colby in 1975 before the Church and Pike committees in separate sessions is instructive.[64] The subject on these occasions was the extent of Agency involvement in the Angolan civil war. Colby's testimony before the Pike committee was much more extensive, detailed, and informative than before the Church committee. This was not because the DCI had more confidence in the Pike panel; on the contrary. Rather, the Pike members (in this particular instance) were better prepared; they had studied the case more thoroughly, and—most important—they were sufficiently interested and patient to ask Colby a long series of detailed questions. The Church committee lectured Colby for forty-five minutes; the Pike committee cross-examined him for over two hours. In the end, the Pike committee knew much more about the original question—the extent of CIA involvement in Angola—than its Senate counterpart. The moral: CIA officials are no more responsive than their overseers demand.

Experience

Whether a CIA director came up through the ranks or was recruited laterally seems to matter little in how he reacts to Congress. One of the most responsive directors has been William Colby; yet his career pattern was similar to one of the most elusive, Richard Helms. Both rose through the ranks of the organization's clandestine service. It is hard to judge the responsiveness of the early CIA directors, for they were rarely asked to respond; but, since the Year of Intelligence (1975), George Bush (from the Congress and the diplomatic corps) and Turner (from the Navy) were both relatively open and cooperative with oversight committees, and Casey (with an OSS, business, and political background) proved to be the least cooperative DCI.

Structure

The most impermeable organizations in the federal government are the intelligence agencies. Few are the journalists who learn what happens behind their doors and fewer still the private citizens. And so the agencies prefer the situation to remain. (If a reduction in the number of oversight committees is a primary goal of the CIA, complete exemption from the Freedom of Information Act has been a close second.) The reason is obvious and in many ways reasonable: these agencies house many of the nation's top secrets that must be protected from the unauthorized intruder—notably, hostile foreign intelligence agents.

While this argument is legitimate, the end result is a collection of highly isolated organizations (which, moreover, are heavily segmented internally). Since the press, interest groups, and private citizens have limited access to the intelligence agencies (unlike most government entities), important eyes and ears are lost to overseers in the Congress. As Rourke has noted, even a tradition of disclosure in bureaucratic agencies can "wither in the shade of administrative evasion or inertia were it not for the continued exercise of outside vigilance."[65] Legislators are left to fare largely for themselves (with the occasional and important exception of insights from investigative reporters like Seymour Hersh and Bob Woodward) and, as a result, the oversight task becomes more difficult.

Environment

Never had the domestic environment been more hostile to the CIA than in 1975. The decision of the Agency, nonetheless, was to cooperate that year with the investigating committees (a controversial position that divided the Ford administration and the Agency). The cooperation, though, was always within distinct limits: relations with the Pike committee steadily deteriorated and finally collapsed, and even with the Church committee the administration flatly refused to let any executive personnel appear at public hearings on covert action.

After the Intelligence Wars, the environment turned more benign—until soured by the Iran-contra affair. Fifty-nine percent of the American public, for example, agreed in a November 1978 poll that the CIA should work inside other countries to weaken forces inimical to the interests of the United States—a policy of covert action sharply criticized by the investigating committees in 1975. Ironically, only 43 percent of the public supported this policy in 1974 before the investigations.[66] The hostage-taking in Iran and the Soviet invasion of Afghanistan (both initiated in 1979) contributed to an environment where opinion favored the immediate removal of "shackles" from the CIA. In his first news conference of the 1980 legislative session, Senate majority leader Robert C. Byrd predicted that "this will be a security-minded Congress" and left no doubt that he favored loosening some of the restraints on the Agency.[67] This view was encouraged further by the spate of foreign spies apprehended in the United States during the 1980s.

Then, under DCI Casey, the tables turned again. Provoked by his surly approach to Congress and—the clincher—the Iran-contra operation, legislators have shifted from increasing skepticism about a CIA-legislative partnership to outright distrust. The CIA now faces a renewed crisis of confidence similar to the Year of Intelligence (though legislative passions were even more inflamed in 1975 because of the CIA's involvement in domestic spying).

The differences in attitudes toward oversight between a Colby and a Casey can be considerable; public views on intelligence issues can fluctuate from year to year; and a sensational news story (like the one on CHAOS in 1974) can create a great stir on Capitol Hill. But more important than the views toward Congress or the past experience of CIA directors, more important than the ebb and flow of public opinion on intelligence matters, has been this single reality: the CIA is largely immune to regular scrutiny from anybody other than insiders and a few interested, energetic congressional overseers.

The isolation of the CIA (and the other intelligence agencies) allows it unusual self-control with respect to oversight responsiveness, for the Agency is relatively free from critiques by the press and the few interested citizen groups who try to follow its activities from outside thick walls. For those who cherish Madison's concept of "auxiliary precautions" (*The Federalist*, No. 51), the remedy lies not in breaking down these walls, of course, since where the general public goes so can hostile foreign intelligence agents. Instead, the time-honored concepts of checks and balances rest here, as always, on the willingness of elected representatives to monitor the intelligence agencies with dedication and seriousness of purpose. In this case, however, their burden assumes greater importance; since these agencies are closed to the public, citizens must depend more than usual upon the eyes of their representatives to pierce the dark side of government.

ELEVEN

Controlling the CIA: A Critique of Current Safeguards

Hidden Power and Democratic Safeguards

Is it possible to have an effective secret service in an open society? That is the central question of this book. The weight of the evidence suggests a positive answer—indeed, democracies in this perilous world must have a secret service. The United States could well perish at the hands of foreign enemies without the protection afforded its citizens by the eyes and ears of the intelligence community. But an important caveat has to be added immediately: democracies must also maintain strong safeguards to shield their citizens against the possible misuse of secret power at home or abroad. Here is the central paradox: Those who protect us from dangers abroad can themselves become a danger. Various forms of self-protection have been tried in this country, with imperfect results. The purpose of this chapter is to sum up the findings presented in the preceeding chapters by offering a succinct critique of the chief safeguards established as a check against abuses by the CIA.

The National Security Council

At the center of the control system for intelligence lies the NSC, or, more specifically, its subcommittees dealing with intelligence. These subcommittees have had responsibility for national policy on covert action, collection and analysis, and, least likely to be discussed at the NSC level, counterintelligence. The purpose of the NSC is to provide a forum for the coordination of foreign-policy information prepared for the president and for the discussion of policy alternatives. The council was not created to make decisions; that is the president's job.

Since the Eisenhower administration, the NSC staff has been expected to follow through on policy initiatives to ensure that the president's decisions are carried out properly.

The importance of the NSC resides in its statutory membership: the president, the vice president, the secretary of state, and the secretary of defense. The panel's coordinating secretary and staff director is referred to as the president's assistant for national security affairs or, simply, the NSC director. This position became particularly significant for American foreign affairs during the tenure of two recent incumbents: Henry A. Kissinger (under Nixon and Ford) and Zbigniew Brzezinski (Carter), men who clearly carried the job description far beyond the original conception of "neutral policy coordination." Also usually attending NSC meetings over the years have been the director of central intelligence (during the Carter and Reagan years, the DCI also met frequently with the president outside the NSC framework); the chairman of the Joint Chiefs of Staff; and, less frequently, the director of the Office of Management and Budget (OMB), the attorney general (AG), and other officials—though rarely members of Congress.

Precisely who is invited to attend NSC meetings, and who is influential during these deliberations, varies from administration to administration, as well as according to the issues before the panel. During the early years of the Reagan administration, the White House counsel (and later attorney general), Edwin Meese III, often reigned over the NSC advisory system because of the confidence the president placed in the judgment of this long-time political ally—an odd situation, for Meese had virtually no foreign policy experience.[1] Neither did another political confidant (and attorney general) who played a prominent role in key NSC meetings during the Kennedy years: Robert F. Kennedy, the president's brother.

Despite its shifting configuration of dominant personalities, the NSC has managed to provide reasonably close control over intelligence activities in recent years—with some striking exceptions, such as the Iran-contra affair of 1985–86. Since 1975, few important intelligence operations have taken place without prior review by the NSC and its stamp of approval—including, for covert actions, the president's authorizing signature. The secret sale of arms to Iran, however, was based initially on a more flimsy oral approval by the president to his assistant for national security affairs, Vice Adm. John M. Poindexter, which provided no paper trail of accountability leading to the Oval Office. More flimsy still, if President Reagan's public statements and Poindexter's congressional testimony are accurate, was the authority relied upon by the NSC staff and the CIA for the subsequent diversion of the arms-sale profits to the contras: Poindexter's approval alone, without the knowledge of the president.

The CIA assassination plots appear to represent another conspicuous exception to the rule of NSC control (though at least these schemes—carried out in the 1960s—have the excuse of preceding the enactment of the strong intelligence oversight statutes passed in 1974 and 1980). The evidence regarding the plots remains murky, but the NSC was evidently unaware that the CIA had hired the Mafia for attempts against Fidel Castro's life and had resorted to other questionable modus operandi, including the shipment of murder instruments in a diplomatic pouch to the U.S. embassy in the Congo (now Zaire) for use against Patrice Lamumba. Indeed, the details of operations aside, not a single living NSC principal recalled—under oath before Congress—any approval whatsoever, oral or written, from the White House for the assassination attempts.

The Church and Pike committees uncovered other instances when the CIA and other intelligence agencies acted like "rogue elephants," but the list was short—at least in recent years. In earlier times, NSC controls over the CIA were much more lax. During the tenure of Allen Dulles as DCI (1953–61), the CIA seems to have had broad freedom of discretion, with Dulles running the Agency (according to a *New York Times* assessment in 1966) "largely as he saw fit."[2] The Church committee found in 1975 that only a small percentage of the total number of covert actions had been sent to the NSC for prior approval during the early days of the CIA.

The debacle at the Bag of Pigs, however, stimulated more serious attention to the question of control at the NSC level, and passage of the Hughes-Ryan Act in December 1974 represented a giant stride toward tighter NSC supervision of covert action by stipulating the requirements of presidential authorization (implicitly in written form, a practice honored by presidents Ford and Carter) and timely reports to Congress. Since Hughes-Ryan and the further reforms stemming from the ensuing Year of Intelligence (which drew Congress more deeply into supervision of the NSC control system—watchdogs watching watchdogs), the NSC as a safeguard against intelligence abuse has grown in reliability. This control system is hardly foolproof; the disclosures regarding the roles of DCI William J. Casey (1981–87), Adm. John M. Poindexter, and Lt. Col. Oliver L. North in the Iran-contra operations are reminder enough of that.

Other White House Overseers

The NSC is assisted by two other White House entities established to monitor CIA activities: the President's Foreign Intelligence Advisory Board, set up during the Eisenhower administration (based on the model of an earlier Eisenhower panel called the Board of Consultants on Foreign Intelligence Activities) and temporarily dismantled during the Carter administration; and the Intelligence Oversight Board (IOB), established during the Ford administration in a response to the congressional intelligence investigations of 1975. In its early days, PFIAB provided some useful guidance to the president and the CIA, especially on technical matters related to intelligence collection via reconnaissance airplane and satellite, since among its members (numbering from nine to twenty-one in most administrations, all civilians and retired military officers) were high-tech specialists like Edwin H. Land of the Polaroid Corporation. More recently, the board has been less influential, though occasionally it offers a useful observation, even if ignored, as in the case of its warnings about possible CI weaknesses at the U.S. embassy in Moscow—well before U.S. Marine guards there were charged with security breaches in 1987.

On the whole, both PFIAB and the three civilian members of the IOB—"three blind mice," concluded one savvy observer[3]—have been feckless participants in the intelligence-control process. Several members of both boards have been distinguished Americans, including over the years Clark Clifford, Henry Kissinger, William Scranton, and Albert Gore, Sr. Often, however, members have owed their appointments more to their political ties with successful presidential candidates than to any intelligence expertise they might claim. The boards have become, in too many instances, sinecures offered as rewards for presidential campaign support.

The chairman of PFIAB during the Reagan administration, Anne Legende Armstrong of Texas, provides a classic example. A big gun in Texas Republican politics, Armstrong served as U.S. ambassador to Great Britain from 1976 to 1977 and was urged upon Reagan by Vice President Bush in 1981. Her knowledge of intelligence matters was reportedly weak, but her ties with Bush and fellow Texan and White House Chief of Staff James Baker, 3d, were strong and that was more important.

One of the more peculiar "qualifications" on the curriculum vita of an IOB member was his former experience as an officer in the CIA, where among other things he was in charge of starting up the controversial covert operation to fund the National Student Association. This considerable blemish for a would-be intelligence overseer apparently faded into insignificance in light of what most endeared him to his White House benefactors: political help—notably proven acumen at fundraising—during the presidential election. While "hands on" expertise can be useful for meaningful oversight, greater sensitivity should be exercised in the selection of overseers. Professional, nonpartisan experience would improve the credibility of the IOB and PFIAB watchdogs.

As the reader may recall from chapter 3, PFIAB and IOB meet only bimonthly. Moreover, the members have accepted a narrow view of their responsibilities. Neither, for example, routinely examines CA proposals. The members of the boards receive occasional, perfunctory briefings from the CIA and seem content to be "participants" in White House affairs, with all the prestige that offices in the Old Executive Office Building confer—even if the boards deal more with shadow than substance. Staff assistance has consisted usually of only one person for each board (though PFIAB's staff grew to a half-dozen during the Reagan administration), competent aides for the most part but unable to achieve meaningful supervision with limited resources and an agenda tied to board members with the short attention span of distracted dillettantes.

This harsh indictment is directed less against particular individuals on the boards (some, like Clifford, have been eloquent advocates of intelligence oversight), than against oversight as a part-time hobby for itinerant dignitaries. Needless to say, PFIAB never uncovered any of the CIA abuses revealed by the Church committee and the Rockefeller commission in 1975. When PFIAB asked to examine a copy of the Huston spy plan, the FBI and the attorney general refused (according to Church committee findings)—just as it had been denied access to 40 Committee minutes, which it sought in order to study the extent of CIA covert action in Chile.

The IOB was established after the Church committee inquiry was well underway and the Rockefeller panel already disbanded; but its record since then has been lackluster at best and most recently, during the Iran-contra affair, flatly pathetic. The IOB staff attorney during the Reagan years had the distinction of failing his bar exams four times before finally passing, and he had never written a single legal opinion until the White House asked him to appraise the relevance of the Boland amendment (prohibiting military aid for the contras) to the activities of the NSC. None, he opined, in spite of weighty legal precedence to the contrary.[4] Learning by chance of the Iran-contra operation, the IOB attorney began an inquiry. According to his testimony before the Inouye-Hamilton investigative panels in 1987, this probe consisted of a five-minute conversation with Lieutenant Colonel North and a thirty-minute meeting with the NSC legal counsel. Assured by these individuals that there was nothing to worry about, he ended his "investigation." The Intelligence Oversight Board: three blind mice and a lamb.

Arguably (and the Church committee settled on this view), PFIAB should strictly limit its role to advising the president on the quality and effectiveness of intelligence, abandoning altogether any "watchdog" pretensions—a task it would leave to the IOB. The series of abuses uncovered by the Church, Pike, Rockefeller, Tower, and Inouye-Hamilton panels suggest, however, that two watchdogs—real ones, not pottery imitations—could be useful. At a minimum, the staff and personnel of the IOB require new talent and a fresh sense of seriousness, including individuals with proven investigative skills and an air of curiosity—not to say skepticism. The same is true of PFIAB—doubly so if its job is to include oversight; but, even if it restricts itself to questions of quality control, this panel needs a better mix of experts (especially on technical matters) and thoughtful generalists unafraid to offer candid criticism.

Executive Budget Reviews

The CIA undergoes further review, of a more serious nature than that offered by PFIAB and the IOB, at the hands of various budget panels whose members poke into the intricacies of Agency spending. As well as sending a representative to some NSC meetings, the OMB has an office for national security affairs that is dedicated to the examination of CIA and other intelligence agency budgets. These reviews by the OMB are real, not make-believe, and the intelligence agencies are required to justify their spending plans before gimlet-eyed budget examiners. Some improper operations in the past have slipped easily through this net, however (among them, the CIA domestic spying program code-named CHAOS), for the simple reason that great mischief can be accomplished at little expense and, if necessary, funds can be shifted over from legitimate programs. Moreover, the CIA has resisted OMB visitations to its stations overseas.

The CIA also has its own budget auditors, along with those attached to the DCI's Intelligence Community Staff—the "IC Staff" or ICS, a community-wide group of assistants assigned to the DCI to help him coordinate the sprawling intelligence empire. Both sets of auditors scrutinize the books at CIA Headquarters and at Agency stations around the world to monitor the flow of cash. The raising of CA funds by Lieutenant Colonel North from private channels outside the government—wealthy American citizens and foreign heads of state—held an added attraction beyond circumventing the Congress; here was a way to avoid the green eyeshades within the executive branch as well.

Internal CIA Overseers

The Agency itself conducts continuous inside inquiries over and above its budget audits. The Directorate for Administration sends inspectors (the dreaded "Admin") to every station abroad and every office at home to check on the proper functioning of the Agency. The CI Staff and the Office of Security also have their own investigators who examine security problems and constantly test the CIA's defenses against hostile penetrations. The Office of General Counsel churns out documents for the Agency on the propriety of planned operations. The Office of Legislative Liaison reminds Agency personnel of various legislative strictures, as well as the personal views of leading legislators and staff on proposed intelligence operations. The CIA Inspector General (IG) has an open door to hear from employees about alleged wrongdoing in their bureaus, and the 1980 Intelligence Accountability Act explicitly requires the CIA to report to the Congress "any illegal intelligence activity."

The Agency, in short, has several in-house safeguards against abuse. It is sobering, though, to recall that Operation CHAOS was run out of the CIA's Office of Security itself and that the CIA's attorneys, budget examiners, and inspectors general apparently never knew about the Iran-contra caper. Further, neither at the time of their occurrence—or even very soon after—did these in-house entities know about the CIA wrongdoings disclosed by the Church committee. One indication of the low status conferred on the IG's office was DCI Colby's reduction of its staff from fourteen to five and his appointment of a person as IG who possessed limited prior training in oversight. Agency directors and other senior officials have simply bypassed these multiple checks when they wished—though always at the risk of another important internal bureaucratic check against malfeasance: the leak, like the one to *Times* reporter Seymour M. Hersh on CHAOS that led to the intelligence investigations of 1975.

The Station

The CIA has controls in the field, too, beyond periodic inspections by teams sent out from the CI Staff, the IG, and other Headquarters offices. The Agency's chief organizational entity in the field, the station (a few small countries have bases and a few large ones have both), is supervised by the chief of station (COS). He or she is responsible for ensuring that directives from Headquarters are properly interpreted and carried out; that CIA officers assigned to the station (case officers) and their local agents (often called assets, if they are paid) perform their duties within the bounds of legislative and presidential guidelines; and that a reasonably harmonious relationship is maintained with the overall chief of the American mission within the country, the U.S. ambassador.

The most difficult of these chores evidently is to control the assets and, sometimes, the case officers themselves. Here is the business end of the CIA, the men and women who secretly plant the newspaper stories, pass the bribes, steal the government documents, smuggle the guns and ammunition, and, if necessary, pull the trigger. The more unsavory the deed, the more unsavory the asset the CIA might have to recruit for its execution. One Agency asset, code-named WI/ROGUE, a potential recruit for the death plot against Patrice Lamumba of the Congo, was described by his case officer as "aware of the precepts of right and wrong, but if he is given an assignment which may be morally wrong in the eyes of the world, but necessary because his case officer ordered him to carry it out, then it is right, and he will dutifully undertake appropriate action for its execution without pangs of conscience. In a word, he can rationalize all actions."[5] Such individuals do not worry excessively about oversight legislation or executive orders.

Nor are many guerrilla "freedom fighters" known for their devotion to the American Constitution. The contras have been intent on overthrowing the Sandinistas in Nicaragua by practically any means available—even when, in 1984, a majority of the Congress formally voted (in an early version of the Boland amendment) to limit the CIA-funded operations of the rebels strictly to the interdiction of weapons bound for left-wing insurgents in El Salvador supported by the Sandinistas. Rebels by definition are zealots who are not easily controlled and who have their own agenda and their own deeply held beliefs, cultures, mores, and allegiances—none of which are necessarily in tune with instructions from their American overseers.

Case officers, too, have sometimes been difficult for Headquarters and the COS to contain. These individuals are not thugs, but rather well-educated individuals with long exposure to American cultural norms and the rule of law; yet some have displayed a remarkable insensitivity to legal constraints. In 1984, it was a CIA case officer who wrote and distributed the assassination manuals in Nicaragua, and during the Kennedy years, it was an Agency case officer who initiated the ties with the Mafia in order to carry out the Castro murder schemes—by all accounts without the knowledge of the White House, the DCI, or other senior CIA officials.[6]

Nor have the station chiefs themselves been immune from lawlessness. During the Iran-contra affair, for instance, the COS in Costa Rica facilitated airdrops of military supplies to the Nicaraguan rebels on the southern front—in clear defiance of the Boland amendment.[7] Generally, the COS has been an effective overseer and most covert actions have been properly supervised. The record, though, reveals several examples of loose or nonexistent supervision at the field level—probably the most consistently weak link in the hierarchy of controls.

The Ambassador

The American "chief of mission"—the ambassador—is also expected to keep tabs on the COS and his operations within a country, providing yet another check upon the propriety of CIA activities abroad. Since little has been published on the relationship between the ambassador and the COS, this safeguard warrants a more extensive examination here.

Personal interviews and public statements on the subject indicate that, by and large, the relationship has been smooth, with the COS accepting leadership and control from the ambassador. Most ambassadors have little interest in digging into the details of CIA plans; as long as its operations do not become an embarrassment in the local newspapers, they are content to give the COS considerable rein. And most COSs have no interest in blindsiding their ambassadors; an angry chief of mission can force the reassignment of a COS and bring about career demerits.

Sometimes, though, the teamwork breaks down—usually when the ambassador fails to establish his or her authority forcefully, or when the COS has limited confidence in the experience and discretion of the ambassador (as when some misguided political appointees view the job of ambassador as a social plum more than a serious responsibility for protecting American interests abroad). On occasion, a strong-willed COS with backing from Headquarters will ignore ambassadorial instructions that the Agency finds disagreeable. The CIA continued to back Chinese nationalists in Burma in 1954, despite objections from the U.S. ambassador (who eventually resigned in protest); supported Laotian Premier Phoumi Nosavan in 1960 over the instructions of the ambassador; and reportedly carried out a sensitive operation in Malaysia in 1960, which backfired, without bothering to clear the project with the ambassador.[8] When James Angleton served as CIA chief of counterintelligence, he conducted various aggressive operations around the world without informing ambassadors—or, often, even station chiefs.[9]

"The State Department through U.S. embassies and consulates offers the only external check upon CIA's overseas activities," stated the Church committee report on State Department–CIA relations abroad. Yet, the committee discovered, this check was far from satisfactory: "uneven" was the best the investigators could say.[10]

The recent history of efforts to define the relationships between the ambassador and the COS begins with a May 29, 1961 letter from President John F. Kennedy addressed to each chief of mission. The President told his emissaries that he expected them "to oversee and coordinate *all* activities of the United States Government" in their assigned countries.[11] The Kennedy letter remained in effect until superseded by a similar letter from President Richard M. Nixon on December 9, 1970. "As Chief of the United States Diplomatic Mission," the Nixon letter said,

> you have full responsibility to direct and coordinate the activities and operations of *all* of its elements. You will exercise this mandate not only by providing policy leadership and guidance, but also by assuring positive program direction to the end that *all* United States activities in [the host country] are relevant to current realities, are efficiently and economically administered, and are effectively interrelated so that they will make a maximum contribution to the United States interests in that country as well as to our regional and international objectives.[12]

Supplementing this message was a classified State Department–CIA communique further explaining the president's intent (State Department Circular Airgram 6693 of December 1970). In essence, the communique told the ambassador that any access he might have to information on intelligence sources and methods would be subject to the approval of the COS; if an argument over access arose, the disagreement would be reported to the secretary of state and the director of central intelligence for arbitration. According to the Church committee, this communique "may well . . . have had the effect of inhibiting ambassadors in seeking to inform themselves fully in this area."[13]

The seemingly comprehensive authority given to the ambassador in the Kennedy-Nixon letters (diluted to a degree by the State Department–CIA communique) was set in legislative concrete four years later (1974) with Public Law 93-475.[14] According to this statute (emphasis added):

> (1) the United States Ambassador of a foreign country shall have full responsibility for the direction, coordination, and supervision of *all* United States Government officers and employees in that country *except* for personnel under the command of a United States area military commander;
>
> (2) the Ambassador shall keep himself fully and currently informed with respect to *all* activities and operations of the United States Government within that country, and shall insure that all government officers and employees in that country, except for personnel under the command of a United States area military commander, comply fully with his directives; and
>
> (3) any department or agency having officers or employees in a country shall keep the United States Ambassador to that country *fully* and currently informed with respect to *all* activities and operations of its officers and employees in that country, and shall insure that *all* of its officers and employees, except for personnel under the command of a United States area military commander, comply fully with all applicable directives of the Ambassador.

A comparison of this law with the language of the Nixon letter shows that P.L. 93-475 went beyond even his considerable effort to strengthen the arm of the ambassador in the field; yet neither the White House, the State Department, nor the CIA issued directives to implement this statute. The law, devoid of implementing regulations, remained (in the words of one ambassador testifying before the Church committee) "suspended."[15]

It became clear through the testimony of its officials before the Church committee, that the CIA, in the words of the committee's report, "opposes giving the Ambassador the unrestricted access to communications and other operational information that the law would appear to authorize."[16] A CIA deputy director for operations (DDO) told the committee that, for example, "individual agent recruitments are not cleared with either the ambassadors or the Secretary of State."[17]

The CIA argues that the COS cannot freely provide the ambassador with all intelligence-related information, since the National Security Act of 1947 charges the DCI with responsibility "for protecting intelligence sources and methods from unauthorized disclosure."[18] This seems to be a gross overreaction on behalf of the CIA, for seldom have American ambassadors engaged in micromanagement to the degree where they felt they had to know the specifics of sources and methods. In one of the few documented cases where this occurred (in Portugal; see below), the ambassador probably had good reason—in a country in the midst of revolution—to know the details of what the CIA was doing. While the debate over the meaning of "unauthorized" remains lively today, the Church committee concluded unequivocally in 1976 that, in its opinion, P.L. 93-475 "resolves any doubts as to whether disclosure to the Ambassador is authorized."[19] Moreover, added the committee, "for CIA operations conducted within his country of assignment, the Ambassador should be a good judge of the risks of such operations, and of their possible usefulness to the U.S."[20]

Three years following the adoption of P.L. 93-475 and seven years after the Nixon letter, a new president—Jimmy Carter—sent out to America's embassies a now familiar refrain giving U.S. ambassadors around the world authority to supervise "*all* United States Government officers and employees in their countries" (emphasis added). The Carter letter, dated October 25, 1977, stated to each chief of mission: "[You] have the authority to review message traffic to and from *all* personnel under your jurisdiction" (emphasis added).

The Carter initiative immediately triggered the release of a series of State Department and CIA explanatory directives to the field (just as the Nixon letter had done in 1969). According to one correspondent, President Carter's letter, and the ensuing directives from the executive branch, reflected "widely divergent interpretations" as to what the proper relationship ought to be between the chief of mission and the chief of station.[21]

The directives following the president's letter included the following:

1. A joint communiqué from Secretary of State Cyrus Vance and DCI Stansfield Turner to each ambassador and COS.[22] (This document is referred to privately in the intelligence services as the State Department–CIA "treaty" on the role of the ambassador.)
2. A cable from Admiral Turner to all CIA stations, supplementing the Vance-Turner "treaty."
3. Two more CIA cables to all stations, supplementing the Turner cable.

Traveling from the Carter letter to the last two directives resembles a journey through a cave. First, with the president's letter, the ambassador stands at the entrance, which is wide and ample and filled with light. Then, with successive messages, the passage narrows, the space for maneuvering grows cramped and the light turns to gray, becoming darker and more uncertain. Finally, the ambassador finds his pathway blocked altogether. While President Carter stated that "all" U.S. government employees in the ambassador's assigned country are under his supervision, the Vance-Turner communique began the process of "amplifying," or, more accurately, narrowing the meaning of the president's message. The ambassador would be notified in advance by the COS before certain—but not all—CIA activities.

The difference between the unclassified letter from President Carter and the increasingly restrictive directives from the CIA led to controversy. Learning about these differences through intelligence sources, an experienced *New York Times* correspondent concluded that the CIA apparently sought to "undercut" the Vance-Turner treaty. According to one Department of State source quoted by this correspondent, the CIA directives following in the wake of the Carter letter "in effect . . . stated that the President's letter and the State Department guidelines do not apply to the CIA." In truth, claimed the source, ambassadors had greater leeway to monitor CIA clandestine operations before the Carter pronouncement. A CIA officer concurred, suggesting that the kind of information the U.S. ambassador to Portugal had reportedly been able to pry loose from his COS in 1975 (regarding CIA assets in the Portuguese government) would no longer be forthcoming; with the new directives, the COS would no longer feel obligated to name his assets for an ambassador. The correspondent concluded, with obvious understatement: "[S]everal ambassadors have indicated unhappiness with the new arrangement."[23]

Some observers suggest that, ultimately, the ambassador-COS relationship is reduced to personality. If the ambassador is tough-minded, aggressive, and interested in intelligence matters, he or she will receive most or all of what is requested from the COS—even when their views are at variance. While the COS may complain about micromanagement by the ambassador (as he did in the Portuguese case mentioned above), he will usually cooperate. If the ambassador would just as soon remain blindly ignorant (perhaps to avoid blame and embarrassment if an operation is exposed) or is more interested in the social and cultural activities of the United States Information Agency (USIA), he or she will receive minimal information from the COS.

In hearings before the Oversight Subcommittee of the House Permanent Select Committee on Intelligence, former ambassador William Porter agreed with the suggestion of Chairman Les Aspin (D, Wisconsin) that a successful chief of mission who wants to know what the CIA is doing must be "strong [so] he can impose himself on the system."[24] Former ambassador L. Dean Brown pointed to another ingredient for success: an ambassador should possess "a very suspicious mind."[25] Despite the ongoing tensions between the two offices, Ambassador Porter arrived at an optimistic conclusion. With P.L. 93-475, he testified, the ambassador "has got the wherewithal today if he wants to use it."[26]

While even the most perfectly written presidential directives would be of little use in the face of a passive or uninterested ambassador, those chiefs of mission who are energetic and dedicated to the full exercise of their duties nonetheless face an uphill struggle—notwithstanding Ambassador Porter's optimism. The CIA has clearly sought in the past to circumscribe the spirit of White House directives designed to strengthen the hand of the ambassador. By invoking its obligation to protect—in that slippery phrase—"unauthorized disclosures" and through secret internal directives that skirt the ambassador, the CIA has been known to conduct its own foreign policy outside the vision of the nominal chief of mission. Testimony by Lieutenant Colonel North during the Inouye-Hamilton hearings gave the impression that DCI Casey hoped to achieve sufficient private funding for covert action that auditors, legislators, ambassadors, and other official nuisances could be bypassed altogether.

Pressure Groups and Parties

One of the more venerable propositions in the literature of political science is that government is strongly influenced by a tripartite alliance of executive agencies, outside interest groups, and congressional committees—the "iron triangles" or "subgovernments" of textbook fame. Much evidence suggests that these three organizations develop a symbiotic relationship to enhance their respective goals: swelling budgets for the bureaucrats, federal largess for the interest groups, and reelection for the legislators. When the relationship operates in this mutually beneficial way, interest groups are unlikely to provide a check on agencies; the agencies have been bought off—or, more often, the interest group has colonized or "co-opted" the agency.

In myriad interest groups that exist in America's pluralist democracy (that is what "pluralism" means), competing groups often become dissatisfied with the response of an agency to their demands and they make their dissatisfactions known to the public. In this public debate—or, less grandly, squabbling—over an issue and an agency's handling of it, a further check is placed on the bureaucracy; an alerted public and its representatives have an opportunity to take corrective measures in an effort to mollify the group conflict. This tension, and the useful publicity it engenders, can be seen dramatically in the periodic clashes of corporate and environmental groups over such policies as clear-cutting and strip-mining.

Such group dynamics are largely irrelevant, however, as a democratic control over intelligence policy. Few outside groups exist in this policy domain, and those that do are weak. Moreover, most aspects of intelligence policy have been too concealed (and some have been genuinely too sensitive) to serve as subjects of public debate. This condition has changed somewhat since 1975, when the legislative investigations spawned both a pro-CIA advocacy group of retired intelligence officers (AFIO) and a civil-liberties group devoted to the prevention of further intelligence abuses (the Center for National Security Studies, CNSS). Other centers of intelligence inquiry and advocacy have cropped up, too, including the right-wing American Security Council (ASC), with Angleton among its founding officers, and the more scholarly, if still rightward-leaning, National Intelligence Study Center (NISC), with former DDI, now professor, Dr. Ray S. Cline, among its founders, and the National Strategy Information Center (NSIC)—all in Washington, D.C.

These groups provide expert testimony at congressional hearings and for the media, send out mailings, publish treatises on intelligence, and the like. The CNSS, for instance, prodded the newly formed House Intelligence Committee in 1978 to undertake extensive hearings on the relationship between the CIA and U.S. domestic organizations. But these groups have little influence over the shaping of intelligence policy—at least compared to the clout wielded over portions of the government by such groups as the AFL-CIO, the American Medical Association, and other giants in the constellation of Washington lobbyists.

The political party stands as another entity in the American political system that sometimes attempts to set boundaries for the making of intelligence policy. Now and then, the parties have become involved in debates over intelligence controversies. Partisan wrangling over missile or bomber gaps (for which the GOP attacked the Kennedy administration), over scandals like the CIA involvement in Watergate—however peripheral—and the Iran-contra affair, over unsuccessful coups and battlefield reversals, and other foreign-policy mishaps (not to mention incidents of illegal domestic spying), often carry at least an implicit allegation that the party in charge of the White House failed to make correct use of the intelligence agencies. The threat is always present that misuse of intelligence might be discovered and decried for partisan advantage by the "loyal opposition."

With the aggressive and more open resort to covert action by the Reagan administration (reportedly the backbone of the so-called Reagan Doctrine[27]), partisan fissures over intelligence have become more conspicuous than ever, with clear-cut party votes occurring with some regularity on the congressional intelligence committees. The Boland amendment votes from 1982 to 1986 are prominent examples. Though a source of acute discomfort for those who believe in "bipartisan foreign policy" (a beguiling slogan, often with a hidden agenda: defer to the president), the new tension between the two parties over intelligence policy may have had the salutary effect of encouraging closer legislative scrutiny over selected CIA operations.

The Media

In the modern history of American intelligence, the media has consistently provided the public with the most information on the abuse of power by the CIA. The National Student Association scandal, the CIA connection to Watergate (which proved to be slight), Operation CHAOS, even early reports (discounted) on assassination plots, came to the attention of U.S. citizens initially through the media. Whether from leaks or from skillful investigative reporting (often both), the public has benefited from a free press able to warn Americans of transgressions and to prod the government into corrective measures.

Dependence on the media, though, is a less than fully reliable means to control the intelligence establishment. Correspondents have limited access to this secret world and must await, for the most part, tips from insiders (with all the biases they may carry). The media has been indispensible as a safeguard, but its reporters have hardly been a timely and infallible deterrent to abuse—or even a reliable chronicler of most abuses. The intelligence community is surrounded by too impenetrable a wall of secrecy for outsiders in the press corps to break through at will (a wall designed appropriately to keep the KGB and other U.S. adversaries away from sensitive American secrets). Among "outsiders" only the Congress has the formal powers (foremost among them the subpoena, budget review, and a capacity to focus public attention during controversial hearings) to wedge through this barrier—if its members have the motivation to use that power. Journalist Charles Peters sensibly recommends that the media concentrate more on stimulating the Congress to fulfill this obligation: "What journalists could do is make the public aware of how little attention Congress devotes to what is called 'oversight,' i.e., finding out what the programs it has authorized are actually doing. If the press would publicize the nonperformance of this function, it is at least possible that the public would begin to reward the Congressmen who perform it consistently and punish those who ignore it by not reelecting them."[28]

The Congress

Ideally Congress should provide one of the main checks against intelligence abuses. Indeed many believe that oversight ought to be the primary focus of the Congress. Legislators, though, largely ignored their responsibilities for intelligence oversight until recently. As illustrated in earlier chapters, from 1947 until 1975 the CIA received precious little attention from those legislative subcommittees supposedly responsible for intelligence oversight—probably less than a total of twenty-four hours each year, and this time was spent mainly in passive listening to briefings that carefully avoided issues of controversy.[29]

With the establishment of the two permanent intelligence committees in 1976–77, legislative oversight took on new vigor. Of particular importance has been the independent check on CIA spending provided by budget specialists on these two committees, augmented by the intelligence subcommittees of the two appropriations committees. Each of these units goes through the CIA annual budget line by line and supplements this audit with a series of closed hearings on whatever money requests the members decide to probe further, including special releases from the Contingency Reserve Fund outside the normal appropriations cycle.

Since 1976, the budget of the CIA (and other intelligence agencies) has risen steadily—indeed, by a factor of three. This might lead some to conclude that the intelligence committees have actually failed to provide much of a check. Interviews with members and staff (current and former) of the committees suggest, however, that the funding increases have been merited by the need to improve the intelligence collection capabilities of the United States and have been the subject of thorough examination by the legislative panels (though a handful of legislators worry that too much money is being spent on satellites and other hardware at the expense of HUMINT). While sometimes legislators have actually *increased* the CIA's budget requests (as in the case of CA monies for Afghanistan), the various Boland amendments, designed to set limits on PM funding for the contras during the 1980s, illustrate that the power of the purse can be used by the Congress against the CIA if legislators strongly oppose an intelligence activity. Forcing the executive to honor the legal prohibitions, though, is another matter—the crux of the Iran-contra scandal.

Still, knowledgeable observers continue to question the steadiness and staying power of legislative oversight, and some have discounted its effectiveness all together—even before the Iran-contra affair. For one senior intelligence official, oversight had descended by 1984 to a level of "anarchy."[30] Rep. George E. Brown, Jr. (D, California) concluded baldly in October 1985 that the new oversight "is not working."[31] A top staffer attached to the CNSS called the state of oversight in 1985 "absymal."[32] A former Senate Intelligence Committee staffer noted, also in 1985, that at best oversight had been "uneven."[33]

Former DCI Stansfield Turner predicted—*before* the Iran-contra scandal came to light—that the new oversight process would fail unless, on the one hand, President Reagan discarded his "indifference about whether his use of the CIA is supported by Congress" and, on the other hand, Congress "[took] the bit in its teeth and exercise[d] the latent authority it has." Turner was especially critical of the relatively mild legislative reaction to the Nicaraguan assassination manual; the CIA's mining of harbors in Nicaragua; CIA association with uncontrollable factions in Lebanon in 1984; and the use of NSC staffer North to guide the contras, despite the Boland amendment prohibition to halt the support, "directly or indirectly, [of] military or paramilitary operations in Nicaragua. . . ."[34] Then, when the full force of the Iran-contra scandal struck the nation in December 1986, CIA-congressional relations reeled further backwards. The reader may recall the judgment reached by the vice chairman of the Senate Intelligence Committee (reported in chap. 6): the oversight process was now "fractured."[35]

Yet if a comparison is made with oversight in the pre-1975 period, when the congressional watchdogs were fast asleep and the CIA was content to let them lie, one would have to acknowledge the presence of greater watchfulness now—even counting the failure to prevent the unfortunate excesses of some NSC staffers and their CIA cohorts during the Iran-contra episode. One must avoid drawing general conclusions from exceptional, even if important, cases like this scandal. Though certainly imperfect, as revealed vividly during the televised hearings in 1987 of the Inouye-Hamilton committees, the experiment in genuine oversight for intelligence policy only began, after all, in 1976. It seems premature to write this experiment off as a total failure already—especially when the record suggests many useful oversight results during this period, including (among several other examples) the Senate Intelligence Committee's thoughtful critique of CIA estimates on Soviet oil production; the House Intelligence Committee's demands (though not always consistent) for the details—sources and methods aside—of CA plans in order to evaluate them properly; and, within the CIA itself, the "A-team, B-team" critique of a key intelligence estimate on the Soviet Union in which outside academic experts were included in the process—a review that, though biased toward an extremely bleak "worse-case" view, at least had the merit of tapping expertise beyond the walls of CIA Headquarters.

Another comparison is relevant: the state of intelligence oversight in other democracies. Though a systematic comparative analysis lies beyond the scope of this book (not to mention the competence of the author), clearly most allied intelligence services have remained free of the restraints that have evolved within the United States. In Great Britain, the Official Secrets Acts sharply limit reliable monitoring of British intelligence by the press or even the Parliament. The French approach is equally slack from the viewpoint of maintaining safeguards against potential intelligence abuses. In 1985, for example, a French intelligence officer sank the vessel "Rainbow Warrior" (an antinuclear protest ship belonging to a group called Greenpeace) in the harbor of Auckland, New Zealand, killing a person aboard. The French government set free the intelligence officer, choosing instead to arrest five individuals for leaking information about the incident to the press. Under the present system of controls in the United States, one suspects that such a mission would never have been approved in the first place and, if approved and discovered, never so easily dismissed. As one CIA official has put it (with some dismay), "We are the most carefully scrutinized intelligence agency in the world."[36] Only Canada's new system of intelligence oversight, guided by the Security Intelligence Review Committee (SIRC), seems to keep comparably close tabs on its intelligence establishment. The SIRC even has the authority, which it has exercised, to conduct surprise, random program audits ("We want to look at files X, Y, and Z").

As the preceding chapter illustrated, the meaningful practice of legislative oversight in the United States will probably have to be sustained by a few dedicated members on the two congressional intelligence committees. In the first decade of the new oversight (1976–86), several legislators on the intelligence committees have provided the necessary leadership and integrity for oversight to work. One must hope that others will step forward to assume this responsibility in the future. The special rotation system adopted by the Congress for the intelligence committees, whereby no member serves for longer than eight years on the Senate panel or six on the House, means a constant replenishment of personnel (at least above the staff level). This ought to help avoid the co-optation that often occurs when committee members serve too long and grow too close to bureaucrats—though the unfortunate trade off is the loss of corporate memory among the members.

It will take another decade or so to appraise how well this experiment in oversight is working. In its first decade, the intelligence partnership between Congress and an executive branch has been stormy—especially during the Iran-contra affair—but far more healthy for democracy than the earlier ostrichlike posture of Congress. Former DCI William Colby has observed (with reference to covert action) that, in this new era, intelligence mistakes "will be American mistakes. They will not be CIA mistakes, but mistakes of the administration *and* the Congress in power."[37] While this proved to be untrue for the Iran-contra operation, intelligence policy for the most part did become more of a partnership between the branches than ever before. This, one presumes, is how democracy in the United States is meant to work—not through reliance on the executive branch alone to determine the destiny of the nation.

The Courts

One can be more succinct about the role of the courts as intelligence overseers: it has been significant, but within a limited range. Occasionally the courts will adjudicate espionage charges or suits involving the public disclosure of classified information. The most celebrated instance of the latter is the Pentagon Papers case, in which the Supreme Court stood against prior restraint on the grounds that disclosure—in this instance at least—would not lead (in the words of Mr. Justice Stewart) to "direct, immediate, and irreparable damage to the Nation or its people."

The courts have also been involved in controversial cases based on contractual secrecy, in which the intelligence agencies have successfully sought to prevent former employees from writing about intelligence policy without submitting their manuscripts to a CIA prepublication review board. This censorship process, though, has been arbitrary; former officers critical of the CIA have been hounded, while favored individuals—including some retired senior officials who write newspaper columns without any clearance—have been left alone. In a landmark case, *Knopf v. Colby* (1975), the Agency brought suit against one of its former senior administrators (Victor Marchetti) who had submitted his book manuscript to the Agency's clearance process but was unwilling to accept the extensive excisions requested by the CIA. The book (*The CIA and the Cult of Intelligence*, coauthored by John D. Marks) was eventually published, but with conspicuous gaps throughout where the CIA censors had worked their scissors—quite fascinating to the curious reader and no doubt a boon to sales.

In one important case where the CIA was unaware of a forthcoming publication by a former employee (intelligence analyst Frank Snepp), the Agency moved after publication to block payment of royalties to the author. The CIA was successful before the Supreme Court in establishing this precedent and in confirming its rights to prepublication review (*U.S. v. Snepp*, 1980). While the intelligence agencies seem to have a legitimate need to prevent former employees (like CIA defector Philip Agee) from revealing sensitive sources and methods, the method of prepublication review currently in place has been widely criticized. Snepp persuasively argues: "Congress should establish an independent review board to keep the C.I.A. and other intelligence agencies from overcensoring; clarify how, and under what circumstances, a censored author can challenge excisions in court; and set a time limit on post-employment censorship. . . ."[38]

Beyond these occasional suits, the systematic use of the court system for intelligence oversight is limited chiefly to its issuance of warrants for electronic surveillance by the intelligence agencies, as a result of the 1978 Electronic Surveillance Act. The special court of district judges established for this purpose rarely refuses warrants; but interviews with officials in Congress and within the intelligence agencies indicate that the court does actually serve as a serious check against poorly justified wiretaps by the federal government.

Conscience and Opinion

The most important checks on the abuse of power in a democracy are, in the end, the attitudes held by people in office and across the land. The Huston Plan, Operations CHAOS, covert actions in Chile and Cuba, and other controversial intelligence operations occurred because the climate of opinion in the country seemed to allow the use of extreme measures—even assassination and the violation of domestic law—to combat Communist-Marxist rivals. Bureaucrats are influenced by the example of their leaders in the executive branch, from agency directors to the president. These leaders in turn must take into account the views of Congress and public opinion. If such views are unformulated or permissive, the executive acts accordingly and proceeds as it wishes; if Congress and the public set limits, the chances increase that limits will be honored, especially if it is clear that Congress intends to monitor the executive branch with serious intent. To tame the bureaucracy, then, the president, the Congress, and the public are obliged to make the boundaries of probity clear and then to ride the fences. This clarity, as a former secretary of state has noted, ought to be made "by insistence, not necessarily by legislation."[39]

Prevailing opinion seems to oppose assassination as an instrument of American foreign policy (executive orders explicitly prohibit it, although, as the F-111 bombing of Libyan leader Col. Muammar el-Qaddafi's home in 1986 demonstrates, at least one administration has been prepared to overlook such legal niceties). It also seems to oppose U.S. involvement in large-scale PM operations of the magnitude conducted in Indochina during the Kennedy-Johnson years, as well as, it goes without saying, spying on American citizens. The CIA, as a result, has curtailed such activities and is unlikely to acquire authority to resume them unless public opinion changes dramatically. The exception has been the CIA covert action in Afghanistan, which spiraled steadily toward the "large-scale" category during the Reagan years, enjoyed widespread support in the Congress, and likely spurred the Soviet decision in 1988 to withdraw its army from this civil war.

Changes in public opinion toward a broader support for major covert actions may occur if, for example, the Soviet Union attempted more direct, aggressive intervention in parts of the world considered vital to American interests, or if terrorism became a greater immediate problem within the United States. Former secretary of state Dean Rusk has commented on the relationship between threats to the country and the legal protections afforded American citizens against the intelligence agencies: "If a president received what he thought was reliable information that a suitcase nuclear bomb had been hidden away in an American city, our constitutional provisions with respect to search and seizure, and wiretapping, and all of our freedoms would go out the window. Both authorities and citizens would turn that city upside down trying to locate such a device."[40]

Following America's unhappy experience with Operation CHAOS and other transgressions, however, one would hope that a return to extreme intelligence measures would be preceded by debate (behind closed doors if necessary), meaningful prior consultation with Congress, the issuance of the appropriate written orders and warrants, and then careful monitoring to help assure fealty to the new boundaries—unless, as in the case of, say, the threat of nuclear terrorism, the president had no time whatsoever to consult and follow the letter of every procedure. Everyone recognizes, in the popular law-school phrase, that the Constitution is not a suicide pact. But for 99 percent of the cases, prior debate, consultation, established procedures, and close monitoring can be honored. Indeed this is the democratic way. It will work, though, only if the media, the Congress, and the people refuse to allow—even when society is under pressure—a return to the pre-1975 era of permissiveness when intelligence policy lay in the hands of a few figures hidden in the shadows of government.

The Church committee documented the relationship between the attitudes of intelligence officers toward their mission and the attitudes prevalent in the wider society. FBI official William C. Sullivan told the committee how contemporary intelligence officials had grown up "topsy-turvy" during World War II, when sensitivity to the law was secondary to defeating the Nazi war machine.[41] President Franklin Roosevelt himself stretched constitutional restraints to the breaking point, forcing Japanese-Americans into camplike prisons and ignoring congressional prerogatives in the early stages of the war. Obviously this rules-be-damned approach spilled over, for some, into the subsequent cold war with the Soviet Union.

Of overmastering importance, then, is the necessity to remind (or freshly educate) intelligence officers—and the broader public to whom they must answer—of the central role oversight and accountability must play in a democracy. A significant number of professional intelligence officers still seem to reject the idea of legislative oversight, seeing it more as a hinderance to America's war against hostile foreign threats than a safeguard against the abuse of power at home. "Do you know why we all enjoyed reading [the 1984 Tom Clancy novel] *The Hunt for Red October*?" asked an intelligence officer of a visitor at a CIA seminar for senior intelligence managers held at Headquarters recently. His answer: "Because in the book a senator and his aide are the villains and their culpability is uncovered by the Agency!" While said in partial jest, the point was made during a discussion of legislative oversight and was meant to convey the group's sense that legislators were a burden to their mission.

At another Agency seminar for senior officials, in 1987, a young officer displayed a brooding hostility toward a former legislative overseer participating in the session. The officer attributed leaks over recent CA operations, notably in Afghanistan, to the involvement of Congress in the oversight process and was bitter about the danger that leaks presented to case officers and their assets in the field. Concern for leaks and the lives of intelligence personnel is warranted—and shared by legislators and their staffs, who take such matters as seriously as officials in the executive branch (widely recognized as the source of most leaks); what this officer failed to appreciate is the importance of checks and balances in a democracy. Only when U.S. intelligence officials accept within their own minds the legitimacy of executive and legislative supervision will the many safeguards against abuse truly work with reliability.

Officials in the CIA often find this Madisonian principle difficult to accept. They are driven each day by the vital mission to identify, describe, and help thwart external threats to the United States. "Was it possible to lose the nation and yet preserve the Constitution?" asked Lincoln, and the question resonates well in the halls of the CIA today. Yet if the peril to the nation is less clear and immediate than in Lincoln's day, is it appropriate to disregard the Constitution and the nation's laws?

Part of the dilemma for the intelligence officer resides in the changing nature of American foreign policy since the unambivalent confrontations with the Soviets during the Truman-Eisenhower-Kennedy years. Until recently a foreign-policy consensus existed in the United States: the nation was in, if not a zero-sum conflict with the USSR, then at least a dangerous competition. Détente, however, illustrated the possibilities for tempering in some degree the steady enmity between the superpowers which has been the hallmark of the cold war. Ransom offers a persuasive hypothesis linking threat and accountability: "The greater the hostility between the United States and Soviet Union and the greater the consensus about the security threat, the less the public demand for accountability for secret intelligence agencies."[42] As the consensus over the urgency of the Soviet threat diminishes (at least compared to earlier periods of the cold war), the new public mood seems to place a greater emphasis on intelligence accountability.

Hence, just as legislators represent an aggravation to intelligence professionals, so, too, does the present uncertainty about American foreign-policy objectives. Life was once much simpler, with no congressional oversight to worry about and a monolithic, aggressive enemy to combat as the executive branch saw fit, few holds barred. Today, the modern intelligence officer must continue to provide the best warnings possible to protect the nation in a world that is more complicated and risky than in 1947, but must maintain as well a heightened appreciation and respect for the democratic process—from regulations in the executive branch to the advice of representatives in Congress. This is the difficult challenge confronting the U.S. secret service. Can it honor democracy while protecting it from antidemocratic forces beyond these shores? Will the CIA live up to this twin obligation? The American people cannot afford for it to fail.

Intelligence and Accountability

"We must strive to assure the people that their intelligence agencies will not be turned against them," said Griffin Bell in 1979, in the first address to CIA employees ever given by an attorney general.[43] This assurance depends upon redoubled efforts, at all the control points, to guard against abuse. The motivations of the men and women in key positions—their judgment, their respect for the law, their honesty—will continue to be of central importance. The recruitment of good people to positions of responsibility, in the CIA and throughout the government, has been and always will be a sine qua non for successful democracy.

In addition, members of Congress must continue to have and, as the Iran-contra affair has shown, to improve access to information about CIA operations through full and advanced briefings in all but the most extraordinary circumstances—and, even then, with at least a warning to key legislators (if not the Gang of Eight, then the Gang of Four: the top party leaders in both chambers) that a finding has been signed and a full report will have to be delayed for a short period.

The former staff director of the Church committee, William G. Miller, has emphasized recently the importance of the provision for *prior* notice in the 1980 Intelligence Oversight Act.[44] This law also uses the phrase "in a timely fashion" as a reporting deadline, which seems to open the door to CIA lawyers for *ex post facto* briefings to Congress; but Miller stresses that this phrase was meant to apply only narrowly to the most exceptional cases.

While an absolute insistence on prior reporting—without exception—has the virtue of clarity, critics maintain that it flies in the face of real-world complications. The reader will recall from the previous chapter the example of the six American diplomats in Iran, rescued with the indispensable help of the Canadian embassy in Teheran. The Canadians were concerned about the security of the operation and demanded that the number of individuals who knew the details—the "witting circle," in intelligence parlance—be kept sharply limited (only about ten American officials were allowed in this "compartment" of information), or else they would not participate. Given this choice, President Carter elected to cooperate with the Canadians.[45]

This centimeter of space allowed to presidents could obviously be misused. To avoid mischief, Lloyd Cutler, who served as President Carter's legal adviser, suggests that a president might simply inform the intelligence committees straightaway, "I have commenced an action" [i.e., signed a finding] and tell them that, because of an extraordinary circumstance, a full reporting will have to be delayed.[46] Under this procedure, Congress would at least know that an operation was underway and that it could expect an accounting soon. Morton H. Halperin, a national security expert, would go further, requiring a slightly more detailed initial report to Congress, yet not so detailed as to jeopardize security, for example: "I have made a finding to help release hostages." This brief description would then be followed by a full accounting to the intelligence committees after the operation.[47] Halperin also advocates for most covert actions a time lag of at least a few days between the time of the finding and its implementation; this would allow the intelligence committees an opportunity to digest the proposal before offering an evaluation to the president and the DCI.[48]

Sen. Arlen Specter (R, Pennsylvania), among others, has reluctantly endorsed a forty-eight hour reporting requirement, pointing out (as noted in chap. 6) that prior notice simply has not worked in the past. He encourages prior notice, but as an "extra safeguard" he would permit an extra two days for reporting in extraordinary circumstances. A forty-eight-hour rule, he argues, satisfies a standard of clarity: "Everyone can understand it"—the same logic that led drafters of the 1988 Intelligence Oversight bill to insert the two-day provision in their legislation (see chap. 10).[49]

Beyond clarifying the reporting expectations, Congress must further insist on regular hearings, random program audits, and overseas inspections of CIA operations (without examining fine operational details that might jeopardize sources and methods); written, prior approval by the president for all important covert actions (with criminal penalties for those who disobey[50]); and an end to the privatization of intelligence operations and the establishment of secret funds outside the appropriations process—both of which have the effect of removing intelligence policy from proper constitutional checks and balances (see chap. 12). The Reagan administration's new National Security Decision Directive on Special Activities (NSDD 286), issued in the wake of the Iran-contra revelations, attempted to address some of these problems, prohibiting oral and retroactive findings. Yet, one should recall that its reporting provision allows a finding to be kept from legislators indefinitely, depending upon the whim of NSC officials.[51]

Urgently needed, too, are better criteria for determining whether the CIA is operating within the bounds of acceptability. This is a difficult assignment. What is *acceptable* covert action? Domestic surveillance? Intelligence failure? CIA relations with the media, academe, and other groups? Legislators, professional associations, scholars, and others interested in intelligence policy must attempt greater definition—short of pursuing suffocating detail (micromanagement) that proscribes reasonable CIA flexibility and initiative—if the CIA is to have proper benchmarks. Existing executive orders and laws, as well as guidelines established by nongovernmental organizations, have helped in this regard, but many unnecessary ambiguities remain.

Foremost among them are the limits to CIA (and FBI) use of surveillance techniques at home. President Reagan's omnibus executive order on intelligence, issued early in his administration, seems to permit investigations of domestic groups, for example, under the guise that they may be tools of foreign powers or possess information that could be relevant to U.S. interests abroad. This authority, though, remains sealed in secret internal regulations; the exact standards for the use of such techniques remains unknown—a situation all too reminiscent of the Huston Plan.[52] In a democracy, one would expect the criteria to be legislated, not signed secretly within the executive branch and implemented without debate or review. Warrantless physical searches and the surveillance of Americans abroad are other examples of operations whose guidelines still need to be codified into law.

All are necessary: dedicated individuals aware of the necessary balance between security and civil liberty; a steady flow of information to the congressional intelligence committees on all important operations; renewed efforts to define societal expectations more precisely through executive orders, laws, and other guideposts; and, always so vital, a spirit of cooperation ("good faith," in the words of the Inouye-Hamilton investigative committees) between intelligence leaders in the executive and legislative branches. With this mixture—a challenging but reachable goal—the United States can have both democracy and a CIA.

TWELVE

Pathways to Reform

"If it ain't broke, don't fix it" states a sensible adage popular in the south. Most of the CIA ain't broke and don't need fixen. The Agency has served the nation well, from high competence in the monitoring of Soviet weapons developments to sound appraisals of global economic trends. Sometimes its officers have risked, and lost, their lives while carrying out duties in hostile regions of the world. Yet like every organization, the CIA has also erred—grievously so in its periodic violations of American law. Regardless of how good most of the Agency's work has been, the use of its secret power against U.S. citizens or in unauthorized ways against foreign nations is intolerable in a democracy. The Iran-contra affair has further reminded Americans that the efforts initiated in the 1970s to establish greater accountability over the intelligence agencies remain flawed. In light of the abuses and mistakes the CIA has made, improvements in its methods of conduct and oversight are warranted, despite the Agency's overall record of admirable achievement—especially given the grave risks to democracy that attend its secret misuse of power.

One must be cautious, however, about recommendations for policy reform. Though CIA officials are too quick to cry "micromanagement!" at the sight of any attempt whatsoever to improve the supervision of intelligence, excessive control can indeed so encumber a government agency that—like a ship weighted down with barnacles—steerage becomes sluggish and inefficient, even dangerous. This condition must be avoided if the United States is to maintain a reliable intelligence capability in a world of imposing threats.

Employing a more terrestrial metaphor, a former senior intelligence official once declared—in anguish over the new oversight and its possible retarding effects on CIA prowess: "The intelligence agencies need *horsepower!*"[1] Certainly

TABLE 12.1. Primary Influences Contributing to the Sins of Strategic Intelligence

Sins of Strategic Intelligence	*Primary Influences*		
	Psychological	*Bureaucratic*	*Historical*
1. Biased intelligence	X		
2. Disregarded intelligence	X		
3. Indiscriminate covert action			X
4. Inadequate protection		X	
5. Misused counterintelligence		X	
6. Inadequate accountability		X	
7. Indiscriminate collection		X	

the United States requires a CIA with adequate horsepower, but a vehicle with great horsepower and no brakes can be as perilous as one with no horsepower at all. The objective should be a good combination of power and restraint. The brief agenda for reform that follows has this ideal in mind.

The Seven Sins Revisited

Early in these reflections, I introduced seven criticisms (or "sins") of strategic intelligence (see the summary in table 12.1). Each is affected by the bureaucratic behavior of men and women in office, their personalities, and their interpretation of history. Combinations of these influences (and others) are often so intertwined that it becomes virtually impossible to separate one from another. The failure of DCI Richard Helms to provide uninhibited intelligence on occasion (see chap. 3) appears to have been, for example, a function of his own cautious personality and years of bureaucratic accommodation. Similarly, a disregard of objective intelligence by policymakers may be attributable to bureaucratic forces at play in the White House ("Nothing permeates the Cabinet Room more strongly than the smell of hierarchy. . . ." observes a recent study[2]), along with wishful thinking by the president himself, or, especially for ideologues, a rigid interpretation of history blocking out information that fails to reinforce an existing worldview.

One is left with what gives the social sciences their fascination, and their frustration: a rich, complex tapestry of interwoven variables. While the figure in the carpet may remain elusive, one can at least point to some conspicuous features. With respect to the seven sins explored in this book, the first two—biased and disregarded intelligence—could be ameliorated to some extent by paying better attention to the personalities of individuals under consideration for key intelligence and policy posts. Surely a necessary condition for coherent and effective decision making is the ability to speak and hear the truth—a question of character, without which the most finely tuned laws, sanctions, and organizational reforms will be to little avail.

Intelligence scholar Richard K. Betts emphasizes the virtue of nagging—dissenting and lobbying to be heard—on the part of the analyst; and being skeptical and questioning on the part of the policymaker: all in all, as healthy a debate between experts and decision makers as time will allow.[3] Yet extensive research over the years reveals that this process of "objective appraisal" is anathema to some personality types.[4] To a considerable extent, then, the ancient problem of judicious leadership selection and personnel recruitment lies at the center of improvements in intelligence policy—especially the careful screening out of ideologues.

William J. Casey provides an illustration. As DCI under President Reagan, Casey established an excellent reputation for improving some aspects of the National Intelligence Estimate (NIE) process (discussed in chap. 5). The NIEs went from about twelve a year in the last throes of the Carter administration to over one hundred after a year under Casey; moreover, their focus was more precisely oriented toward the topical interests of policymakers.[5] Further, compared to any of his predecessors, Casey enjoyed unparalleled access to the Oval Office—a much valued attribute for an Agency that has sometimes been almost completely ignored by presidents (Nixon for one).

Yet, according to many observers, Casey brought far too much political and ideological baggage to this sensitive post. As a close friend to the president and his former national campaign manager, he sometimes—certainly not always—seemed to be more attuned to pursuing the administration's political fortunes and his own cold-war ideology (widely shared in the White House), than to gathering the purest form of objective analysis the intelligence community could muster. "On Mexico," recalls one senior intelligence official, "Casey had such a fixed idea that the country was about to collapse that he turned the Agency upside down to find the bad news."[6]

The chairman of the House Intelligence Committee, Lee Hamilton, has also complained that, under Casey, intelligence was "often selectively used to influence public debate over policy."[7] His examples included CIA criticism of the Marxist regime in Nicaragua (the Sandinistas) for its violations of human rights and its use of propaganda against the United States, without comparable criticism of the excesses perpetrated by the Reagan-supported contras; "incontrovertible" charges of terrorism against Libya, without solid evidence; and claims that a "flood" of arms flowed from the Sandinistas to the Marxist rebels of El Salvador, again without persuasive documentation. In addition, Casey's behavior during the Iran-contra affair seems to have gone far beyond the bounds of prudence, not to mention the law.

Casey's attention to NIEs and his large welcome mat at the White House were properly valued, but they have been overshadowed by his apparent willingness to draw the CIA—still in trauma from the "Year of Intelligence"—once more into the realm of "political intelligence" and improper operations. A person like Casey with close partisan ties to the White House is, in a word, unlikely to be the best choice to head an intelligence community whose primary mission is to provide the unvarnished truth—as hard as that task is without the insertion of political and ideological filters.

Even with greater sensitivity to personnel selection, however, the United States will continue to fall short of an intelligence and decision process of crystalline perfection. Structural "sluggishness" will remain a problem, and so will miscalculations about world events—as long as human beings continue to be fallible.[8] As a consequence, innovative suggestions about how to make America's governmental institutions work more effectively ought to remain a primary concentration for political scientists.[9] But even with the necessary structural improvements—an important one would be to increase the DCI's authority over the far-flung agencies within his "community"—the nation's first requirement for good intelligence policy will always be individuals in key positions who are capable of a dispassionate assessment of information, individuals who will ask difficult, probing questions and solicit (as well as appreciate) hard facts, as jarring as they might be: a CIA director with an unalloyed affinity for the truth; a president committed to a search for information, broad and deep, and an unyielding allegiance to the law; lower-level Agency analysts with the nerve to report unpopular findings at CIA, interagency, and NSC meetings; a White House willing to hear from "lowly" analysts recently immersed in the day-to-day cable traffic from Patagonia. Even then mistakes will be made, but they will be fewer.[10]

As for the indiscriminate use of covert action, it seems that this sin is driven primarily by the prevailing interpretation of Soviet-American relations in essentially zero-sum terms. The adherence of the United States to an anti-Marxist doctrine has been so strong that it has led policymakers to intervene almost everywhere for fear a local revolution might lead to a Soviet military or ideological foothold. While certainly contributing to this covert intervention, bureaucrats have largely been servants to cold-war ideology. Yet by most accounts the usefulness of covert action has been modest at best. "Covert action is a cop-out," concludes former secretary of state Alexander Haig, complaining about the tendency of the United States to engage in secret foreign policy rather than to confront problems head on and decide as a nation which direction to take.[11] While acknowledging that occasionally minor covert actions can be useful as supplements to broad policy initiatives, former DCI Turner, looking back on the Carter years, concludes that "on the whole, major covert actions were not terribly successful. . . . Only two worked."[12]

Henry Kissinger, secretary of state during the Nixon and Ford years, is a supporter of covert action as one instrument of American foreign policy, but he too remains wary about its misuse and overuse. During the Church committee investigation, he expressed his opposition to large-scale PM operations like the kind pursued by the CIA in Laos during the 1960s.[13] He found it ill-advised to use covert action for what was essentially a "jet-fighter war." If it became impossible to keep a CIA PM operation strictly limited in scope, the United States would be better off to declare its intentions and openly arm its side. He altogether opposed the use of the military for PM operations ("special operations" or "special ops"; see chap. 2). Whenever covert actions were employed, Kissinger recommended that they be closely tied to a diplomatic "track" [objective]. For America's other most celebrated statesman, George F. Kennan—a strong supporter of covert action in 1947–48—the quiet option had become more harmful than helpful. "We will have more respect in the world by being honest," he told the

Church committee, "than by following the fancy [covert] tricks of the Soviet Union."[14]

According to one experienced hand, the "most valuable covert actions over the years have been political and psychological [propaganda]," including financial assistance to friendly politicians overseas and Western literature smuggled into Soviet society.[15] Others point to (among others) successes in the Philippines against Huk rebels, support to Portugese moderates during the Ford administration, and, most recently, help for the Afghan resistance.[16] The rising CA budget during the Reagan years (see chap. 6) attests to the broad support that covert action continues to enjoy in the government. Indeed, as one savvy former U.S. senator has stated, once in the White House, "*every* president—regardless of his party or his ideology, or what he may have said in criticism of covert action during his presidential campaign—will turn to covert action to one degree or another."[17] Nevertheless, the observations of thoughtful and experienced men like Kissinger and Kennan, coupled with the excesses of the Iran-contra experience, suggest that the "quiet option" deserves closer scrutiny and a more discriminating use in the future.

The other sins—inadequate protection of intelligence officers and assets abroad, misused and mismanaged counterintelligence, inadequate accountability, and too much intelligence collection—appear to be matters susceptible in part to organizational adjustments. At least the government can require better security abroad, improved planning for the relocation of endangered assets, more centralized counterintelligence (and no counterintelligence at all directed against American citizens without strict adherence to proper judicial warrants), and closer supervision of intelligence activities from the lofty perches of the NSC staff down to the grassroots of the infrastructure. Extensive collection of intelligence may be less sinful than suspected—except when it focuses on the USSR at the expense of other global perils. As a former CIA analyst notes, what might once have appeared as "obscure research" has often "later proved useful when events broke in parts of the world where activity was unexpected."[18]

Again, though, even more important than organizational improvements will be the quality and attitudes of officials in the executive and legislative branches. With respect to the analytic side of intelligence, a former CIA analyst asks and appropriately answers: "Can one legislate integrity, honesty and professionalism? The strength of U.S. intelligence production mechanisms lies in the quality of its people, not in the bureaucracy in which they work."[19] This conclusion applies to all aspects of intelligence policy and is to be read between the lines of each recommendation outlined below.

A Checklist of Prescriptions

Intelligence Collection

1. More contact should take place between the intelligence producer and consumer. An important way to improve the CIA's primary mission—the production of useful information for policymakers—is to increase the opportunities for

analysts and decision makers to interact. This means that both groups should take the time to engage in face-to-face discussions about intelligence needs and possibilities. The danger here is that the analyst might become caught up in the political requirements of the policymaker, but well-trained, professional analysts should be able to resist the temptation of "political intelligence."

2. Intelligence officials—the managers, the analysts, the case officers—must be carefully recruited to avoid political and ideological zealots.

3. More open dialogue is necessary between intelligence officials (especially analysts) and thoughtful outside experts—that is, more "A team, B team" exercises (though with a better balance of views, not just a ploy to counter politically unpopular internal estimates with skewed outside sources), more visiting scholars at the CIA, more participation by CIA people at scholarly conferences, and the like, all designed to open up the Agency to views outside the Washington Beltway.

4. Foreign language training at the CIA must be further improved, and analysts should make more frequent trips abroad to the countries (if travel is possible there) they follow from their desks at Headquarters.

5. Less fascination with intelligence hardware and more emphasis on HUMINT is in order, as well as a greater concentration on global problems other than the Soviet threat (though the USSR, with its potent military capabilities, must obviously continue to be a main focus).

Covert Action

6. The United States should avoid large-scale PM operations, and resort to smaller ones only after diplomacy by itself has failed, and only after an exhaustive calculation of the risks to the United States and to the recipients of American weapons. To what extent did the United States ultimately help the Kurds, the Meo tribesmen, and so many others who have received PM assistance in the past?

7. The United States should lessen its reliance on all forms of covert action, so that the quiet option does not again become the tail that wags the foreign-policy dog, as sometimes seemed to be the case during the 1960s. Some covert actions, like the destruction of crops, the use of biological substances to cause epidemics, and assassinations should never be instruments of America's secret foreign policy. Some covert, truthful propaganda might continue to be useful in countries that otherwise would have exposure to only a perverted view of the United States, and some political covert actions may be helpful when a democratic friend is in danger. President Corazon C. Aquino of the Philippines is an example of an individual the United States might well help quietly and honorably, although here, as in most cases, the best assistance would be an open and united encouragement from all democratic nations, with spirited public rhetoric channeled overtly to the Philippines along with money and arms through institutions dedicated to the advancement of democracy around the world—far more than President Aquino has been provided in her time of continuing danger.

8. Strict limits must be placed on the privatization of covert action (and all other intelligence activities). The existing intelligence oversight statutes must be

honored, not bypassed through unconstitutional fundraising procedures outside the view of legislative overseers. Covert actions should be reported to Congress (or, in times of extreme emergency, to its top leaders) in advance of implementation. Moreover, in most cases, covert action must be funded by Congress. If some other nation (or a business corporation) wishes to pay for U.S. foreign-policy objectives, the Congress—or its surrogates, the House and the Senate intelligence committees—may wish to consider this generosity (and the inevitable strings attached); but this approach must not take place without consultation with the Congress, or else U.S. intelligence will move outside a constitutional framework. This does not mean that private U.S. citizens could not be employed for CA operations, say, for a PM flying mission; it does mean, however, that the covert action would have to be approved and reported according to existing statutory guidelines.

9. Throughout the process of approving and reporting covert action, overseers must check more closely for adherence to extant guidelines and must insist on greater care in selecting and supervising personnel, especially at the infrastructure level (though as the Iran-contra affair reminded us, at no level can the integrity of individuals be taken for granted).

Counterintelligence

10. The CIA must not conduct counterintelligence against any U.S. person, other than suspected spies within its own organization and—after obtaining the appropriate warrants—outside individuals who have demonstrated beyond reasonable doubt that they are in the employment of foreign intelligence agencies or other hostile entities. Even then, the CIA's role (beyond protecting its own facilities within the United States) lies overseas, and espionage cases inside the United States rightfully become the responsibility of the FBI.

11. Counterintelligence leadership at the CIA must be more centralized (short of the Angleton excesses), and CI training throughout the Agency and with its private contractors must be given higher priority.

12. The counterintelligence chief—and all high-ranking CIA officials—should have a fixed limit of service (say, ten years maximum, the current provision for the FBI director) to avoid the empire building and ossification that came into the CI division under Angleton (for all his good work in many instances of ferreting out dangers to CIA and U.S. security). On behalf of a six- or seven-year limitation for DCI, Richard Helms told the Church committee on January 30, 1976: "[M]en get tired of these kinds of jobs. They are very demanding and they're twenty-four hours a day and you're never without a telephone, either beside your bed or in your car . . . [A]fter six or seven years, you've had a whack at the problems, you've had a chance to make your mark for good or bad and you're beginning to get tired, and I must say also there's a little tendency to get more sure of yourself. So I see no problem [to] this limitation; as a matter of fact, I think it might be constructive."

13. The quality of protection and cover for CIA officers and assets abroad—a CI responsibility—requires much more attention.

14. The CIA must resist improper CI operations pushed upon it by the White House, like the Huston Plan, just as the Agency more successfully rejected requests to assist in the Watergate cover-up in 1973–74.

Organization

15. The intelligence community cries out for better management. "It is a loose confederation of equals with no real leader," observed a chairman of the House Intelligence Committee.[20] The authority of the DCI must be upgraded; this does not mean, however, that he or she should be made a member of the cabinet, which holds the risk of turning the DCI into a political figure. A key place to begin would be to give the DCI increased authority over tasking for strategic collection, increased budget authority for the entire community, and statutory membership on the NSC. The DDCIA's in-house authority should be elevated at the same time, within limits, in order to free the DCI from day-to-day CIA management details—though not to the point of emasculating the DCI's stature within the Agency, his key base of support within the government.

16. According to a DDCIA, "only four to five out of 160 U.S. embassies around the world are supportive of the CIA mission."[21] If accurate, this situation is unacceptable. The DCI and the secretary of state must do more to improve the working cooperation of ambassadors, foreign service officers, and case officers abroad.

Relations with U.S. Groups

17. The CIA should not engage U.S. scholars and media personnel in covert operations. It is primarily up to these groups (and others, like missionary organizations, who may wish to be excluded from CIA activities) to protect themselves through strong professional guidelines and penalties for those who violate them. The congressional intelligence committees can help by insisting that the expressed views of these citizens be honored by the intelligence community.

18. Covert recruitment and secret research should be banished from the nation's campuses, for openness lies at the heart of the university's purpose and reason for existence.

Oversight

19. The two congressional intelligence committees are apt to provide more comprehensive oversight than one joint committee, so both should be retained. The two committees, however, ought to engage in more joint hearings and discussions now and then. Up until now their members have met together infrequently, but the occasional sharing of witnesses, documents, ideas, and frustrations could be useful for both chambers.

20. The intelligence committees need to use surprise "spot" audits to double-check the CIA's allegiance to existing procedures of accountability.

21. Committee members need to spend more time in discussions, mostly informal, with intelligence officials at all levels (including personnel overseas)—at

CIA Headquarters, over lunch, in their offices. This is what it takes to understand an Agency's problems and to educate the CIA on the importance of the legislative viewpoint.

22. Committee members need to spend more time preparing for oversight hearings. The tough questions, with follow-ups, are too seldom asked by members; nor is attendance at these sessions as good as it should be. Indeed, getting the Congress to concentrate on anything for very long is difficult. "The House is pure chaos," observes a senior staffer on the House Intelligence Committee. "Getting members to conduct Committee business is like herding chickens."[22] If the public were to express a greater appreciation for the importance of oversight, legislators might be inclined to take this part of their job more seriously.

23. On the other side of the coin, intelligence officials could be much more forthcoming. Their failure to be straightforward with the intelligence committees has been a primary source of conflict over intelligence policy. William Casey's "briefing" on the Nicaraguan mining, consisting of twenty-seven words, was hardly a model of candor, not to mention his retreat to the technicality that "piers"—and therefore not "harbors"—had been mined. A central problem for legislative overseers is revealed in an exchange that occurred in 1972 between the chairman of the Senate Foreign Relations Committee, J. William Fulbright (D, Arkansas) and the U.S. ambassador to Laos during the war in Indochina:

> FULBRIGHT: Why has the administration not informed Congress of these military operations in Laos?
>
> AMBASSADOR: We felt we had Congress' tacit approval . . . and because we had not been asked any *direct* questions about U.S. air operations in northern Laos, we naturally concluded . . .
>
> FULBRIGHT: There is no way for us to ask you questions about things we don't know you are doing, Ambassador.[23]

As the Inouye-Hamilton panels concluded after their investigation into the Iran-contra affair, "a spirit of good faith with the Congress" was badly needed among the officials of the intelligence community.[24]

24. The oversight role of the President's Foreign Intelligence Advisory Board (PFIAB) and the Intelligence Oversight Board (IOB) needs to be upgraded.

Laws and Executive Regulations

25. Ideally, the CIA's original charter—left purposefully vague in 1947 because the United States at that time remained inexperienced and hesitant about the proper use of secret agencies in peacetime—should be rewritten based on what we now know over forty years later. The attempts to write a Grand Intelligence Charter in 1978 failed, bogged down by excessive detail. A new effort would have to aim for a degree of specificity halfway between the 1947 and the proposed 1978 charters. If officials lack the will to redraft the National Security Act of 1947 in a fundamental way, then at least they should support some corrections in existing intelligence law. The Intelligence Oversight Act proposed in 1988 represented a commendable step in this direction and warrants passage—

with the exception of its provision that abandons prior notice for all covert actions.

The Supreme Court once noted that good law "has a generality and an adaptability comparable to that found to be desirable in Constitutional provisions. It does not go into detailed definition which might either work injury to legitimate enterprises or through particularization defeat its purposes by providing loopholes for escape."[25] Within these sensible limits, the current statutes should be amended to emphasize the need for prior notice (at least to the Gang of Eight or Four and, if necessary in rare cases, in broad language to be elaborated upon later when the operation is less sensitive), and to tighten congressional controls over any privatization of covert action or other intelligence activities.

26. The CI portions of the Reagan executive order on intelligence should be revised and clarified to strengthen the protection of U.S. persons against unwarranted (in the literal and figurative sense) intelligence activities; then the improved language should be enacted into law.

27. Clear sanctions against intelligence abuses need to be adopted by Congress in order to instil a greater sense of respect for the nation's laws among those few intelligence officials who, once again, might be tempted to snub their legislative overseers. The Iran-contra plotters, for example, faced no direct penalties in the Boland amendments. To remedy this, the legislative language proposed by Sen. Arlen Specter (R, Pennsylvania) is on the mark:

> Whoever being an officer or employee of the United States, in any matter within the jurisdiction of the Senate or the House of Representatives of the United States, or any committee or subcommittee thereof, knowingly and willfully falsifies, conceals, or covers up by any trick, scheme, or device a material fact, or makes any false, fictitious or fraudulent statement or misrepresentation, or makes or uses any false writing or document knowing the same to contain any false, fictitious, or fraudulent statement of entry shall be imprisoned for not less than one year nor more than five years and may be fined not more than $10,000. . . .[26]

High officials in the Reagan administration apparently failed to understand the importance of obeying the law—even after the embarrassing lessons of the Iran-contra scandal. When DCI William H. Webster reprimanded a senior CIA official for lying to the intelligence committees in their preliminary inquiries into the Iran-contra events, Elliott Abrams, the assistant secretary of state for Inter-American Affairs, opined that this treatment of a CIA official "would send exactly the wrong signal to young officers at the Agency."[27] Abrams had gotten it 180 degrees backward. A reprimand for lying was just the right signal. The wrong signal was his own lie to the same panels[28] and the Reagan administration's failure to dismiss him from this important government post for unprofessional conduct.

Attitudes

28. As the Abrams example illustrates, more important than any of the proposals offered above is the need for a better attitude among government officials

toward the relationship between intelligence and democracy. "It's the *tone* you set in office," emphasizes former DCI Turner, critical of the permissiveness allowed during the Reagan administration.[29] "An atmosphere was created," says former secretary of state Haig, offering an explanation as to how the Iran-contra excesses could have occurred.[30] Thomas Jefferson once argued that, on rare occasions, laws might have to be disobeyed if the safety of the nation required it.[31] Short of a extraordinary calamity, though, widespread allegiance to the law and its protections remains the greatest single strength of American democracy. If the United States is to remain free, the laws must be obeyed. When intelligence officials and their policy bosses lose sight of this, they have lost sight of what counts the most.

Most intelligence officers understand this. Discussions of intelligence reform should keep in mind that the people who work for the CIA and the other intelligence agencies are among the best in the government. With few exceptions, they are well-educated, thoughtful, and dedicated. Mistakes have been made; but, almost always, the CIA has made them at the direction of elected political officials and their aides. The men and women of the CIA have served their country with distinction most of the time—and sometimes in dangerous settings. As the nation continues to define their role in this democracy, an eighth sin of strategic intelligence would be forget this fact.

Yet, as Madison, Lord Acton, and many others have reminded us over the years, human beings are not angels. Power is ever liable to abuse, and so the constant need to stand guard. In the Japanese Police Academy, officers are taught floral arranging to develop sensitivity, as well as judo and fencing for strength and discipline. How well this approach works is unclear, but the spirit is commendable. American intelligence officers (some of them at least) also need to become more sensitive to the value—and the fragility—of democracy, through education, edict, and example.

This checklist is by no means exhaustive; other proposals (and the justification for the ones listed above) appear throughout this volume. The list, though, does suggest some pathways for the CIA that would lead toward a better balance between intelligence and democracy. The United States is in an enviable position. It has the strength and stature of a world leader. In the spirit of the nation's founders, its first obligation in this capacity is, as George F. Kennan has observed, "to set a good example. . . . to be a teacher for all of humanity." For Kennan, the United States is "no place for secret [covert-action] operations . . . it doesn't fit our national character."[32] Few officials seem prepared to accept this view, as wise as the advice may be, but surely the United States can reject the past practices of plotting assassinations, mining harbors, relying on assets from the utter dregs of society, bribing foreign politicians, and discrediting (or worse) every person or nation that criticizes American policy.

"I do not think we have any call to appoint ourselves as God's avenging angels," Speaker of the House Jim Wright once said, "and to reform by force governments with whose policies we disagree."[33] If the leaders of the United

States will only remember this, and if they can remember as well the essential purpose of the CIA—to gather and analyze information from abroad, not to spy on Americans—the union of democracy and intelligence will endure and even prosper.

Appendix

The 1974 Hughes–Ryan Act

Also known as Section 662 (a) of the Foreign Assistance Act of 1974, the amendment states:

> "No funds appropriated under the authority of this or any other Act may be expended by or on behalf of the Central Intelligence Agency for operations in foreign countries other than activities intended solely for obtaining necessary intelligence unless and until the President finds that each such operation is important to the national security of the United States and reports, in a timely fashion, a description and scope of each operation to the appropriate committees of Congress."

The 1980 Intelligence Oversight Act

SEC. 407. (a) Section 662 of the Foreign Assistance Act of 1961 (22 U.S.C. 2422) is amended—

(1) by striking out "(a)" before "No funds";

(2) by striking out "and reports, in a timely fashion" and all that follows in subsection (a) and inserting in lieu thereof a period and the following: "Each such operation shall be considered a significant anticipated intelligence activity for the purpose of section 501 of the National Security Act of 1947."; and

(3) by striking out subsection (b).

(b)(1) The National Security Act of 1947 (50 U.S.C. 401 et seq.) is amended by adding at the end thereof the following new title:

"*Title V—Accountability for Intelligence Activities*

"CONGRESSIONAL OVERSIGHT

"SEC. 501. (a) To the extent consistent with all applicable authorities and duties, including those conferred by the Constitution upon the executive and legislative branches of the Government, and to the extent consistent with due regard for the protection from unauthorized disclosure of classified information and information relating to intelligence sources and methods, the Director of Central Intelligence and the heads of all departments, agencies, and other entities of the United States involved in intelligence activities shall—

"(1) keep the Select Committee on Intelligence of the Senate and the Permanent Select Committee on Intelligence of the House of Representatives (hereinafter in this section referred to as the 'intelligence committees') fully and currently informed of all intelligence activities which are the responsibility of, are engaged in by, or are carried out for or on behalf of, any department, agency, or entity of the United States, including any significant anticipated intelligence activity, except that (A) the foregoing provision shall not require approval of the intelligence committees as a condition precedent to the initiation of any such anticipated intelligence activity, and (B) if the President determines it is essential to limit prior notice to meet extraordinary circumstances affecting vital interests of the United States, such notice shall be limited to the chairman and ranking minority members of the intelligence committees, the Speaker and minority leader of the House of Representatives, and the majority and minority leaders of the Senate;

"(2) furnish any information or material concerning intelligence activities which is in the possession, custody, or control of any department, agency, or entity of the United States and which is requested by either of the intelligence committees in order to carry out its authorized responsibilities; and

"(3) report in a timely fashion to the intelligence committees any illegal intelligence activity or significant intelligence failure and any corrective action that has been taken or is planned to be taken in connection with such illegal activity or failure.

"(b) The President shall fully inform the intelligence committees in a timely fashion of intelligence operations in foreign countries, other than activities intended solely for obtaining necessary intelligence, for which prior notice was not given under subsection (a) and shall provide a statement of the reasons for not giving prior notice.

"(c) The President and the intelligence committees shall each establish such procedures as may be necessary to carry out the provisions of subsections (a) and (b).

"(d) The House of Representatives and the Senate, in consultation with the Director of Central Intelligence, shall each establish, by rule or resolution of such House, procedures to protect from unauthorized disclosure all classified information and all information relating to intelligence sources and methods furnished to the intelligence committees or to Members of the Congress under this section. In accordance with such procedures, each of the intelligence committees shall promptly call to the attention of its respective House, or to any appropriate committee or committees of its respective House, any matter relating to intelligence activities requiring the attention of such House or such committee or committees.

"(e) Nothing in this Act shall be construed as authority to withhold information from the intelligence committees on the grounds that providing the information to the intelligence committees would constitute the unauthorized disclosure of classified information or information relating to intelligence sources and methods."

(2) The table of contents at the beginning of such Act is amended by adding at the end thereof the following:

"TITLE V—ACCOUNTABILITY FOR INTELLIGENCE ACTIVITIES

"Sec. 501. Congressional oversight."

Directors, Central Intelligence Agency

1947–50	Rear Adm. Roscoe H. Hillenkoetter
1950–53	Gen. Walter Bedell Smith
1953–61	Allen W. Dulles
1961–65	John A. McCone
1965–66	Vice Adm. William F. Raborn, Jr.
1966–73	Richard Helms
1973	James R. Schlesinger
1973–76	William E. Colby
1976–77	George Bush
1977–81	Adm. Stansfield Turner
1981–87	William J. Casey
1987–	William H. Webster

Chairmen, Senate Select Committee on Intelligence

1976–77	Daniel K. Inouye, Democrat, Hawaii
1977–81	Birch Bayh, Democrat, Indiana
1981–85	Barry Goldwater, Republican, Arizona
1985–87	David Durenberger, Republican, Minnesota
1987–	David L. Boren, Democrat, Oklahoma

Chairmen, House Permanent Select Committee on Intelligence

1977–85	Edward P. Boland, Democrat, Massachusetts
1985–87	Lee H. Hamilton, Democrat, Indiana
1987–89	Louis Stokes, Democrat, Ohio
1989–	Anthony C. Beilenson, Democrat, California

Notes

Preface

1. For an account of this investigation, see Loch K. Johnson, *A Season of Inquiry: Congress and Intelligence,* 2d ed. (Chicago: Dorsey, 1988).

2. Informal remarks to staff aides on the Senate Select Committee to Study Governmental Operations with Respect to Intelligence Activities (hereafter the Church committee, named for its chairman, Sen. Frank Church [D, Idaho]), Washington, D.C., September 1975.

3. Richard K. Betts, "Intelligence and Policy," statement to the House Subcommittee on International Security and Scientific Affairs, Committee on Foreign Affairs, Washington, D.C., January 28, 1980. More recently, another leading intelligence scholar, Glenn Hastedt, has noted: "Rather than having little or no information on which to make judgments about the place of intelligence in foreign policy making, there was [as a result of the flood of publications on intelligence since 1976] a growing abundance of information whose accuracy and relevance were not easily weighed." Paper presented at the annual meeting of the American Political Science Association, Washington, D.C., September 2, 1988.

4. See *Report of the Congressional Committees Investigating the Iran-Contra Affair,* Senate Select Committee on Secret Military Assistance to Iran and the Nicaraguan Opposition and House Select Committee to Investigate Covert Arms Transactions with Iran (the Inouye-Hamilton committees, chaired by Sen. Daniel Inouye [D, Hawaii] and Rep. Lee Hamilton [D, Indiana]), November 1987, S. Rept. No. 100–216 and H. Rept. No. 100–433 (hereafter cited as Inouye-Hamilton committees, *Report*).

5. Peter St. John, "Canada's Accession to the Allied Intelligence Community, 1940–45," *Conflict Quarterly* 4 (Fall 1984), 5.

Chapter 1

1. 50 U.S.C. (United States Code) 403 and 50 U.S.C. 403 a–p.

2. See the almost daily reports from December 22 through 31. For the phrase "massive

spying" (which critics maintained was a massive exaggeration on the reporter's behalf), see Seymour M. Hersh, "Underground for the C.I.A. in New York: An Ex-Agent Tells of Spying on Students," *New York Times,* December 29, 1974, p. 1.

3. Quoted in Seymour M. Hersh, "Hunt Tells of Early Work for a C.I.A. Domestic Unit," *New York Times,* December 31, 1974, p. 1.

4. For a detailed examination of the Senate investigation, see Johnson, *A Season of Inquiry.*

5. On these specific cases, see, respectively, Sol Stern, "NSA and the CIA," *Ramparts* 5 (March 1967), 29–38; Peter Wyden, *Bay of Pigs: The Untold Story* (New York: Simon and Schuster, 1979); William Colby and Peter Forbath, *Honorable Men: My Life in the CIA* (New York: Simon and Schuster, 1978); Drew Pearson and Jack Anderson, "House Lauds Rivers, Rejects Censure," *Washington Post,* March 3, 1967, p. D15; Rockefeller commission (Commission on CIA Activities within the United States), *Report to the President,* Washington, D.C.: Government Printing Office, June 1975, chap. 14 (hereafter cited as the Rockefeller Commission Report); and, generally, Thomas Powers, *The Man Who Kept the Secrets: Richard Helms and the CIA* (New York: Knopf, 1979).

6. Colby and Forbath, *Honorable Men,* 391.

7. The findings of the panels may be found in the following sources: the Rockefeller Commission Report; *Final Report,* Church committee, S. Rept. No. 94–755, 94th Cong., 2d Sess., April 23 and 26, 1976 (6 vols.: vol. 1, *Foreign and Military Intelligence;* vol. 2, *Intelligence Activities and the Rights of Americans;* vol. 3, *Supplementary Detailed Staff Reports on Intelligence Activities and the Rights of Americans;* vol. 4, *Supplementary Detailed Staff Reports on Foreign and Military Intelligence,* including Anne Karalekas, "History of the Central Intelligence Agency," 1–108; vol. 5, *The Investigation of the Assassination of President John F. Kennedy: Performance of the Intelligence Agencies;* vol. 6, *Supplementary Reports on Intelligence Activities,* including Harold C. Relyea, "The Evolution and Organization of the Federal Investigative Function: A Brief Overview, 1776–1975," 1–353 (hereafter cited as Church committee *Final Report* with volume, and page numbers); and "The CIA Report the President Doesn't Want You to Read," *The Village Voice,* February 16 and 23, 1976, pp. 69–92, an unauthorized publication (via a leak) of the Pike committee findings (hereafter cited as the Pike Committee *Report*). The *Village Voice* version of the Pike committee findings is, according to Otis Pike, accurate and almost complete (interview with Mr. Pike, Athens, Georgia, April 20, 1983; unless otherwise indicated, all interviews cited in notes were conducted by me). The Pike committee officially released its twenty recommendations at a press conference in Washington, D.C., on February 10, 1976, given extensive coverage in the next day's newspapers.

8. Charles Douglas Lummis, "The Radicalism of Democracy," *democracy* 2 (Fall 1982), 9–11; emphasis in original.

9. E. H. Carr, *The New Society* (London: Macmillan, 1951), 77.

10. Ibid., 79.

11. Arend Liphart, *Democracies* (New Haven, Conn.: Yale University Press, 1984), 8.

12. Alexander Hamilton (or possibly the writer was James Madison), from the New York Packet, February 5, 1788, *Federalist Paper* No. 51; rpt. in *The Federalist* (New York: Modern Library, 1937), 337.

13. Roger Hilsman, *To Move a Nation: The Politics of Foreign Policy in the Administration of John F. Kennedy* (New York: Dell, 1964), 7.

14. For an introduction to oversight as an analytic concept, see: Joel D. Aberbach, "Changes in Congressional Oversight," *American Behavioral Scientist* 22 (1979), 394–515; Morris S. Ogul, *Congress Oversees the Bureaucracy: Studies in Legislative Supervision* (Pittsburgh, Pa.: University of Pittsburgh Press, 1976): and Bert A. Rockman, "Executive-Legislative Relations and Legislative Oversight," *Legislative Studies Quarterly* 9 (August 1984), 387–440.

15. Title VI of the National Security Act of 1947 (P.L. 97–200; 50 U.S.C. 403).

16. In 1971 CIA Director Richard Helms advised that "the nation must to a degree take it on faith that we too are honorable men devoted to its service" (cited by Victor

Marchetti and John D. Marks, *The CIA and the Cult of Intelligence* [New York: Knopf, 1974], 205), a viewpoint almost uniformly accepted by Washington officialdom throughout the first era. Colby entitled his memoirs *Honorable Men.*

17. See, among many, the criticisms expressed by President Ronald Reagan, *New York Times,* September 28, 1984, p. 1.

18. John F. Blake, deputy director for administration, affidavit, *Nathan Gardels v. Central Intelligence Agency,* Civil Action No. 78-0330, U.S. District Court for the District of Columbia, June 7, 1978.

19. Cited by Marchetti and Marks, *The CIA,* 167.

20. From the Dolittle Report, quoted in Church committee *Final Report* 1:9.

21. Thomas Paine, *The American Crisis,* 1, December 23, 1776.

22. Edward Gibbon, *The Decline and Fall of the Roman Empire,* D. M. Low ed. (New York: Harcourt, Brace, 1960), 838.

Chapter 2

1. Lewis Carroll, *Alice's Adventures in Wonderland* (New York: Knopf, 1983), 138.

2. See Central Intelligence Agency, *Intelligence in the War of Independence* (Washington, D.C.: CIA, 1975). Also, William E. Colby's statement to the Church committee, reprinted in the *Congressional Record,* October 28, 1975, p. 33927; Church committee, *Final Report* 6:9–14.

3. For a useful lexicon of modern intelligence terms, see Church committee, *Final Report* 1:617–29.

4. Observations by Sen. John Tower (R, Texas), Church committee hearings, May 21, 1975, Washington, D.C., based on notes provided by the CIA.

5. See Robert Wallace, "The Barbary Wars," *Smithsonian* 5 (January 1975), 87.

6. CIA, *Intelligence in the War of Independence,* 11.

7. See William R. Corson, *The Armies of Ignorance: The Rise of the American Intelligence Empire* (New York: Dial, 1977), chap. 2; Powers, *The Man Who Kept the Secrets;* John Ranelagh, *The Agency: The Rise and Decline of the CIA* (New York: Simon and Schuster, 1986); the Church committee, *Final Report,* vol. 4:1–108, and vol. 6:1–353.

8. On avoiding surprise attacks (so-called "warning intelligence"), former DCI Richard Helms has observed: "It's everything, and underline everything" [quoted by Bob Woodward, *Veil: The CIA's Secret Wars, 1981–87* (New York: Simon & Schuster, 1987), p. 49]. On the Pearl Harbor intelligence failure, see Roberta Wohlstetter, *Pearl Harbor: Warning and Decision* (Stanford, Calif.: Stanford University Press, 1962); Rear Adm. Edwin T. Layton, with Capt. Roger Pineau and John Costello, *"And I Was There": Pearl Harbor and Midway—Breaking the Secrets* (New York: Morrow, 1985).

9. The necessary fin adjustment was devised by the Japanese less than two months before the Pearl Harbor attack. See Wohlstetter, *Pearl Harbor: Warning and Decision,* 369; also, John Deane Potter, *Yamamoto: The Man Who Menaced America,* (New York: Viking, 1965), 53.

10. See Seth W. Richardson (general counsel for the Joint Congressional Investigating Committee on Pearl Harbor), "Why Were We Caught Napping at Pearl Harbor?" *Saturday Evening Post,* May 24, 1947, 79–80.

11. See Wohlstetter, *Pearl Harbor: Warning and Decision,* 55–56, 225, 387.

12. See Alvin D. Coox, "Pearl Harbor," in Noble Frankland and Christopher Dowling, eds., *Decisive Battles of the Twentieth Century* (New York: McKay, 1976), 148; and Wohlstetter, *Pearl Harbor: Warning and Decision,* 394.

13. Richard C. Synder and Edgar S. Furniss, Jr., *American Foreign Policy* (New York: Rinehart, 1954), 229.

14. Ibid.

15. On the OSS, see Richard Harris Smith, *OSS: The Secret History of America's First Central Intelligence Agency* (Berkeley: University of California Press, 1972); Ray S. Cline,

Secrets, Spies, and Scholars: Blueprint of the Essential CIA (Washington, D.C.: Acropolis, 1976); Stewart Alsop and Thomas Braden, *Sub Rosa: The OSS and American Espionage* (New York: Reynal and Hitchcock, 1946). Christopher Andrew captures the absurdity to which American intelligence descended during World War II. The Navy's OP-20-G Signal Intelligence Corps and the Army's Signal Intelligence Service spent much of the war struggling against one another. Finally, the two resorted to an arbitrary compromise: traffic intercepted on even days would go to the Army for analysis, on odd days to the Navy. See Christopher Andrew, "Codebreakers and Foreign Offices: The French, British and American Experience," in Christopher Andrew and David Dilks, eds., *The Missing Dimension: Governments and Intelligence Communities in the Twentieth Century* (Urbana: University of Illinois, 1984), 52.

16. Cited by Merle Miller, *Plain Speaking: An Oral Biography of Harry S. Truman* (New York: Berkeley, 1973), 420n. Intelligence reorganization had been one of the subjects on President Roosevelt's mind, too, shortly before he died. On April 5, 1945, he wrote to Maj. Gen. William S. Donovan, head of the OSS: ". . . I should appreciate your calling together the chiefs of the foreign intelligence and internal security units in the various executive agencies, so that a consensus of opinion can be secured [about the establishment of a central intelligence service]." Memorandum from Franklin D. Roosevelt to Major General Donovan. One week later the president died. General Donovan often complained during World War II about the inefficiencies of American intelligence. In one memo to President Roosevelt, dated November 22, 1944, Donovan wrote:

> The immediate revision and coordination of our present intelligence system would effect substantial economies and aid in the more efficient and speedy termination of the war.
>
> Information important to the national defense, being gathered now by certain Departments and agencies, is not being used to full advantage in the war. Coordination at the strategy level would prevent waste, and avoid the present confusion that leads to waste and unnecessary duplication. . . .
>
> We have now in the Government the trained and specialized personnel needed for the task. This talent should not be dispersed.

The Roosevelt and Donovan memos are from the National Archives Collection.

17. See the account in Harry Howe Ransom, *The Intelligence Establishment* (Cambridge, Mass.: Harvard University Press, 1970), 79–80.

18. On the NSC, see Karl F. Inderfurth and Loch K. Johnson, *Decisions of the Highest Order: Perspectives on the National Security Council* (Pacific Grove, Calif.: Brooks/Cole, 1988).

19. Snyder and Furniss, *American Foreign Policy*, 230.

20. Ransom, *Intelligence Establishment*, 81.

21. Quoted by Marchetti and Marks, *The CIA*, 70.

22. See the account in the Church committee, *Final Report* 1:170.

23. Quoted by Sen. Frank Church (D, Idaho), *Congressional Record*, January 27, 1976, p. 1165.

24. National Security Act of 1947, signed on July 26, 1947 (P.L. 97-200; 50 U.S.C. 403, Sec. 102). For a useful collection of intelligence laws and regulations, see House Permanent Select Committee on Intelligence, *Compilation of Intelligence Laws and Related Laws and Executive Orders Of Interest to the National Intelligence Community* (Washington, D.C.: Government Printing Office, April 1983).

25. "Special Message to the Congress on Greece and Turkey: the Truman Doctrine" (March 12, 1947), *Public Papers of the Presidents of the United States: Harry S. Truman, 1947* (Washington, D.C.: Government Printing Office, 1963), 178.

26. See Wyden, *Bay of Pigs*.

27. NBC, "Evening News," January 13, 1978. Speaking in favor of covert action, an assistant to the president for national security affairs in the Reagan administration argued: "In the late part of the twentieth century we are going to face in many countries, not only in Central America, a determined effort by the Soviet Union to subvert friendly govern-

ments. Now when they do that, using great violence, do the American people really want their president, faced with the question of whether a friend like Salvador or Korea or Israel is being attacked, to have no other options than to war or to do nothing? I don't think so." Robert McFarlane, NBC, "Meet the Press," May 13, 1984.

28. See *Public Papers of the Presidents of the United States: Jimmy Carter, 1978* (Washington, D.C.: Government Printing Office, 1979), 1:194–214.

29. Respectively, 88 Stat. 1804, Sec. 662, 22 U.S.C. 2422; and 94 Stat. 1981, Title IV, Sec. 501, 50 U.S.C. 413. In 1988, legislators seeking passage of the 1988 Intelligence Oversight Act combined the language of the Carter executive order and the Hughes-Ryan Act into an omnibus definition of special activities. See *Report No. 100-276,* U.S. Senate (January 27, 1988), 46. The purpose was to try to make more clear what was already accepted practice in the government, namely, that the CIA and the other intelligence agencies were not expected to obtain a finding for all operations other than collection (even though Hughes-Ryan states this). Examples of overseas operations that are secret, but which do not require a finding are counterintelligence activities (discussed later in this chapter) and assistance to the Department of State and Defense in routine diplomatic or military programs (*Report No. 100-276,* 38–40). Clark Clifford, an original author of the 1947 National Security Act, settles on a simpler definition: "[secret] active efforts to alter political conditions in foreign countries through financial, paramilitary, and other means" (Testimony, House Permanent Select Committee on Intelligence, *Hearings,* February 24, 1988, 2).

30. B. Hugh Tovar, "Strengths and Weaknesses in Past U.S. Covert Action," in Roy Godson, ed., *Intelligence Requirements for the 1980s: Covert Action* (Washington, D.C.: National Strategy Information Center, 1981), 194–95. Looking back on the use of covert action in Chile, a senior CIA official argued before the Church committee: "Was our role in Chile bad and anti-democratic? I think not. The United States was acting within the broad mainstream of traditional U.S. policy in Latin America. That policy has been to resist the establishment of governments in Latin America with close ties to European powers—in this case the Soviet Union. That policy, around the world, has also been to oppose the attempt by communist and radical Marxist parties to take over governments in the knowledge that once in power these forces ultimately destroy the elements of democracy and diversity that enabled them to gain power. What has been preserved in Chile is the chance to begin again" (October 23, 1975).

31. Observation by Professor Roy Godson, Panel on "The Future of Intelligence Research," at the Annual Convention of the International Studies Association, Anaheim, California, March 17, 1986.

32. See Harry A. Rositzke, *The CIA's Secret Operations: Espionage, Counterespionage, and Covert Action* (Pleasantville, N.Y.: Reader's Digest, 1977); Gregory F. Treverton, *Covert Action: The Limits of Intervention in the Postwar World* (New York: Basic Books, 1987); Woodward, *The CIA's Secret Wars.*

33. "Gesprach mit William E. Colby," *Der Spiegel* 4 (January 23, 1978), 101, my translation.

34. Pacem in Terris IV Convocation, Washington, D.C., December 4, 1975.

35. CIA official, interview, Washington, D.C., December 16, 1975. The official title of the Clark amendment (sponsored by Sen. Dick Clark [D, Iowa]) is Sec. 118 of the International Security and Development Cooperation Act of 1980 (22 U.S.C. 2293, note).

36. Church committee, "Covert Action," *Hearings,* October 23, 1975. The official name of the Murphy commission (chaired by Robert D. Murphy) was the Commission on the Organization of the Government for the Conduct of Foreign Policy. The government published its multivolume report in June 1975 (Washington, D.C.: Government Printing Office). In discussion with the Church committee about covert action in Chile, Colby argued that the Agency had gotten "a bum rap." The operations had been appropriately designed, in his view, to support the democratic forces against a threat to Chile and, primarily, to the United States by a Marxist faction.

37. Church committee, "Covert Action" Clifford and Vance testified on December 4, 1975.

38. Ibid., December 4, 1975. Halperin advocated open support for U.S. allies overseas. Two organizations established recently for the purpose of aiding America's friends abroad openly with federal funds (Panamanian democrats, for example) are the National Democratic Institute for International Affairs and the National Endowment for Democracy.

39. Pacem in Terris IV Convocation.

40. For the Gates confirmation hearings, see "Hearings on Robert M. Gates," Senate Select Committee on Intelligence, January 1987.

41. The data in this section and the observations that follow are based upon interviews with CIA managers in 1975, 1980, 1984, and 1987, as well as on the scholarly sources cited in nn. 42 and 58 below. (See also the sources in n. 1 of chap. 6.) On the affinity of the Reagan administration for covert action, see Joseph Lelyveld, "The Director: Running the C.I.A.," *New York Times Magazine,* January 20, 1985, 25. With reference to CA trends, the Lelyveld article leaves a misleading impression in one important respect. The Carter administration did not, as asserted, have "a higher level [of covert action] than at any time since [the] Kennedy [administration], when covert operations were at their peak" (p. 25). Just the opposite: the frequency of covert action reached a lower point during the Carter years than at any time since Kennedy, and CA levels were substantially higher under President Johnson than under his predecessor. Furthermore, which administration—Carter's or Reagan's—was more extensively involved in covert action depends upon the measure of involvement employed. Lelyveld concludes that the Carter administration used covert action more extensively. This is accurate in terms of the frequency of operations, as a result of the sudden surge of covert actions resorted to in the last two years of the Carter term (largely in response to the Afghanistan invasion, which shattered that administration's experiment in detente). In terms of money expended, however—a more significant index than the frequency of often minor operations—the Reagan administration surpassed its immediate predecessor, though remained below the levels of the Kennedy and Johnson administrations. As a second in command at the CIA has put it, "Covert actions in the Reagan administration have been fewer, but 'noisier' and more expensive" (John McMahon, deputy director of the CIA, remarks, June 12, 1984). These conclusions are drawn from interviews with several CIA officers in June 1984 and April 1987, Washington, D.C. See Church committee, *Final Report,* vol. 1, which notes that CA projects "reached an all-time high" in fiscal year 1964 in terms of *numbers* of operations. Interviews I have conducted suggest that in terms of *money* expended, the high point is calendar year 1968 because of the cost of paramilitary operations in the Far East (see fig. 6.1 in chap. 6). Misleading, too, is a recent comment by McMahon's replacement as second in command at the CIA, who attempts to downplay the importance of covert action with the observation that "over 95 percent of the national intelligence budget is devoted to the collection and analysis of information. Only about three percent of the CIA's people are involved in covert action." The national intelligence budget includes Department of Defense intelligence, which amounts to some 85 percent of the entire intelligence budget—most of which goes toward collection (especially via costly technical collection methods). The assertion, then, badly underestimates the importance of covert action within the CIA's budget. So does the reference to "the CIA's people" dedicated to covert action, since money is a better index; moreover, most of the people involved in covert action are indigenous assets abroad—not CIA officers. This is the sort of legerdemain that has led to soured relations between Congress and the CIA in recent years (see chaps. 6 and 10).

42. See Marchetti and Marks, *The CIA;* Godson, *Covert Action;* Church committee, *Final Report,* vol. 1; Colby and Forbath, *Honorable Men;* Stansfield Turner, *Secrecy and Democracy: The CIA in Transition* (Boston: Houghton Mifflin, 1985); Theodore Shackley, *The Third Option: An American View of Counterinsurgency* (Pleasantville, N.Y.: Reader's Digest, 1981); and Rositzke, *The CIA's Secret Operations.*

43. On CIA propaganda, see House Permanent Select Committee on Intelligence, Subcommittee on Oversight, "The CIA and the Media," *Hearings,* December 28, 1977 (hereafter cited as Aspin Hearings). For several examples, see chap. 7 of this volume, and, especially with respect to Asia and Latin America, Joseph Burkholder Smith, *Portrait of a Cold Warrior* (New York: Putnam, 1976).

44. See Church committee, "Covert Action in Chile: 1963–1973," *Staff Report,* December 1975; and Treverton, *Lessons of Covert Action.*

45. "Covert Action in Chile," 15.

46. Interview with former CIA-CAS officer, Washington, D.C., February 21, 1976.

47. For a fascinating account, see David Wise and Thomas B. Ross, *The Invisible Government* (New York: Random House, 1964).

48. See Stuart H. Loory, "The CIA's Use of the Press: 'A Mighty Wurlitzer,' " *Columbia Journalism Review* (September/October 1974), 8–18; and *Final Report,* vol. 1, chap. 10.

49. For one account, see Wise and Ross, *The Invisible Government.* The CIA actually tried to buy out all the copies of the first printing of the Wise and Ross book in order to get it out of circulation; the publisher simply printed more—an author's dream come true. (See also n. 51.)

50. Aspin Hearings, 5.

51. See Marchetti and Marks, *The CIA,* 145. Some CIA disinformation operations have been aimed at U.S. citizens. The Agency, for instances, sent cables abroad to its stations requesting personnel to plant unfavorable book reviews of David Wise and Thomas B. Ross, *The Invisible Government,* a critical account of the CIA's covert actions. According to an internal Agency memorandum obtained by Wise through a Freedom of Information Act (FOIA) request, the stations were to "secure unfavorable books reviews . . . to lessen the book's impact and to cast doubt on the validity of its claims" (letter from David Wise to this author, November 24, 1978, with a copy of the Agency memo). For an insider's account of this effort to besmirch Wise and Ross, see Smith, *Portrait of a Cold Warrior,* 432–33. The Soviet Union is probably more involved in the practice of disinformation than any other country; the KGB, the Soviet equivalent of the CIA and the FBI combined, has an entire Department of Disinformation. The latest Soviet euphemism for this art form is "active measures." For examples of KGB handiwork, see Aspin Hearings, 531–627. Also, see Richard H. Shultz and Roy Godson, *Dezinformatsia: Active Measures in Soviet Strategy* (New York: Pergamon-Brassey, 1984).

52. Interviews with CIA officials experienced in propaganda operations, Washington, D.C., June 1984.

53. See the sources in n. 42 above; on the Hussein case, see *Newsweek,* March 7, 1977, 16 and the *Washington Post,* February 18, 1977, 1.

54. See Tom Wicker et al., "C.I.A. Operations: A Plot Scuttled," *New York Times,* April 28, 1966, p. 1.

55. Church committee, "Covert Action in Chile," 15.

56. Interviews with former CIA officials, Washington D.C., October 1980 and July 1984, and Boston, February 1988; on CIA covert actions in Nicaragua, see Woodward, *The Secret Wars of the CIA,* 174, 276, 281, and 430. See also the *Washington Post,* January 9, 1977, for a claim that the CIA may have been involved in the spreading of a swine virus in Cuba during 1971.

57. See Shackley, *The Third Option.*

58. In addition to the sources in n. 42 above, see "Should the CIA Fight Secret Wars?" *Harper's,* September 1984, 33–47; *New York Times,* June 1, 1978; Nicholas Bethell, *Betrayed* (New York: Times Books, 1985); and Philip Taubman, "Casey and His C.I.A. On the Rebound," *New York Times Magazine,* January 16, 1983, 21 ff.

59. See Marchetti and Marks, *The CIA,* and Shackley, *The Third Option.*

60. Church committee, "Alleged Assassination Plots Involving Foreign Leaders," *Interim Report,* Rept. No. 94-465, November 20, 1975.

61. *Joint Hearings,* Senate Select Committee on Secret Military Assistance to Iran and the Nicaraguan Opposition and House Select Committee to Investigate Covert Arms

Transactions with Iran (the Inouye-Hamilton committees, named after Senator Daniel Inouye [D, Hawaii] and Representative Lee Hamilton [D, Indiana]), U.S. Congress, May–August 1987 (hereafter cited as Inouye-Hamilton committees, *Hearings*).

62. See Shackley, *The Third Option;* on CIA support of the Kurds, see the Pike Committee *Report.*

63. Church committee, "Alleged Assassination Plots," 181n.

64. See Colby and Forbath, *Honorable Men,* 272; and "Gesprach mit William E. Colby," 103, 106.

65. *New York Times,* December 6, 1984. The manual, entitled "Psychological Operations in Guerilla Warfare," has been subsequently published by Random House, New York, 1985 (see p. 57 for the "neutralize" terminology).

66. See, for instance, Church committee, "Alleged Assassination Plots," 41.

67. Quoted by the *New York Times,* December 4, 1984, p. 3. In 1986, when U.S. fighter bombers dropped their payloads on the home of Libyan leader Muammar Qaddafi (where he was expected to be at the time, though it turned out he was not), President Reagan denied that the attack represented a violation of the prohibition against assassination attempts against foreign leaders. Former DCI Stansfield Turner has called the bombing "a deliberate assassination attempt" against Qaddafi (interview with Ted Koppel, ABC, "Nightline," April 21, 1988). For an indication that the Reagan administration authorized covert antiterrorist operations in the Middle East that may have involved assassinations, despite President Reagan's executive order banning assassinations (No. 12333, signed on December 4, 1981), see Woodward, *The CIA's Secret Wars,* 393–97.

This is a book on the CIA, not on Soviet intelligence, but the reader should keep in mind that whatever the CIA has done so has Soviet intelligence in spades. On the covert-action front, among the most barbaric operations to come to light have been the use of booby traps designed to look like toys, which have exploded to kill or maim mujahideen children as part of the Soviet Union's paramilitary effort against the resistance in Afghanistan during the mid-1980s (see Egbal Ahmad and Richard J. Barnet, "A Reporter at Large: Bloody Games," *New Yorker* [April 11, 1988], 65). The Marxist regime in Nicaragua once similarly charged the CIA with placing plastic explosives in Mickey Mouse dolls and other toys as part of its secret paramilitary campaign in Central America; *Washington Post,* December 25, 1982, A15. The Soviets have also been accused of using lethal gases in Afghanistan and Cambodia, but the evidence has been less conclusive.

68. Interview with FBI counterintelligence specialist, May 8, 1975, Washington, D.C. Most of the statistics and observations in this section on counterintelligence draw upon my research (with John Elliff) conducted for the Church committee, *Final Report* 1:163–73; the annual reports of the congressional intelligence committees also usually provide current data on the Soviet CI challenge (e.g., *S. Report* 98-665, October 10, 1984, 22–23).

69. FBI memorandum, "Intelligence Activities within the United States by Foreign Governments," March 20, 1975, cited in Church committee, *Final Report* 1:164.

70. James Angleton, interview, July 30, 1975, Washington D.C. This phrase is originally from T. S. Eliot's poem "Gerontion."

71. Raymond Rocca, CIA-CI specialist, interview, November 25, 1975, Washington, D.C.

72. Memorandum from John McCone to chairman, President's Foreign Intelligence Advisory Board, October 8, 1963, cited in Church committee, *Final Report* 1:167.

73. Cited in ibid., 167.

74. Interview with CIA-CI specialist, November 1, 1975, Washington, D.C.

75. Interview with CIA-CI specialist, October 17, 1975, Washington, D.C.

76. On all four, Dunlop, Philby, Penkovsky, and Popov, see David C. Martin, *Wilderness of Mirrors* (New York: Harper and Row, 1980); and John Ranelagh, *The Agency: The Rise and Decline of the CIA* (New York: Simon and Schuster, 1987).

77. "Klop" Ustinov, quoted by Peter Wright, *Spycatcher* (New York: Viking, 1987), 87.

78. Rocca, interview, op. cit.

79. See Sir John Masterman, *Double Cross System of the War of 1939–45*, (New Haven, Conn.: Yale University Press, 1972).

80. Interview with Bruce Solie, former CIA director of the Office of Security, November 25, 1975, Washington, D.C.

81. Interviews with James Angleton, November and December 1975, Washington, D.C.

82. "Yuri Nosenko, KGB," *Foreign Intelligence Literary Scene* 5 (November/December 1986), 11; see also *Washington Star*, September 22, 1978, pp. A1–A5. The CIA has been charged from time to time with using, or teaching foreign intelligence services to use, CI interrogation techniques crueler than even those experienced by the hapless Nosenko. According to one recent charge, CIA officials trained Honduran soldiers in CI psychological torture techniques—including the feeding of dead rats to prisoners, depriving them of sleep, giving them rotten food, dousing them with icy water, and placing cockroaches and other vermin in their cells—but advised against physical torture (James Le Moyne, "Testify to Torture," *New York Times Magazine*, June 5, 1988, 44). According to one knowledgeable former CIA-CI officer, physical torture as a technique is foolish, since subjects will say *anything* to stop the pain. Psychological approaches—no toothpaste, blankets, watches, toilet paper, decent food, walking room, or someone to talk with—are more apt to elicit useful information when the CI interrogator arrives. (See William R. Johnson, "Tricks of the Trade: Counterintelligence Interrogation, *International Journal of Intelligence and Counterintelligence* 1, 1986, 103–33.) The trouble with allegations like the ones reported in the *Times* is that it seems unlikely such foul deeds really need to be taught. Those who wish to torture psychologically or physically can no doubt think of many ingenious approaches without special tutoring from the government of the United States. The former Argentine dictatorship did not require a torture hotline to CIA Headquarters in order to learn that electrode shocks to genitalia can bring confessions even from the innocent.

83. Interview with CIA-CI specialist, January 15, 1976, Washington, D.C.

84. Henry Brandon, "The Spy Who Came and Then Told," *Washington Post* (National Weekly Edition), August 24, 1987, p. 36. Brandon was the *London Sunday Times* chief correspondent in the United States for thirty-five years.

85. *Hearings*, House Committee on Expenditures in the Executive Departments, June 17, 1947, 454.

86. *Hearings*, Senate Committee on Armed Services, April 30, 1947, 497.

87. The language reads: "[T]he [CIA] shall have no police, subpoena, law-enforcement powers, or internal-security functions. . . ." (50 U.S.C. 403, Sec. 102).

88. Harry Howe Ransom, "Don't Make the C.I.A. a K.G.B.," *New York Times*, December 24, 1981; for the precise language of the statute, see 50 U.S.C. 403 a–n.

89. See the *Congressional Record*, May 27, 1949.

90. See Church committee, *Final Report* 1:144 and 4:25–41.

Chapter 3

1. On the structure, staffing, and funding of the intelligence community, see John Hamer, "Intelligence Community," *Editorial Research Reports* 2 (July 25, 1973), Congressional Quarterly, Washington, D.C., pp. 559–61; Wise and Ross, *The Invisible Government* (New York: Random House, 1964) and *The Espionage Establishment* (New York: Random House, 1967); Frank J. Donner, "The Theory and Practice of American Intelligence," *New York Review of Books*, April 22, 1971, 27–39; Paul W. Blackstock, "The Intelligence Community under the Nixon Administration," *Armed Forces and Society* 1 (February 1975), 231–50; Church committee, *Final Report*, vol. 1; Jeffrey Richelson, *The U.S. Intelligence Community*, (Cambridge, Mass.: Ballinger, 1985); Stephen J. Flanagan, "The Coordination of National Intelligence," in Duncan L. Clarke, ed., *Public Policy and Political Insti-*

tution. United States Defense and Foreign Policy: Coordination and Integration (Greenwich, Conn.: JAI, 1985), 157–196; Marchetti and Marks, *The CIA*, 49–98; Ransom, *The Intelligence Establishment*, 82–146; Central Intelligence Agency, *Factbook on Intelligence* (Washington, D.C.: CIA, April 1983) and *Intelligence: The Acme of Skill* (Langley, Va.: CIA, no date); Bonner Day, "The Battle Over U.S. Intelligence," *Air Force Magazine* 61 (May 1978), 42–7; Murphy commission, (Commission on the Organization of the Government for the Conduct of Foreign Policy) *Report to the President* (Washington, D.C.: Government Printing Office, June 1975), 96; Harry F. Eutace, "Special Report: Changing Intelligence Priorities," *Electronic Warfare/Defense Electronics* 28 (November 1978), 35–37. This last reference is the source of the estimate; see also George Lardner Jr., "Missing Intelligence Charters," *Nation* 227 (September 2, 1978), 169. According to the ranking minority member of the Senate Intelligence Committee in 1984, Daniel Patrick Moynihan (D, New York), the CIA's budget declined dramatically during the Nixon and Ford years because of cutbacks in Vietnam, but funding has increased each year since 1977 (CBS, "Evening News," September 27, 1984). For a comparative analysis of intelligence organization in the English-speaking democracies, see Jeffrey T. Richelson and Desmond Ball, *The Ties That Bind: Intelligence Cooperation between the UKUSA Countries* (Boston: Allen and Unwin, 1985).

2. William J. Barnds, "Intelligence and Foreign Policy: Dilemma of a Democracy," *Foreign Affairs* 48 (January 1969), 286.

3. See, for example, Lee Lescaze, "Pentagon vs. CIA," *Washington Post*, June 10, 1977, p. 1.

4. Dean Rusk, interview, January 9, 1985, Athens, Georgia.

5. William Scranton, former Pennsylvania governor and U.S. senator (R), remark to me, December 16, 1979, Washington, D.C. On PFIAB, see the *New York Times*, April 28, 1966, p. 28. Harry Howe Ransom once described PFIAB as more of a "polite alumni visiting committee than a vigorous watchdog." See his "Secret Mission in an Open Society," *New York Times Magazine*, May 21, 1961, 80.

A former member of PFIAB claims that it has had its vigorous moments, as in the early days of the Reagan administration, but that by 1986 it had become little more than (in a profusion of mixed metaphors) "a celebrity board" and "a purring lap cat" with "very few boat rockers" (Martin Anderson, *Revolution* [New York: Harcourt Brace Jovanovich, 1988], 361–62). For an alternative view that PFIAB was engaged in some "serious projects" throughout the Reagan years, see the account of another member (who opines as well that it is an act of "cowardice" for intelligence officials "to regard congressional oversight as of equal if not greater importance than White House oversight"), University of California, Berkeley, political science professor Paul Seabury, "A Massacre Revisited," *Foreign Intelligence Literary Scene* 7 (May–June 1988), 1, 2.

6. Marchetti and Marks, *The CIA*, 96. On the CIA's share of the intelligence budget, see Eutace, "Special Report: Changing Intelligence Priorities."

7. This observation is reported by former DCI Admiral Stansfield Turner, *Secrecy and Democracy*, 185. One method used by DCIs to gain leverage over the intelligence community has been to place ("detail") CIA officers on the National Security Council where they serve as intelligence specialists on the staff and, presumably, keep an eye out for the DCI and the Agency. Other intelligence agencies also engage in the practice, offsetting, to some extent, the CIA's attempt at high-level advantage. These "detailees" perform a valuable service for the NSC, providing national security expertise without drawing down the White House budget. Yet the presence of intelligence officers so close to the center of power has its risks. The Reagan administration provides a recent example. Its DCI, William J. Casey, evidently detailed his top covert propaganda specialist to the NSC in order to direct a "public diplomacy" project. The plan was to sway U.S. public opinion against the regime in Nicaragua (the Sandinistas)—in essence a CIA propaganda operation directed toward American citizens, which CIA officers housed within the Agency are prohibited from carrying out. (See Robert Parry and Peter Kornbluh, "Iran-Contra's Untold Story," *Foreign Policy* 72 [Fall 1988], 3–30.) Beyond detailees, some DCIs and other intel-

ligence officials also try to persuade regular NSC and White House staffers to support the objectives of their agencies. The most extreme cases involved Tom Charles Huston of the Nixon administration and Lieut. Col. Oliver L. North of the Reagan administration (see chap. 7). Casey apparently looked upon his NSC detailee and elements of the regular NSC staff (especially North) as a kind of surrogate CIA, free from the legal restrictions placed upon the intelligence agencies.

8. Marchetti and Marks, *The CIA*, 85; see also Richelson, *The U.S. Intelligence Community*, 21; and the Church committee, *Final Report*, vol. 1.

9. DDO remarks, Conference on U.S. Intelligence: The Organization and the Profession, Central Intelligence Agency, Langley, Va., June 11, 1984.

10. Operations Directorate official, interview, September 12, 1980, Washington D.C. A former second in command at the CIA has explained the mission of the case officer in this way: "Anyone seeking to act covertly must select key influential personalities in each country who may be useful. Then they must study the desires, ambitions, prejudices and vulnerabilities of such people to see how they may be manipulated. Some, by a misguided idealism, some by the lust for power, others for money or other material ambitions. Some are motivated by a desire for revenge for some real or imagined hurt. Some may be manipulated through blackmail because of past misconduct or irregularities in their sexual lives." Vernon A. Walters, "The Uses of Political and Propaganda Covert Action in the 1980s," in Godson, *Covert Action*, 121.

11. Rositzke, *The CIA's Secret Operations*, 208. The number of assets comprising the CIA's infrastructure has ranged over the years from some six hundred individuals to as many as seven thousand (interviews with CIA officials, November 1980 and June 1984, Washington, D.C.).

12. OTA director "Conference on Intelligence," June 11, 1984. Of course, CIA involvement in the Iran-contra operation two years later—though not OTA—raises doubts about who in the Agency learned what.

13. Marchetti and Marks, *The CIA*, 90.

14. See, respectively, Alfred W. McCoy, *The Politics of Heroin in Southeast Asia* (New York: Harper and Row, 1972), esp. 144–45, 263, 360; *New York Times*, May 21, 1988, pp. 1, 4, 8; and Ahmad and Barnet, "Bloody Games," 44–86.

15. Sanford J. Ungar, *FBI* (Boston: Atlantic Monthly, 1975), 125.

16. Donner, "American Intelligence," 27. See also Frank J. Donner, *The Age of Surveillance* (New York: Knopf, 1980).

17. On these disputes and liaison difficulties, see Martin, *Wilderness of Mirrors;* Unger, *FBI*, 107; Church committee, *Final Report* 1:170.

18. Church committee, *Final Report* 1:163; also, John Barren, *KGB: The Secret Work of Soviet Agents* (Pleasantville, N.Y.: Reader's Digest, 1974).

19. See John T. Elliff, *The Problem of FBI Intelligence Operations* (Princeton, N.J.: Princeton University Press, 1979); Donner, *Age of Surveillance*, and Unger, *FBI*, as well as the Church committee, *Final Report*, vol. 2.

20. Dean Rusk, interview, op. cit.

21. Donald Gregg, assistant to the vice president for National Security Affairs, Conference on U.S. Intelligence, June 13, 1984.

22. On the NSA, see David Kahn, *The Codebreakers: The Story of Secret Writing* (New York: Macmillan, 1967), chap. 19, and "Big Ear or Big Brother?" *New York Times Magazine*, May 16, 1976, pp. 13, 62–72; James Bamford, *The Puzzle Palace* (New York: Houghton Mifflin, 1984); Nicholas M. Horrock, "East or West, Spy Central Is a Whirring Big Computer," *New York Times*, July 17, 1977, E6; Deborah Shapley, "Who's Listening," *Washington Post*, July 9, 1978, p. B–7.

23. For an example of one NSA operation directed against the Soviets from Turkish soil, see the *New York Times*, May 15, 1977, p. 14; see also "Unveiling the Secret NSA," *Newsweek*, September 6, 1982, 20–28.

24. Gayler, interview, June 19, 1975, Washington, D.C.

25. The Fitzhugh Report, cited in the Church committee, *Final Report* 1:342.

26. See the discussion of Operations Minaret and Shamrock in Johnson, *A Season of Inquiry.*

27. See the testimony of Secretary of Defense Casper Weinberger, *Hearings,* Inouye-Hamilton committees.

28. See Maj. Gen. Daniel O. Graham, "Estimating the Threat: A Soldiers' Job," *Army,* April 1973, 14–18. See, as well, Church committee, *Final Report* 1:349–54, 365–66.

29. The Fitzhugh Report, cited in the Church committee, *Final Report* 1:350.

30. See Lawrence M. Baskir, "Reflections on the Senate Investigation of Army Surveillance," *Indiana Law Journal* 49 (Summer 1974), 618–53.

31. The Schlesinger Report, cited in the Church committee, *Final Report* 1:351.

32. The Pike committee in the House of Representatives; see the *New York Times,* February 4, 1976.

33. The memorandum and statistics are cited in the *Washington Post,* February 2, 1975; the figure of 47,000 Secret Service names is from the *Washington Post,* September 9, 1975.

34. For illustrations (the FBI COINTELPRO Operation, among others), see Johnson, *A Season of Inquiry.*

35. See, for example, John Marks, "The CIA's Corporate Shell Game," *Washington Post,* July 11, 1976, p. C–1. A proprietary is an ostensibly private business enterprise that also serves as an instrument of the CIA (or other intelligence agency) to advance its operations around the world.

36. Ransom, *The Intelligence Establishment,* 146.

37. For this and the other comments from Turner in these paragraphs, see his foreward to David D. Newsom, *The Soviet Brigade in Cuba: A Study in Political Diplomacy* (Bloomington: Indiana University Press, 1987), vii–xiii. On this incident, see also Richard E. Neustadt and Ernest R. May, *Thinking in Time: The Uses of History for Decision-Makers* (New York: Free Press, 1986), 91–97, and Gloria Duffy, "Crisis Mangling and the Cuban Brigade," *International Security* 8 (Summer 1983), 67–87.

Chapter 4

1. William E. Colby, "Understanding the Realities," *New York Times Book Review,* June 26, 1983, 3.

2. Senior CIA analyst, interview, Washington, D.C., December 1978.

3. See House Permanent Select Committee on Intelligence, Subcommittee on Evaluation, "Iran: Evaluation of U.S. Intelligence Performance Prior to November 1978," *Staff Report,* January 1979.

4. In Godson, *Analysis and Estimates,* 213.

5. Thomas L. Hughes, "The Power to Speak and the Power to Listen: Reflections in Bureaucratic Politics and a Recommendation on Information Flows," in Thomas M. Franck and Edward Weisband, eds., *Secrecy and Foreign Policy* (New York: Oxford, 1974), 19.

6. Accounts of the Westmoreland libel suit brought against CBS may be found in the *New York Times* throughout December 1984 and January and February 1985.

7. "The Tet Intelligence Flap: One Out of Step, or Many?" *Washington Star,* November 16, 1975, p. E3.

8. Dean Rusk, interview, Athens, Georgia, January 9, 1985. Rusk argues further that "false official statements" are viewed as contemptible by the military—"not the kind of thing an officer of Westmoreland's high character would consider." Another former secretary of state who knew Westmoreland well concurs. "Westy was too straight-arrow, too much of a Boy Scout, to lie about this," says Alexander Haig (interview, Athens, Georgia, October 15, 1985). Another veteran of the Vietnam War (a CIA officer with six years experience there and periodic contact with Westmoreland), suggests that the general—

however honest he might be—did not follow the intricacies of intelligence and was manipulated by subordinates to authorize misleading figures (interview, Cambridge, Massachusetts, February 27, 1988).

9. *New York Times,* December 5, 1984. For McNamara's views, see the *New York Times,* December 7, 1984. McNamara, like Rusk and Haig, spoke of Westmoreland's "integrity" and, like Nitze and Komar, concluded that what had arisen was "an honest disagreement between people putting forth their best figures."

10. "The Tet Intelligence Flap."

11. See the *New York Times,* January 8, 1985. For Adams's initial public criticism, see Samuel Adams, "Vietnam Cover-Up: Playing War With Numbers," *Harper's* 250, May 1975, 41–44.

12. See the Pike Committee *Report.*

13. *New York Times,* January 7, 1985.

14. Interviews with two senior NIOs, Washington, D.C., June 16, 1984; for Allen's statement, see the *New York Times,* January 23, 1985, p. 11.

15. "The Tet Intelligence Flap."

16. Powers, *The Man Who Kept the Secrets,* 240.

17. See the account in Church committee, *Final Report* 1:81.

18. Interviews with House and Senate staff aides, June 1983, Washington, D.C. For additional, recent charges that CIA officials have "cooked" intelligence for policymakers, see the listing of Lee H. Hamilton (chairman of the House Intelligence Committee), "View from the Hill," *Extracts from Studies in Intelligence* (Langley, Va.: Central Intelligence Agency, September 1987), 68; and the testimony of Secretary of State George P. Shultz during the Iran-contra congressional hearings, Inouye-Hamilton Committees, *Hearings,* July 24, 1987.

19. Tuchman, *The Zimmerman Telegram* (New York: Viking, 1958), 26; Albert Speer, *Inside the Third Reich* (New York: Macmillan, 1970), 243. See also a vivid account of Reich Marshall Hermann Goering's dismissal of an intelligence report that might have upset the Führer (at p. 290). President Kennedy's advisers seemed to have been timid about speaking to him plainly during the planning for a covert invasion of Cuba at the Bay of Pigs in 1961. "There is a time when you can't advise by innuendoes and suggestions," concluded retired army chief of staff Maxwell D. Taylor, who investigated the fiasco for the president after its collapse. "You have to look at him in the eye and say, 'I think it's a lousy idea, Mr. President. The chances of our succeeding are about one in ten.' And nobody said that." From an interview cited in Wyden, *Bay of Pigs,* 317. On this phenomenon of "mindguarding" generally, see Irving Janis, *Groupthink,* 2d ed. (Boston: Houghton Mifflin, 1982), 40–41. For an excellent collection of case studies on intelligence performance, see Ernest R. May, ed., *Knowing One's Enemies: Intelligence Assessments Before the World Wars* (Princeton, N.J.: Princeton University Press, 1985).

20. Robert M. Gates, "The CIA and American Foreign Policy," *Foreign Affairs* 66 (Winter 1987/88), 227.

21. Hughes, "The Power to Speak," 28–37, quote on 28.

22. Godson, *Analysis and Estimates,* 79.

23. Interview with senior CIA official, Washington, D.C., June 1983. John Huizenga, former chair of the BNE, testified before the Church committee that "[w]hen intelligence people are told, as happened in recent years, that they were expected to get on the team, then a sound intelligence-policy relationship has in effect broken down." Church committee, *Final Report* 1:75. For another alleged example of pressure brought to bear on analysts to provide data in support of policy views, see Ralph McGehee, "The C.I.A. and the White Paper on El Salvador," *Nation,* April 11, 1981, 423–25. In 1984, CIA analyst John Horan further charged that he had been pressed to redraft an NIE to buttress the Latin American policies of the Reagan administration. The House Committee on Intelligence investigated the allegation, but found that in fact the NIE had presented dissenting views. See House Permanent Select Committee on Intelligence, *Annual Report,* H. Rept.

No. 98-1196, January 2, 1985, p. 6 (hereafter cited as House Intelligence Committee, *Annual Report*).

24. Church committee, *Final Report* 1:78.

25. Ibid., 78. The paragraph deprecated the likelihood of Soviet preparations for a first-strike capability against the United States.

26. See Wyden, *Bay of Pigs*, 99.

27. See ibid. on this point and for the following characterization of Bissell's part in the episode. According to Powers, Bissell never consulted his Agency's Cuban analysts either (*The Man Who Kept the Secrets*, p. 145). Former secretary of state Dean Rusk remembers that the Bay of Pigs operation was closely held for security purposes; to consult widely would have been, in his view, to risk leaks on the operation. Interview, Athens, Georgia, January 9, 1985.

28. See Janis, *Groupthink*, 41.

29. Hughes, "The Power to Speak," 24.

30. Richard K. Betts, in Godson, *Analysis and Estimates*, 179.

31. Colby, "Understanding the Realities," 3.

32. Church committee, *Final Report* 1:274, summarizing "A Review of the Intelligence Community" (the Schlesinger Report), Office of Management and Budget, March 10, 1971, 11.

33. On the Noriega connection, see the *New York Times*, February 6 (p. 5), 7 (p. 18), 12 (p. A6), and 13 (p. 1), April 5 (p. A1), and May 21, 1988 (p. 8); and the *Washington Post*, May 21, 1988, p. A14. The "hard information" quote is from the Director of DEA's predecessor, the Federal Bureau of Narcotics and Dangerous Drugs, cited by the *New York Times*, February 13, 1988, p. 1. For a look at a Senate Foreign Relations Subcommittee's hearings on Noriega's drug dealings, see David Corn, "Can He [Subcommittee Chairman John Kerry, D, Massachusetts] Lift the CIA Veil?" *Nation* 246, April 30, 1988, p. 589.

Intelligence professionals are inclined to view discussions of morality and espionage as more than faintly naive. (See, for instance, the *London Observer* interview with former DCI Richard Helms, reprinted in the *Washington Star*, January 20, 1980, pp. G1, G4.) Sleazy characters like Manucher Ghorbanifar, an arms middleman used by the Reagan administration in the Iran-contra affair, are sometimes indispensible, they maintain, to the success of covert operations in parts of the world where the United States has few contacts. A passage from the pen of novelist John le Carré, in which an embittered British intelligence officer describes his colleagues, captures a widespread attitude in the CIA about the real world in which intelligence officials must operate: "What do you think spies are: priests, saints, and martyrs? They're a squalid procession of vain fools, traitors too, yes; pansies, sadists and drunkards, people who play cowboys and Indians to brighten their rotten lives. Do you think they sit like monks in London, balancing the rights and wrongs?" (*The Spy Who Came in from the Cold*, [London: Pan Books, 1964], 210–11). For a scholarly discussion of intelligence and morality, see E. Drexel Godfrey, Jr., "Ethics and Intelligence," *Foreign Affairs* 56 (April 1978), 624–42.

34. Interview with Thomas L. Hughes, Washington, D.C., June 1983.

35. CIA officials, interviews, Washington, D.C., June 1983. For Church committee criticism of this broad-net approach, see its *Final Report* 1:275. In 1979, it proved useful to have a data bank on Cuba to check on whether or not the Soviets had recently developed a new combat unit of brigade strength, as claimed by the NSA; other intelligence agencies were able to trace—far too slowly in the opinion of critics—the existence of this brigade back to 1962, when agreements between the Kennedy administration and the Soviet Union authorized its presence as part of the negotiations over the removal of Soviet missiles from Cuba. See David D. Newsom, *The Soviet Brigade in Cuba.*

36. Pacem in Terris IV Convocation.

37. CIA official, interview, Washington, D.C., January 1976.

38. The Church committee, "Alleged Assassination Plots Involving Foreign Leaders," 142, n. 2.

39. Ibid., 72. On alleged OSS efforts to create a better world through chemicals, see the plot to make Adolf Hitler's moustache fall off and his voice become soprano by injecting the Führer's carrots with female sex hormones, Stanley Lovell, *Of Spies and Strategems* (New York: Prentice-Hall, 1963), 94.

40. See Loory, "The CIA's Use of the Press," 13.

41. Sen. Frank Church, "Covert Action: Swampland of American Foreign Policy," *Bulletin of the Atomic Scientists* 32 (February 1976), 9. "How impoverished must a country be before it is not a threat to the U.S. government?" asked the prominent West German writer Günter Grass in 1983, in reference to CIA covert action in Nicaragua (*Nation*, March 12, 1983, p. 301).

42. United States Senate, Washington, D.C., October 24, 1975.

43. Quoted by Powers, *The Man Who Kept the Secrets*, 266, emphasis in original.

44. Church committee, *Final Report* 1:9.

45. Deputy director for operations, interview, Washington, D.C., March 1978.

46. Treverton, *Covert Action* 9. For former CIA director Richard Helms's apparent disgust with covert propaganda operations "of no consequence," see Powers, *The Man Who Kept the Secrets*, 101 (for his skepticism about covert action generally, see p. 28).

47. For former senator Church, this remained the central argument against covert action. "If we have gained little [from covert action], what then have we lost?" he once asked. His response was "our good name and reputation." Church, "Covert Action," 11 [n. 49]. See also Frank Church, "Do We Still Plot Murders? Who Will Believe We Don't?" *Los Angeles Times*, pt. 2, June 14, 1983, p. 5.

48. See Stansfield Turner (DCI, 1977–81), "From an Ex-CIA Chief: Stop the 'Covert' Operation in Nicaragua," *Washington Post* (Outlook Section), April 21, 1983, p. C1.

49. Cover is defined by the Church committee as "a protective guise used by a person, organization, or installation to prevent identification with clandestine activities and to conceal the true affiliation of personnel and the true sponsorship of their activities." *Final Report* 1:620.

50. For an account of this tragedy, see Jermiah O'Leary, "Cover Blown, CIA Agent in Athens Killed," *Washington Star*, December 24, 1975, p. A1.

51. See Dan Morgan, "Slain Agent Feared for CIA Lives," *Washington Post*, December 26, 1975, p. A1. The COS in Lebanon, William Buckley, murdered in 1985 by a terrorist group, was also reportedly well known in the region as a probable CIA officer; for him to have stayed on in Lebanon under those conditions seems of questionable judgment—though little is still known about this case. On the visibility of CIA officers abroad, see John Marks, "How to Spot a Spook," *Washington Monthly*, November 1974, 5–11.

52. Aspin Hearings, 8.

53. Though for criticism of proprietaries, see the testimony of Lawrence R. Houston, former CIA general counsel, the Church committee, *Final Report* 1:255.

54. Testimony of Ambassador William Porter, Aspin Hearings, 180.

55. Ferdinand Mount, "Spook's Disease," *National Review* 32 (March 7, 1980), 300. On these causes and peoples, see generally Powers, *The Man Who Kept the Secrets*, and Marchetti and Marks, *The CIA*. On the Kurds, see the Pike Committee *Report*. On the Vietnam debacle, see Frank Snepp, *Decent Interval* (New York: Random House, 1979), especially the "Postscript," pp. 573–80.

56. CIA official, interview, Washington, D.C., June 1983.

57. CIA official, interview, Washington, D.C., June 1983.

58. Angelo Codevilla, "The CIA: What Have Three Decades Wrought," *Strategic Review* (Winter 1980), 68, emphasis in original. On counterintelligence, see Newton S. Miler, "Counterintelligence," in Roy Godson, ed., *Intelligence Requirements for the 1980s: Elements of Intelligence* (Washington, D.C.: National Strategy Information Center, 1979), 47–60; and Roy Godson, ed., *Intelligence Requirements for the 1980s: Counterintelligence* (Washington, D.C.: National Strategy Information Center, 1980). The CIA went to great lengths in 1975 to negotiate with the Church committee for the removal, on security

grounds, of large sections of the committee's CI report. Much of this censured material is freely discussed by former CIA officials in the Godson-edited volumes.

59. Interviews with Angleton, Washington, D.C., June–December 1975. An Angleton protege has written that CI must be "monolithically centralized," Miler, "Counterintelligence," 42.

60. Interviews with Angleton's successor and his staff, Washington, D.C., November 1975, and subsequent interviews with CI personnel.

61. Norman L. Smith, "Counterintelligence Organization and Operational Security in the 1980s," in Godson, *Counterintelligence,* 216.

62. Ibid., 222.

63. Ibid., 254.

64. Press conference, U.S. Capitol, Washington, D.C., July 19, 1975.

65. Church committee, "Intelligence Activities," *Hearings,* September 18, 1975, 1:187. So did another troubling case: the testing by the CIA of LSD, an hallucinogenic drug, on an unwitting Department of the Army scientist. The victim experienced serious side effects and within days took his own life. See the Rockefeller Commission Report, 227.

66. On the lack of knowledge for each of these officials, see Church committee, "Alleged Assassination Plots," esp. pp. 92, 119, 151, and 179.

67. Ibid., 154.

68. Church committee, "Intelligence Activities," *Hearings,* December 4, 1974, 7:51–52. Adm. Stansfield Turner's inside look at the CIA is a startling acknowledgement of the difficulties even an Agency director has in controlling his domain; Turner felt that he never did have the CIA under proper control. See his *Secrecy and Democracy.*

69. On the Iran arms sale, see the Tower commission, *Report of the President's Special Review Board* (Washington, D.C.: Government Printing Office, February 26, 1987) (hereafter cited as the Tower Commission *Report*), and the Inouye-Hamilton committees, *Report.* On the Army's covert-action capability, see Jay Peterzell, "Can Congress Really Check the CIA?" *Washington Post,* April 21, 1983, p. C4; House Intelligence Committee, *Annual Report,* 14–15; Steven Emerson, *Secret Warriors* (New York: Putnam, 1988); and *Time,* August 31, 1987.

71. CIA official, interview, Washington, D.C., June 1983.

72. Intelligence official, interview, Washington, D.C., June 1983.

73. House Intelligence Committee, *Annual Report* (1985), 12.

74. Ibid., 13–14.

75. Ibid., 16.

76. Former CIA official, interview, Washington, D.C., June 1983. See also the multivolume series edited by Roy Godson, *Intelligence Requirements for the 1980s* (Boston: Lexington, 1984, 1985).

77. William B. Bader in a *New York Times* Roundtable, "The Battle over 'Covert' Activities in Central America," *New York Times,* June 12, 1983, p. E3.

Chapter 5

1. Central Intelligence Agency, *Fact Book on Intelligence,* 17.

2. See Arthur S. Hulnick, "The Intelligence Producer-Policy Consumer Linkage: A Theoretical Approach," *Intelligence and National Security* 1 (May 1986), 212–33; and, "Relations between Intelligence Producers and Policy Consumers: A New Way of Looking at an Old Problem," in Stephen J. Cimbala, ed., *Intelligence and Intelligence Policy in a Democratic Society* (Dobbs Ferry, N.Y.: Transnational, 1987), pp. 129–44.

3. Robert M. Gates, remarks, Conference on U.S. Intelligence, June 11, 1984.

4. See John Lewis Gaddis, *Strategies of Containment: A Critical Appraisal of Postwar American National Security Policy* (New York: Oxford, 1982).

5. "Report of the Secretary of Defense to the Congress of the FY 1982 Budget," January 19, 1981.

6. Quoted in Theodore H. White, "Weinberger on the Ramparts", *New York Times Sunday Magazine*, February 6, 1983, p. 19.

7. Written by a senior CIA analyst, February 21, 1974, mimeograph, and provided in a declassified form to the Church committee, September 1975.

8. Gates, Conference on U.S. Intelligence, June 11, 1984.

9. For a similar CIA listing of its interests, "staggering in their diversity," see Robert M. Gates, "Text of Speech at Harvard by Deputy CIA Director Outlining Policy Shifts," *Chronicle of Higher Education*, February 26, 1986, p. 27.

10. Gates, Conference on U.S. Intelligence, June 11, 1984.

11. Carter, press conference, November 30, 1978, answer to question No. 16.

12. Arthur S. Hulnick, Conference on Intelligence, Policy, and Process, U.S. Air Force Academy, Colorado Springs, Colorado, June 6, 1984.

13. See Flanagan, "The Coordination of National Intelligence," 157–96.

14. Quoted by Suzanne Garment, "Casey's Shadows: A Greater Emphasis on CIA Analysis," *Wall Street Journal*, July 16, 1982, p. 16.

15. Gates, Conference on U.S. Intelligence, June 11, 1984.

16. Ibid.

17. Ibid.

18. Ibid.

19. Ransom, *The Intelligence Establishment*, 20. My interviews indicate a figure of 75 percent open-source collection.

20. Comment by senior CIA staffer, Conference on U.S. Intelligence, June 11, 1984.

21. See Douglas Sutherland, *The Great Betrayal: The Definitive Story of Blunt, Philby, Burgess, and Maclean* (New York: Time, 1980); Barrie Penrose and Simon Freeman, *Conspiracy of Silence: The Secret Life of Anthony Blunt* (New York: Farrar, Straus, and Giroux, 1987).

22. See Greville M. Wynne, *The Man from Moscow: The Story of Wynne and Penkovsky* (London: Hutchinson, 1967).

23. See Powers, *The Man Who Kept the Secrets*, 447, n. 6.

24. Interview with senior intelligence analyst, Langley, Virginia, June 11, 1984.

25. Interview with a long-time NIO, Langley, Virginia, June 12, 1984.

26. Gary D. Brewer and Paul Bracken, "Some Missing Pieces of the C^3 Puzzle," *Journal of Conflict Resolution* 28 (September 1984), 453. See also Martin Tolchin, "Pick a Number," *New York Times*, June 5, 1984, p. 10.

27. William E. Burrows, *Deep Black: Space Espionage and National Security* (New York: Random House, 1986), 116.

28. House Intelligence Committee, "Iran: Evaluation of U.S. Intelligence Performance," 7.

29. Carter, press conference, November 30, 1978, in answer to question No. 16.

30. Letter to the Editor, *Washington Post*, December 10, 1978, p. C–6. In a separate interview, Turner has commented: "I am criticized for chopping the espionage section [of the CIA]. I eliminated 805 spaces not by firings, but by not rehiring, and the fact is, many CIA espionage people were near retirement. It was overstaffed and this was blown out of proportion," John Patrick Quirk et al., *The Central Intelligence Agency: A Photographic History* (Guilford, Conn.: Foreign Intelligence Press, 1986), 199.

31. Interview with Les Aspin (D, Wisconsin), then a member of the House Permanent Select on Intelligence, June 10, 1979.

32. On these capabilities, see John Prados, *The Soviet Estimate: U.S. Intelligence Analysis of Russian Military Strength* (New York: Dial, 1982), 276. The *New York Times* reports that U.S. satellites provide constant coverage of the Soviet Union and can detect objects on the ground that are less than six inches wide (February 24, 1986, p. 12). A pioneer in the development of CIA aerial reconnaissance, Richard Bissell (also the chief architect of the Bay of Pigs operation), has said that one reconnaissance satellite "can read a message printed on a fingernail" from its cameras, Quirk et al., *The Central Intelligence Agency*, 212. We apparently have the capability as well to pinpoint surveillance in trouble spots. In

early 1986, U.S. spy satellites reportedly passed over Libya several times a day; the photographs provided an accurate count of runways, missiles, launchers, radar, and antiaircraft sites (ABC, "Evening News," January 28, 1986).

33. For one account of a CIA recruitment pitch, see David Atlee Phillips, *The Night Watch: Twenty-Five Years of Peculiar Service* (New York: Antheneum, 1977), 20–25. According to my interviews with experienced CIA case officers, most agents—probably in the neighborhood of 70 percent—want money, though usually not much more than a modest monthly "stipend." On Soviet inducements to recruit spies in the West, William Colby observes: "Today, nobody is ideologically attracted to the Soviets, anywhere; even the Soviets have lost their belief in the ideological approach. Today they go for the sergeant or the major, or something like that, who's in trouble with money and offer him a little money" (ABC, "Evening News," January 7, 1986).

34. CIA briefing paper (unclassified) prepared for Sen. Frank Church, September 30, 1975. See also Fred Kaplan, *The Wizards of Armageddon* (New York: Simon and Schuster, 1983), 161; and Burrows, *Deep Black*. Roger Hilsman, former director of INR at the Department of State, observes that "the Air Force counted missiles behind every cloud" during this period, further inflating the estimates of the intelligence community, Conference on U.S. Intelligence, June 5, 1984. According to a senior CIA official, the Air Force had a budget motive: the more Soviet missiles in the estimate, the better the Air Force's chances for increased missile funding. Staff interview, Church committee, July 22, 1975, Washington, D.C. (Church committee files). See also the *New York Times*, April 28, 1966, p. 28.

35. Briefing on photographic intelligence, Conference on U.S. Intelligence, June 12, 1984. The U.S. government is disinclined to acknowledge formally the presence of satellite surveillance; officials prefer to keep the subject "black," in intelligence argot, and refer to this capability obliquely as National Technical Means (NTM). The capability, however, is widely known, and every now and then even an official acknowledgment slips by. See, for example, *New York Times*, March 17, 1967; *Department of State Bulletin*, September 1, 1975; *New York Times*, October 1, 1978; and DCI Stansfield Turner, remarks, National Conference of Bar Presidents, Dallas, Texas, August 10, 1979.

36. Unclassified CIA briefing paper for Senator Church, op. cit. The satellite photographs showed that the Air Force count of Soviet missiles had been far too inflated, Hilsman, Conference on U.S. Intelligence, June 5, 1984.

37. Burrows, *Deep Black*, 306.

38. On the *Glomer Explorer*, see the *Washington Post*, March 19, 20, 21, 23 and 30, 1975; *New York Times*, March 20 and 26, 1975.

39. Gates, Conference on U.S. Intelligence, June 11, 1984.

40. Kent, *Strategic Intelligence for American World Policy* (Princeton, N.J.: Princeton University Press, 1949), 64–65.

41. Senior intelligence officials, Conference on U.S. Intelligence, June 11, 1984.

42. NSC aide, interview, November 19, 1984, Washington, D.C.

43. Charles Briggs, CIA executive director, remarks, Conference on U.S. Intelligence, June 11, 1984.

44. See Laurence Stern, "CIA Stops Sending Daily Report to Hill," *Washington Post*, February 4, 1976, A–1.

45. Ibid.

46. Interviews with staff members, congressional committees on intelligence, December 21 and 22, 1980.

47. Kirkpatrick, *Military Review*, May 1961, p. 20, cited by Ransom, *The Intelligence Establishment*, 147.

48. Ibid., 147.

49. See Kaplan, *The Wizards of Armageddon*, and Prados, *The Soviet Estimate*.

50. NIO, interview, Washington, D.C., November 10, 1980.

51. Flanagan, "The Coordination of National Intelligence," 187.

52. CIA briefing paper for Senator Church, op. cit.

53. See *Congressional Record*, September 24, 1986, S13567.

54. Chief, CIA product evaluation staff, Conference on U.S. Intelligence, June 12, 1985.

55. Richard K. Betts, "Analysis, War and Decision: Why Intelligence Failures Are Inevitable," *World Politics* 31 (October 1978), 78. Several observers have suggested that one way to reduce the failures would be to test more NIEs against the rigors of independent analysis by specialists outside the Agency, as occurred in 1976–77 during the now famous "A-Team, B-Team" exercise. Team A comprised a group of CIA analysis who prepared an NIE on Soviet nuclear capabilities; Team B comprised a group of outside experts asked by DCI George Bush (who reportedly found the NIE insufficiently hardline) to critique the NIE. Although Team B had a distinctly conservative bias in the known views of its members, the idea of an outside scrutiny of CIA analytic assumptions and methods remains valuable—though in this instance the Agency's professionals seemed to resent outside "second-guessing" and may be resistant to future A-Team, B-Team critiques. For an account of this experience from the vantage point of a B-Team member, see Richard Pipes, "Team B: The Reality Behind the Myth," *Commentary* 82, October 1986, 25–40.

56. *Congressional Record*, November 11, 1975, p. 35786.

57. Gregg, Conference on U.S. Intelligence, June 13, 1984.

58. CIA officer Arthur S. Hulnick, public lecture, Athens, Georgia, November 20, 1984.

59. Gates, Conference on U.S. Intelligence, June 14, 1984.

60. Ibid.

61. Arthur S. Hulnick, Conference on U.S. Intelligence, June 6, 1984.

62. Hulnick, public lecture, op. cit.

63. Remarks, Roger Hilsman, Conference on Intelligence, Policy, and Process.

64. Donald Gregg, Conference on U.S. Intelligence, June 13, 1984.

65. Hulnick, Conference on Intelligence, Policy, and Process.

66. Directorate of Intelligence official, interview, Washington, D.C., November 16, 1980.

67. Gregg, Conference on U.S. Intelligence, June 13, 1984.

68. Gates, Conference on U.S. Intelligence, June 11, 1984.

69. Donald Gregg's phrase describing the 1975–76 legislative investigations of the intelligence community, Conference on U.S. Intelligence, June 13, 1984.

70. Ibid. Gates sees a strength in this congressional involvement: "[T]he sharing of intelligence with Congress—where members of both parties, with a wide range of views and philosophy all see the information—is one of the surest guarantees of the CIA's independence and objectivity." "The CIA and American Foreign Policy," 229.

71. Gates, Conference on U.S. Intelligence, June 11, 1984.

72. Hulnick, public lecture, op. cit.

73. See Loch K. Johnson, "Decision Costs in the Intelligence Cycle," *Journal of Strategic Studies* 7 (September 1984), 318–35.

74. Gates, Conference on U.S. Intelligence, June 11, 1984, emphasis in original.

75. Ibid.

76. Senior officer, Intelligence Community Staff, Conference on Intelligence, Policy, and Process, June 6, 1984.

77. MacEachin, Conference on U.S. Intelligence, June 11, 1984.

78. Ibid.

79. Interview with CIA Legislative Liaison staffer, September 28, 1980.

80. MacEachin, Conference on U.S. Intelligence, June 11, 1984.

81. Ed Quam, Defense Intelligence Agency, Conference on Intelligence, Policy, and Process, June 6, 1984.

82. Interview with senior CIA official, Langley, Virginia, June 13, 1984. For an illustration of a close working relationship between a president and a DCI, see the account of the John F. Kennedy–John McCone association during the Cuban missile crisis, as recounted by Peter S. Usowski, "An Activist Approach to the Intelligence-Policy Relation-

ship: John McCone and the Cuban Missile Crisis," paper delivered at the Annual Meeting of the International Studies Association, St. Louis, April 2, 1988. The White House valued and listened to McCone's intelligence data and analysis, but kept him at arms length when it came to policy recommendations—an arrangement McCone understood and accepted. When Lyndon Johnson assumed the presidency, however, McCone gradually found himself phased out of frequent White House contact (see Powers, *The Man Who Kept the Secrets*, 209).

83. Briggs, Conference on U.S. Intelligence, June 11, 1984.

84. See the Tower commission *Report*.

85. MacEachin, Conference on U.S. Intelligence, June 11, 1984.

Chapter 6

1. I have been conducting interviews since 1975 with CIA officials knowledgable about covert action (CA) and other intelligence subjects, first as a congressional investigator on the Church committee, then as an aide on the Boland committee (the House Permanent Select Committee on Intelligence) and, since 1979, as a university researcher. The interviews are hard to characterize, for they have been of many forms and lengths. Some have been full-blown, in-depth sessions lasting several hours and probing the nuances of the entire subject: history, definitions, targets, modus operandi, decision paths, accountability, ethics. Others have been of shorter duration, either because the official's time was limited or because the purpose was to probe only a specific aspect of the subject (say, CIA–State Department coordination for propaganda operations).

In the period from 1975 to 1988, I have explored the subject of covert action with 158 CIA, Department of State, and congressional officials. Of these, 74 were affiliated with the CIA: 35 still on active duty and the rest retired. Their positions ranged from DCIs through every major CA position down to case officers. The rule of thumb used to select people to interview was simple: anyone in the business who would speak. I learned of possible individuals to interview at panels on national security during various professional conferences, through my contacts as a former legislative aide, and by referrals from early interviews.

The interview sessions have taken place in private homes, Washington offices and restaurants, CIA Headquarters (where as a scholar I have lectured and attended conferences), and in hotel rooms in between panels at conferences. In each instance, I asked the respondent to comment about the issues that comprise this chapter: what the types and targets of covert action are, how the decision to use covert action is reached, how the operations are supervised, and the like. Intelligence officers will simply not respond to an overly formal set of questions, at least not in any insightful fashion; I discovered early that the best approach was to raise general questions about the decision process and proceed in a conversational tone—staying away from highly sensitive subjects like the details of specific operations abroad (which were of no interest anyway to my purpose, which is a scholarly examination of the decision paths and the extent of accountability for covert action). Though I attempted to be conversational, these sessions were more than "conversations"; I always had in mind and gently directed the interviews toward the subjects covered in this study. I almost always took notes during the interviews, except for three occasions when a taperecorder was permitted; immediately after the interviews, I would find a quiet place to write down further recollections of the responses. Each of the observations presented in this chapter (and throughout this book) has been substantiated by at least three independent interviews with individuals holding relevant institutional positions within the government.

With rare exception, those interviewed insisted on anonymity—an instinct based on a career of hidden identity. Virtually all the respondents were biased—as former or current practitioners—in favor of covert action; to one degree or another, they thought it worked. For the purposes here, though, these biases are largely irrelevant; by cross-checking the various interviews, I believe I have achieved a reliable account of what I set out to understand: how the United States decides upon and supervises the so-called quiet option.

For useful public sources on covert action, see: Philip Agee, *Inside the Company: CIA Diary* (Harmondsworth, Eng.: Penguin, 1975); Paul W. Blackstock, *The Strategy of Subversion* (Chicago: Quadrangle, 1964); Cline, *Secrets, Spies, and Scholars;* Leslie Cockburn, *Out of Control: The Story of the Reagan Administration's Secret War in Nicaragua, the Illegal Arms Pipeline, and the Contra Drug Connection* (New York: Atlantic Monthly, 1988); Colby and Forbath, *Honorable Men;* John M. Collins, *Green Berets, Seals, and Spetsnaz: U.S. and Soviet Special Military Operations* (New York: Pergamon, 1987); William R. Corson, *The Armies of Ignorance;* Godson, *Covert Action;* Inouye-Hamilton committees, *Report;* Loch K. Johnson, "The CIA: Controlling the Quiet Option," *Foreign Policy* 39 (Summer 1980), 143–52, and *A Season of Inquiry;* Church committee, *Final Report,* vols. 1, 4; Hans Moses, *The Clandestine Service of the Central Intelligence Agency,* vol. 1 of the Intelligence Profession Series (McLean, Virginia: Association of Former Intelligence Officers, 1983); Phillips, *The Night Watch;* Powers, *The Man Who Kept the Secrets;* John Prados, *Presidents' Secret Wars: CIA and Pentagon Covert Operations Since World War II* (New York: Morrow, 1986); Ranelagh, *The Rise and Decline of the CIA;* Ransom, *The Intelligence Establishment;* Richelson and Ball, *The Ties That Bind;* Rositzke, *The CIA's Secret Operations;* Shackley, *The Third Option;* Joseph Burkholder Smith, *Portrait of a Cold Warrior;* John Stockwell, *In Search of Enemies: A CIA Story* (New York: Norton, 1978); Tower Commission *Report;* Treverton, *Covert Action;* Turner, *Secrecy and Democracy;* Wise and Ross, *The Invisible Government;* Woodward, *The CIA's Secret Wars;* Peter Wyden, *Bay of Pigs.*

The findings of the Church and Pike panels may be found in the Church committee, *Final Report* (6 vols.) and the Pike Committee *Report.*

2. Church committee, *Final Report* 4:29.

3. Ibid., 31. In 1949, the budget for OPC was $4,700,000; by 1952, it had risen to $82,000,000.

4. Ibid., 67.

5. For the expose, see Stern, "NSA and the CIA," op. cit., 29–38.

6. For Helms's skeptical views on covert action, see Powers, *The Man Who Kept Secrets,* 28.

7. Church committee, *Final Report* 4:69.

8. Ibid.

9. Ibid., 87.

10. On covert action in Chile, see the *New York Times,* December 22–31, 1974; on the assassination plots, see Church committee, "Alleged Assassination Plots."

11. Though Casey's OSS responsibilities were primarily to handle clandestine intelligence collection in Europe, many of these agents were also used for covert action, and evidently Casey ran some of these operations, too. Interviews with senior CIA administrators, June 4, 1984, Washington, D.C., and September 5, 1987, Chicago, Illinois. See also Roger Morris, "William Casey's Past," *Atlanta Constitution,* August 31, 1987, p. A11.

12. Church, "Covert Action," 9.

13. For the relevant wording of the National Security Act and the Hughes-Ryan Act, see respectively 50 U.S.C. 401 and 50 U.S.C. 403. On the question of legal authority for the conduct of covert action, see the *New York Times,* February 8 and 13, 1976; Jay Peterzell, "Legal Constitutional Authority for Covert Operations," *First Principles,* (Washington, D.C.: Center for National Security Studies, Spring 1985), 1–5; and John T. Elliff, "Statutory Limitations on Covert Actions," paper delivered at the Annual Meeting of the International Studies Association, Washington, D.C., March 8, 1985.

14. Church committee, *Final Report* 4:26 and 1:48–49.

15. Church committee, *Final Report* 4:35. These observations are based as well on the Pike Committee *Report* and a series of interviews with CIA officials.

16. Church committee, *Final Report* 1:50, 47.

17. Rositzke, *The CIA's Secret Operations,* 153.

18. Inouye-Hamilton committees, *Hearings,* July 15, 1987; see also the committee's *Report,* 17.

19. Interview with Dean Rusk, February 21, 1985, Athens, Georgia.

20. The memorandum was dated February 21, 1967; see Church committee, *Final Report* 1:56, 57.

21. Church committee, *Final Report,* 4:89.

22. Johnson, *A Season of Inquiry,* 6.

23. Quoted by Bob Wiedrich, "Can Congress Keep a Secret?" *Chicago Tribune,* February 3, 1976.

24. Richard B. Russell Library, Oral History No. 86, taped by Hughes Cates, February 22, 1977, University of Georgia, Athens, Georgia.

25. See Church committee, "Alleged Assassination Plots."

26. On this statute, see Johnson, "Quiet Option," and "Legislative Reform of Intelligence Policy," *Polity* 17 (Spring 1985), 549–73.

27. Turner, *Secrecy and Democracy,* 170.

28. I was a member of the drafting subcommittee for the Democratic party platform in 1976 and witnessed the control of the Carter delegates over these procedures and their interest in an intelligence plank.

29. For the text of the Order, see *Public Papers of the Presidents of the United States: Jimmy Carter, 1978,* 1:194–214.

30. Leslie H. Gelb, "Shift Is Reported on C.I.A. Actions," *New York Times,* June 11, 1984, p. 1.

31. Ibid.

32. Ibid.

33. Stockwell, *In Search of Enemies.*

34. CBS, "60 Minutes," May 14, 1978. According to a knowledgeable senior CIA officer, "The Agency had no *advisers* in Angola at that time, but it did have *liaison officers* [with the pro-Western rebels], plus a radioman—just four or five people" (interview, October 22, 1980, Washington, D.C.; emphasis added). Such distinctions can be rather fine—slippery slopes of terminology upon which covert actions might slide quickly past even the most attentive overseer.

35. Interview with senior DDO official, Washington, D.C., November 16, 1980.

36. Based on interviews with DDO personnel, November 1980 and June 1984, Washington, D.C.

37. On reserve releases from the Contingency Reserve Fund, see Senate Select Committee on Intelligence, *Annual Report to the Senate,* May 18, 1977, S. Rept. No. 95–217, 18–19 (hereafter cited as Senate Intelligence Committee, *Annual Report to the Senate*).

38. Seymour M. Hersh, "Congress Is Accused of Laxity on C.I.A.'s Covert Activity," *New York Times,* June 1, 1978, p. 1. The ties between Gen. Manual Antonio Noriega of Panama and the CIA, which became controversial in 1988, probably find their formal authorization in Ford's generic finding to combat drug trafficking—ironically a service Noriega promised to provide (among others) even as he himself became involved in the trafficking.

39. Interview with senior DDO official, Washington, D.C., October 22, 1978.

40. Turner, *Secrecy and Democracy,* 169. According to the Inouye-Hamilton committees, "while [William J.] Casey was Director of Central Intelligence, CIA personnel attempted to craft Findings in terms so broad that they would not limit the CIA's freedom to act" (*Report,* 379).

41. Interviews with CIA and Department of State officials, Washington, D.C., November 9, 1980; June 22 and 23, 1984; and August 20, 1984.

42. CIA official, interview, September 30, 1980, Washington, D.C.

43. Interviews with CIA officials, June 16 and 17, 1983.

44. Interview with retired DDO official, March 21, 1985, Washington, D.C.

45. Interview with former senior NSC staff aide, December 30, 1985, Washington,

D.C. According to interviews with CIA officials (June 4 and 5, 1984, Washington, D.C.), about 54 percent of all CA proposals in recent years have been rejected at the SCC and NSPG levels. Of those that made it to the Oval Office, about 15 percent were turned down.

46. On the representativeness of the congressional intelligence committees, see Frederick M. Kaiser, "Congressional Roles and Conflict Resolution: Access to Information in the House Select Committee on Intelligence," *Congress and the Presidency* 15 (Spring 1988), 49–73.

47. For Hamilton's views, see his "The Role of Intelligence in the Foreign Policy Process," lecture, University of Virginia, reprinted in *Extracts from Studies in Intelligence* (Langley, Va.: Central Intelligence Agency, September 1987), 69. A chairman of the Senate Intelligence Committee has stated: "After four years of reviewing the covert operations of our intelligence system, I cannot conceive of any circumstances which would require the withholding of prior notice except where the nation is under attack and the president has no time to consult with Congress before responding to save the country" (Sen. Daniel K. Inouye, *Congressional Record,* September 19, 1980, S12691). For more on this debate over prior notice, see also Turner, *Secrecy and Democracy,* 170; *New York Times,* February 29, 1980; "Congressional Oversight of Covert Activities," *Hearings,* House Permanent Select Committee on Intelligence, September 22, 1983, esp. p. 98; and, chap. 10 of this book.

48. See Turner, *Secrecy and Democracy,* 170; and House *Report,* No. 100–705, June 15, 1988, 54.

49. Interview with Senate Intelligence Committee staffers, November 17, 1980; and with Les Aspin (D, Wisconsin), November 18, 1980, Washington, D.C.

50. Interviews with staffers on the Senate and House intelligence committees, June 6 and 7, 1986. The covert mining of harbors was hardly a novel idea; the Agency used this technique in Cuba and Vietnam during the 1960s.

51. Discussions with senior intelligence officials, including DCI Casey, June 11, 1984, Langley, Virginia.

52. The letter was dated April 9, 1984; see the *Washington Post,* April 11, 1984, p. A–17.

53. Discussions with senior intelligence officials, June 11, 1984, op. cit.

54. Turner, *Secrecy and Democracy,* 167, 168.

55. See the Tower Commission *Report,* IV–9, B–60, B–67.

56. See the *New York Times,* January 15, 1987, p. 12.

57. See Steven V. Roberts, "More Lessons in the Secrecy Trade," *New York Times,* April 10, 1986, p. 12; and David B. Ottaway and Patrick E. Tyler, "New Era of Mistrust Marks Congress' Role," *Washington Post,* May 19, 1986, pp. A1, A10.

58. *New York Times,* January 13, 1987 p. 1; see also the Tower Commission *Report,* IV–5, and Inouye-Hamilton Committees, *Report,* 6–7.

59. Interview, David Brinkley, ABC, "This Week with David Brinkley," December 14, 1986.

60. Interviews with senior intelligence officers, November 1980; see also Senate Intelligence Committee, *Annual Report to the Senate,* 2. Former DCI Turner has said that, under congressional pressure, "three times Reagan signed, then cancelled, covert action operations," interview, WGST Radio, July 30, 1985, Atlanta, Georgia. Votes in the House and Senate intelligence committees have tended to be unanimous on most occasions, though on a few controversial votes (like the Boland amendments) divisions have occurred—along party lines in the House Committee, less so in the Senate Committee (see Sen. David L. Boren, chairman, Senate Intelligence Committee, remarks before the Association of Former Intelligence Officers, Ft. Myer, Virginia, March 28, 1988, rpt. in *Periscope* 13, Spring 1988, 7).

61. Interview with staff aide, Senate Select Committee on Intelligence, December 12, 1980, Washington, D.C. Most of the time, though, the intelligence committees have been supportive of the intelligence agencies, whose budgets have increased each year beginning

in 1977 (interviews with CIA and congressional officials, June 11, 1984, March 22, 1986, and September 2, 1986)—indeed, according to one report, have tripled in the past decade (Gelb, "Overseeing of C.I.A. by Congress," p. 1).

62. The Congress passed the Clark amendment in 1975 and then repealed the statute, at the urging of the Reagan administration, a decade later; the Hamilton amendment, an attempt to reinstate the prohibition against covert action in Angola (see House Resolution 4759, *Rept.* No. 99–690, July 30, 1986), failed on September 17, 1986. On the Clark amendment and the Hamilton amendment, see respectively: Section 118 of the International Security and Development Cooperation Act of 1980 (22 U.S.C. 2293, note), and Section 793 of the FY 1983 Defense Appropriation Act (P.L. 97-377). Congress passed a series of six Boland amendments between 1982 and 1986, the last a limit on funding for the contras to a $100 million ceiling for humanitarian and military assistance. For an outline of these various measures, see Henry A. Kissinger, "A Matter of Balance," *Los Angeles Times,* July 26, 1987, p. V-1.

63. See, for example, Gelb, "Overseeing of C.I.A. by Congress," A8; *New York Times,* July 28, 1986, p. 1; and Johnson, *A Season of Inquiry,* 206–7. In 1971, an internal CIA probe concluded that only 5 percent of serious leaks seemed to have come from Capitol Hill (reported by George Lardner, Jr., "Moynihan Unleashes the C.I.A.," *Nation,* February 16, 1980, 177, the same source for the number of personnel with top-secret clearances). [The House Intelligence Committee reports that over five million Americans held security clearances in 1985—a 40 percent increase since 1980 (H. *Report 100–5,* February 4, 1987, 11–12)]. Though few leaks have come from Congress, two in 1987 caused a stir. Patrick J. Leahy (D, Vermont), the former vice-chairman of the Senate Intelligence Committee, presented information from his panel's preliminary Iran-contra inquiry during an NBC "Nightly News" interview on January 8, 1987. Though the information was unclassified, the committee had voted against the release of its findings and therefore the Leahy disclosure represented a breach of the committee's rules. In March 1987, while addressing an audience in Florida, David Durenberger (R, Minnesota), the former chairman of the Senate Intelligence Committee, let slip that during the early 1980s a U.S. intelligence agency had recruited an Israeli military officer—despite the existence of a U.S.-Israeli agreement prohibiting the recruitment of assets in one another's governments. According to one former CIA officer, this incident "sent a shock wave through the professional intelligence community around the world" (George Carver, House *Report No. 100–705,* June 15, 1988, 59). While the Durenberger gaffe—one of a kind—was indeed cause for dismay, even more unsettling have been the some thirty major espionage cases in the past few years involving U.S. intelligence professionals. Here have been the most devastating blows to America's national security interests, not the rare—virtually nonexistent—disclosures of sensitive information from the congressional intelligence committees. (For some reasons why these spy cases have occurred, see "United States Counterintelligence and Security Concerns—1986," House Intelligence Committee *Report No. 100–5,* February 4, 1987.) On the Leahy-Durenberger leaks, see, respectively, the *New York Times,* July 29, 1987, pp. A1, A8, and March 31, 1987, p. 3.

64. *Federalist Paper* No. 51, op. cit. urges this balancing of legislative and executive power. *Federalist Paper* No. 64 (written by John Jay on March 7, 1788) recommends, in what seems to be a contradiction or an exception, that the president "manage the business of intelligence in such a manner as prudence may suggest" (p. 419). It would be prudent, however, to honor the higher democratic principle advanced in No. 51. As historian Dexter Perkins once observed, "Blind acceptance of the views of the executive would not be consistent with the principles of democracy," *The Evolution of American Foreign Policy* (New York: Oxford University Press, 1948), 168.

65. DDO, Interview, November 4, 1978, Washington, D.C.

66. Interviews with CIA officials, November 17 and 18, 1980, Washington, D.C.

67. William Colby claims to have spent 60 percent of his time in his three years as DCI reporting to Congress; he felt that 10 percent would have been more appropriate (cited by Sen. Charles Percy, R, Illinois, *Congressional Record,* May 12, 1976, S13688).

68. Church, press conference, U.S. Capitol, July 19, 1975, Washington, D.C.; Boland, quoted by Don Overdofer, *Washington Post,* August 6, 1983, p. A13.

69. See Stansfield Turner, "Has Reagan Killed CIA Oversight?" *Christian Science Monitor,* September 26, 1985, p. 14; Associate Press report on the contras, *Atlanta Constitution,* June 11, 1986; p. 2; the Tower Commission *Report;* Inouye-Hamilton committees, *Hearings*; and Dan Morgan and Walter Pincus, "Somehow, the Irangate Story Still Doesn't Add Up," *Washington Post National Weekly Edition,* September 28, 1987, pp. 23–25. According to a recent report, the Reagan NSC staff also fabricated "evidence" claiming that the Sandinista regime was running drugs to the United States and that left-wing Salvadoran guerrillas were receiving East-bloc weapons from the Sandinistas—both propaganda operations of the kind run by the CIA's Operations Directorate, only in this case conducted by a surrogate CIA on the NSC staff led by Lieutenant Colonel North (see Parry and Kornbluh, "Untold Story," 12).

70. See the Tower Commission *Report* and *Hearings,* Inouye-Hamilton Joint Committee.

71. Quoted by Norman Kempster, *Washington Star,* November 12, 1975. During hearings before the Church committee on December 4, 1975, Clark Clifford emphasized this problem:

> I believe on a number of occasions a plan for covert action has been presented to the NSC and authority is requested for the CIA to proceed from point A to point B. The authority will be given and the action will be launched. When point B is reached, the persons in charge feel it is necessary to go to point C, and they assume that the original authorization gives them such a right. From point C, they go to D and possibly E, and even further. This has led to some bizarre results, and, when an investigation is started, the excuse is blandly presented that authority was obtained from the NSC before the project was launched.

72. Seymour M. Hersh, "The Angleton Story," *New York Times Magazine,* June 25, 1978, 13 ff.

73. See the *New York Times,* December 1984, and House Intelligence Committee, *Annual Report* (1985).

74. Senior DDO official, interview, June 11, 1984, Washington, D.C.

75. On the Mafia connection, see Church committee, "Alleged Assassination Plots"; on the difficulty of controlling Cuban assets, see Wyden, *Bay of Pigs;* and, on CIA defections and related problems, see Senate Select Committee on Intelligence, "Meeting the Espionage Challenge: A Review of United States Counterintelligence and Security Programs," October 3, 1986, S. Rept. No. 99–522. Specifically on the Wilson case, see Joseph C. Goulden, with Alexander W. Raffio, *The Death Merchant: The Rise and Fall of Edwin P. Wilson* (New York: Simon and Schuster, 1984), and Peter Maas, *Manhunt* (New York: Random House, 1986).

The Christic Institute, a Washington-based public interest group founded in January 1980, brought a suit against various former CIA and other government officials in 1986, charging them with criminal conspiracy involving drug-running, arms sales, and even assassination in Latin America. At the center of the case stood an apparent bombing plot against disgruntled contra leader Eden Pastora. At a press conference called by him on May 30, 1984, a bomb exploded. Pastora was uninjured, but eight people were killed, including a U.S. reporter for the Religious News Service. Tony Avigan, an ABC-TV reporter, was seriously injured. The Christic Institute maintained that, among others, ex-CIA officers Theodore Shackley and Thomas Clines, CA specialists, had established a private covert-action organization resembling Lieutant Colonel North's "Enterprise" (see Inouye-Hamilton committees, *Report*) and with some of the same participants—retired Air Force Maj. Gen. Richard V. Secord, for one, who reportedly paid a former CIA asset $100,000 to smear the Institute. On June 23, 1988, the U.S. Federal District judge for the Southern District of Florida dismissed the suit on grounds of insufficient material facts. A Christic Institute official vowed to appeal the ruling. The defendants, meanwhile, ridiculed the suit as merely a "media stunt" to raise money for the Institute, and the Reagan administration called the Institute's charges a "political fantasy." See the *New York Times,* July 20, 1987, p. 12; March 4, 1988, p. 28; and June 24, 1988, p. 6; and, the *National Review* 40, March 4, 1988, 28.

76. Interview conducted by Church committee staff, Washington, D.C., May 21, 1975 (Church committee files).

77. Church, press conference, Los Angeles, California, February 21, 1976.

78. DDO, interview, February 24, 1978, Washington, D.C.

79. Interview, PBS, "MacNeil-Lehrer Show," July 16, 1987. In contrast, the Reagan administration seemed more bent on removing oversight safeguards—"restrictions," from its perspective—for the CIA. As the eventual GOP vice presidential nominee (and former DCI) George Bush put it early in the 1980 campaign: "It's time to untie the CIA" (CBS, "Evening News," December 7, 1979).

80. Harry How Ransom, "The Politicization of Intelligence" in Stephen J. Cimbala, ed., *Intelligence and Intelligence Policy in Democratic Society* (Dobbs Ferry, N.Y.: Transnational, 1987), 43.

81. "Gesprach mit William E. Colby," 114.

82. Inouye-Hamilton committees, *Report,* 19.

83. Ibid., 383.

84. See the Tower Commission *Report;* Inouye-Hamilton committees, *Hearings;* and Church Committee, "Alleged Assassination Plots." The apparently unauthorized distribution of assassination manuals in Nicaragua by a case officer in 1984 provides a further illustration.

85. See David B. Ottaway, "U.S. Reported Halting Some Afghan Arms," *Washington Post,* March 26, 1988, p. A-1.

86. "Should the CIA Fight Secret Wars?" *Harpers,* September 1984, 39, 44 (a round-table discussion with national security experts).

87. Ibid., 37.

Chapter 7

1. To use Harry Howe Ransom's definition, *The Intelligence Establishment,* 14.

2. Theodore H. White, *Breach of Faith: The Fall of Richard Nixon* (New York: Atheneum, 1975), 133.

3. Interview with William C. Sullivan, assistant director, FBI, June 10, 1975, Boston, Massachusetts. For an overview of FBI operations against domestic radicals, see Sanford J. Ungar, FBI (Boston: Little, Brown, 1976) and Richard Gid Powers, *Secrecy and Power: The Life of J. Edgar Hoover* (New York: Free Press, 1987), especially chaps. 12 and 13. This account of the Huston Plan is based mainly on my investigative findings as a staffer on the Church committee and that committee's hearings on the plan. For a more exhaustive account, see Church committee, *Final Report,* 3:923–86. The hearings are entitled "Huston Plan," *Intelligence Activities,* September 23, 24, and 25, 1975 (hereafter cited as Huston Plan Hearings).

4. Memorandum from C. D. Brennan of the FBI Counterintelligence Branch to William C. Sullivan, dated June 20, 1969 (Huston Plan Hearings, Exhibit 6, September 25, 1975).

5. Testimony of C. D. Brennan, ibid., p. 101; Sullivan interview, op. cit.; and J. Edgar Hoover's handwritten notes on memorandum from William C. Sullivan to Cartha DeLoach, July 19, 1966, p. 3 (Church committee files). As early as 1963, Hoover began to oppose the broad use of domestic wiretaps (memorandum from Sullivan to DeLoach, March 7, 1970, Church committee files). For reference to FBI break-ins into embassies in Washington, D.C., see the answer by former president Richard M. Nixon to Church Committee Interrogatory 17, March 3, 1976, 11. Former CIA CI chief James Angleton has observed that "thousands of man-hours would have been saved if the FBI had been

willing to place taps on phones in embassies," interview, Washington, D.C., September 21, 1980.

6. Deposition of former CIA director Richard Helms, September 10, 1975, 3; deposition of former DIA director Gen. Donald V. Bennett, August 5, 1975, 12; deposition of former NSA director Adm. Noel Gayler June 19, 1975, 6–7; deposition of Tom Charles Huston, May 23, 1975, 36—all in the Church committee files, as well as Sullivan interview, *op. cit.* In the latter part of 1969, Hoover advised the CIA to see the attorney general—not him—if it wanted to expand its intelligence collection on foreigners within the United States (Sullivan memorandum, March 30, 1970, [Church committee files]).

7. Interview with Dr. Louis Tordella, June 16, 1975, Fort Meade, Maryland.

8. Ibid.

9. Huston deposition, op. cit., 34.

10. FBI CI officer, interview, August 20, 1975, Washington, D.C.

11. Col. John Downie, former U.S. Army CI officer, interview, May 13, 1975, Easton, Pennsylvania. On the Army domestic-surveillance scandal, see Lawrence M. Baskir, "Reflections on the Senate Investigation of Army Surveillance," *Indiana Law Journal* 49 (Summer 1974), 618–53.

12. Huston deposition, op. cit., p. 36.

13. Memorandum from Huston to H. R. "Bob" Haldeman, the White House chief of staff, July 1970 (precise day unknown, but during the first two weeks), Huston Plan Hearings, exhibit 2.

14. Attachment to Huston memorandum, ibid.

15. The "and" in the following sentence is probably an original error and should read "a."

16. White, *Breach of Faith,* 133.

17. Attachment to Huston memorandum, op. cit., 2.

18. Ibid., 3. In using the word "burglary," Huston later explained, he sought to "escalate the rhetoric . . . to make it as bold as possible." He thought that, as a staff aide, he should give the president "the worst possible interpretation of what the recommendation would result in" (Huston deposition, op. cit., 69).

19. Ibid., 8.

20. Memorandum from H. R. Haldeman to Huston, July 14, 1970, Huston Plan Hearings, Exhibit 3. Former president Nixon has since stated: "My approval was based largely on the fact that the procedures were consistent with those employed by prior administrations and had been found to be effective by the intelligence agencies" (answer to Church committee Interrogatory 19, March 19, 1970, 13).

21. Huston Plan Hearings, 23–24.

22. Sullivan interview, op. cit.

23. Memorandum for the record from Richard Helms, July 28, 1970, Huston Plan Hearings, Exhibit 20. See also John Mitchell, Huston Plan Hearings, October 24, 1975, p. 123, where he testified that he "made known to the President any disagreement with the concept of the plan and recommended that it be turned down."

24. Helms memorandum for the record, July 28, 1970, Exhibit 20, op. cit.

25. Mitchell testimony, op. cit., 123.

26. Answer by Richard M. Nixon to Church committee Interrogatory 17, March 3, 1976, 11.

27. Ibid.

28. Interview, former chief of the White House Situation Room, July 1, 1975, Washington, D.C.

29. Memorandum from Huston to Haldeman, August 5, 1970 (Church committee files).

30. Memorandum from Huston to Haldeman, August 7, 1970 (Church committee files).

31. Huston, interview, May 22, 1975, Washington, D.C.

32. Ibid.

33. Sullivan, interview, op. cit.

34. Huston deposition, op. cit., 77.

35. John Dean, interview, Beverly Hills, California, August 7, 1975.

36. Huston deposition, op. cit., 50.

37. Responses by Nixon to Church committee interrogatories, op. cit., 1, 4, 5, and 14.

38. Huston deposition, op. cit., 50–51.

39. Testimony of Tom Charles Huston, Huston Plan Hearings, September 23, 1975, 33.

40. Ibid.

41. Ibid., 33–34.

42. See the testimony of Richard Helms, Huston Plan Hearings, October 22, 1975, 89, 96.

43. Testimony of James Angleton, Huston Plan Hearings, September 24, 1975, 54.

44. Ibid., 37.

45. Testimony of Tom Charles Huston, Huston Plan Hearings, September 23, 1975, 16.

46. Ibid., 17.

47. Attachment to Huston memorandum, op. cit., 2–3.

48. Sullivan interviews, op. cit., 92–93. On Operation Minaret and the NSA "Watch List," as well as the companion surveillance operation codenamed SHAMROCK, see Johnson, *A Season of Inquiry,* chap. 10.

49. Response of Richard M. Nixon to Church committee Interrogatory 23, March 9, 1976, 13.

50. Huston testimony, op. cit., 21.

51. Angleton testimony, op. cit., 21.

52. Sullivan interview, op. cit., 92–93.

53. Ibid., 95–96.

54. Ibid., 98.

55. Huston testimony, op. cit., 95–96. In the summer of 1970, Huston held the belief that "the Fourth Amendment did not apply to the President in the exercise of matters relating to internal security or national security." (See, ibid., p. 20, as well as p. 14.)

56. Huston deposition, May 23, 1975, 35.

57. Huston testimony, op. cit., 35; interview with Lt. Gen. Donald V. Bennett, Hilton Head, South Carolina, June 5, 1975.

58. Huston deposition, op. cit., 167. On this dichotomy, see Roy Godson ed., *Intelligence Requirements for the 1980s: Domestic Intelligence* (Boston: Lexington, 1986), esp. 190–91.

59. Huston testimony, op. cit., 45. In 1988 another example of mixing up political with national security considerations came to light. Responses to a Freedom of Information Act lawsuit revealed that the FBI had conducted improper surveillance against the U.S.-based Committee on Solidarity with the People of El Salvador (CISPES). The Bureau's counterterrorism investigation of CISPES was devoid of a signal charge of criminality, yet, according to Rep. Don Edwards (D, California), "through innuendo and guilt by association, CISPES was prosecuted, tried and found guilty of subversion" (*Congressional Record,* March 3, 1988, p. H707). An examination of the FBI investigation of CISPES, prepared by the Center for National Security Studies (CNSS) in Washington, D.C., concluded that the Bureau's activities were "relatively wide-ranging, undefined, unsupervised, and infringed on constitutionally protected rights" (Gary M. Stern, "The FBI's Misguided Probe of CISPES," CNSS *Report* No. 111, Washington, D.C., June 1988, p. 11).

Elements in the FBI even seemed prepared to stray into aggressive operations against CISPES, reminiscent of the darkest pages of the Huston Plan and Operation COINTELPRO. According to a memo from the Bureau's field office in New Orleans: "It is imperative at this time to formulate some plan of attack against CISPES and specifically against [names deleted] who defiantly display their contempt for the U.S. government by making speeches and propagandizing their cause while asking for political asylum" (dated November 10, 1983, cited by Stern, 2: n. 5).

60. Testimony of C. D. Brennan, Huston Plan Hearings, exhibit 6, September 25, 1975, 134.

61. Ibid., 104, 107, and 135.

62. Memorandum from W. R. Wannall to C. D. Brennan, March 23, 1975 (Church committee files).

63. Memorandum for the files by J. Edgar Hoover, April 12, 1971 (Huston Plan Hearings, exhibit 31). Subsequent to the meeting with Mitchell, "the Attorney General reversed the FBI decision" against a proposed CIA electronic surveillance, according to James Angleton, and in May 1971 "all the devices which had been installed . . . were tested and all were working." See Memorandum for the record by Angleton, Huston Plan Hearings, exhibit 61, May 18, 1973, 3.

64. For detailed documentary evidence on these points, see the Church committee reports on the CIA mail program and the NSA and FBI internal security programs, especially Vol 2 of its *Final Report.* See also the Brennan testimony on the extent of the FBI internal security investigation, op. cit., 100.

65. Memorandum from Executives Conference to Clyde Tolson, October 29, 1970 (Church committee files), emphasis added. The Executives Conference was an occasional gathering of senior officials in the FBI.

66. Brennan testimony, op. cit., 138–39.

67. Mitchell testimony, op. cit., 141. On the apparent lack of presidential awareness of the NSA watch-list expansion, see the testimony of NSA director Gen. Lew Allen, Huston Plan Hearings, October 29, 1975, 28–29, and former President Nixon's response to Church committee interrogatories, March 9, 1976, 1.

68. *Presidential Documents,* May 22, 1973, 693–95.

69. Huston Plan Hearings, 70–71.

70. Ibid., 82.

71. Ibid., 83.

Chapter 8

1. CIA personnel officer, interview, August 1, 1979, Washington, D.C.

2. A CIA recruitment officer referring to a professor-source, cited by Philip W. Semas, "How the CIA Kept an Eye on Campus Dissent," 15 *Chronicle of Higher Education,* December 5, 1977, 3.

3. CIA Operations Directorate official, interview, July 26, 1979, Washington, D.C.

4. Warren Hinckle, Sol Stern, and Robert Scheer, "The University on the Make," *Ramparts* 4 (April 1966), 11–22; Sol Stern, "NSA and the CIA," 29–38. Hinckle, who was editor of *Ramparts* at that time, recalled recently that "the classic of the CIA's corruption of independent academia was Michigan State University's project in Vietnam, where the university knowingly provided academic cover for CIA agents and its professors, who helped set up a dictatorship under the guise of advising a democracy" (Warren Hinckle, "CIA Reunion," *San Francisco Examiner,* January 27, 1986).

5. See remarks by Robert M. Gates, "Text of Speech at Harvard by Deputy CIA Director Outlining Policy Shifts," *Chronicle of Higher Education* 31 (February 26, 1986), 26–29, esp. p. 26; and, by Gates as well, "CIA and the University," address given at the National Convention of the Association of Former Intelligence Officers, Tyson's Corner, Va., October 10, 1987, reprinted in *Periscope* (Fall 1987), 17–19. See also Robin Winks, *Cloak and Gown: Scholars in the Secret War,* 1939–1961 (New York: Morrow, 1987). As a university professor, it seems fair and proper for me to state precisely at the outset of this chapter my own connections with the CIA. Before March 1975, I had none. In that month, while on leave of absence, I became assistant to the chairman of the Senate Select Committee to Study Governmental Operations with Respect to Intelligence Activities,

the long-winded official title of the Church committee. For the duration of this inquiry (sixteen months), I had almost daily contact with personnel in the various intelligence agencies, as part of my job as Sen. Frank Church's aide and as a staff investigator on the committee. My tasks were chiefly taking depositions, preparing hearings, writing reports on investigative findings, and helping to prepare Church for his various obligations as committee chairman. In 1976 I resigned from my university position to serve as Church's issues director in his presidential bid and, later, as his aide on the Foreign Relations Committee. My next contact with the CIA was from 1977 to 1978, when I served as staff director for the Subcommittee on Oversight, House Permanent Select Committee on Intelligence (the Boland committee). Here again I had almost daily contact with CIA and other intelligence officials as I prepared hearings, wrote reports, conducted oversight inspections here and abroad, and the like. In January 1979, I left Washington to join the faculty of political science at the University of Georgia. From then until now, my contacts with the CIA have consisted of the following: openly participating in a conference on intelligence at the Agency in June 1984 (with about thirty other faculty members from around the country chosen, presumably, because of their interest in intelligence policy, as demonstrated in published writings or teaching concentrations); hosting openly at the University of Georgia a lecturer from the CIA Office of Public Affairs once each year to discuss intelligence issues with political science classes in my department (just as I have invited representatives of other government entities, as well as speakers outside the government—including individuals quite critical of the CIA; on two occasions (in 1986 and 1987) meeting at the CIA with its Senior Seminar participants (GS15s and above) for discussion of legislative oversight, for which I received travel expenses and a modest stipend; and, finally, conducting many interviews with current and former Agency officials periodically over the telephone and in Washington two or three times a year, as part of my continuing research into the problem of how to balance democracy with the existence of secret intelligence agencies.

6. See Church committee, *Final Report* 1:181–91. The relationship was a two-way street. On its side, the CIA funded research centers (like the Center for International Affairs at MIT), provided travel support and information on foreign countries, and offered related research assistance to friendly academicians. See Nardo Zacchino and Robert Scheer, "CIA Papers Show Long History of UC Contacts," *Los Angeles Times*, February 19, 1977; Marchetti and Marks, *The CIA*, 153; Hinckle, Stern, and Scheer, "The University on the Make." Harvard University historian Richard Pipes has commented on the natural affinity between "spooks" (CIA slang for spies) and scholars: "[The CIA's] analytic staff, filled with American Ph.D.'s in the natural and social sciences along with engineers, inevitably shares the outlook of U.S. academe, with its penchant for philosophical positivism, cultural agnosticism, and political liberalism. The special knowledge which it derives from classified sources is mainly technical; the rest of its knowledge, as well as the intellectual equipment which it brings to bear on the evidence, comes from academia" ("Team B: The Reality Behind the Myth," *Commentary* 82 [October 1986], 29). For examples of academicians who have had successful careers in both the world of spies and scholars, see Andrew Sommer and Marc Cheshire, "The Spy Who Came in from the Campus," *New Times*, October 30, 1978, p. 14; and, "Our Wrong Rush to Unleash the C.I.A.," *New York Times*, February 6, 1980.

7. Katzenbach Report, *Weekly Compilation of Presidential Documents*, April 3, 1967, 556.

8. On these cases, see Angus Paul's two-part series, "CIA Eases Stand on Research Role," *Chronicle of Higher Education* 31 (February 26, 1986), 1, 26; *Chronicle of Higher Education* 31 (March 12, 1986), 29; *New York Times*, January 1 and February 21, 1986, pp. 1 and 11, respectively.

9. *Newsweek*, January 13, 1986, 75.

10. *New York Times*, January 20, 1986, p. 9.

11. *Boston Globe*, December 3, 1985.

12. Hinckle, "CIA Reunion." All the controversy seemed to have no negative effect

on the Harvard-CIA relationship; the next year the Agency publicly awarded the Kennedy School of Government $400,000 for research and training.

13. See Gates, "Text of Speech at Harvard," 26.

14. For these guidelines, see Senate Select Committee on Intelligence [the Bayh committee, after Chairman Birch Bayh (D, Indiana)], "National Intelligence Reorganization and Reform Act of 1978," *Hearings,* July 20, 1978, letters submitted for the record from Admiral Turner to President Bok, dated May 15, 1978, p. 660.

15. Ibid., letter from Turner to Bok, dated June 13, 1977, p. 651.

16. Gates, "Text of Speech at Harvard."

17. *New York Times,* February 18, 1986.

18. DCD director, interview, August 3, 1978, Washington, D.C.

19. Rockefeller Commission Report, 210. Recall, too, the references to campus spying in the Huston Plan (chap. 7).

20. See Church committee, *Final Report* 1:438.

21. See, for example, Hersh, "Underground for the C.I.A."

22. Church committee, *Final Report* 1:438.

23. Gates, "Text of Speech at Harvard."

24. Interview with CIA official, August 28, 1979, Washington, D.C.

25. Interview with director of personnel, July 26, 1979, Washington, D.C.

26. Letter dated June 14, 1976, obtained through the Freedom of Information Act (FOIA).

27. Quoted by William Trombley, "CIA Agents on U.S. Campuses Alleged," *Los Angeles Times,* June 25, 1976.

28. *New York Times,* March 20, 1977.

29. Wolfinger, "For Love or Money," *PS* [a newsletter of the American Political Science Association] 1 (Summer 1978), 336–38.

30. Loch K. Johnson, questionnaire, "CIA and Academe," University of Georgia (March 1980).

31. *New York Times,* March 20, 1977. For a controversy regarding CIA ties with professors at Rutgers University, see the *New York Times,* November 28, 1984, p. B-2; and the *Chronicle of Higher Education* 29 (December 5, 1984), 2, and 29 (January 16, 1985), 3. In contrast, a prominent political scientist has argued for strong ties between the CIA and the academic world: "Do you feel that the United States government does not seem to understand Vietnamese villagers, or Dominican students, or Soviet writers? If you think that Washington could act better if it had a deeper comprehension of the social processes at work around the world then you should be demanding that the CIA hire and write contracts with our best social scientists" (Ithiel de Sola Pool, "The Necessity for Social Scientists Doing Research for Governments," *Background* 10 [August 1966], 114.) Professor de Sola Pool and his colleagues at M.I.T. severed research ties with the CIA in 1965, however, on grounds that the Agency's insistence on keeping the relationship secret was unethical. "We were perfectly willing to do public, published university research for the United States government via any of its departments," he writes, "but we have to say who the sponsor is" (p. 115).

32. See the *New York Times,* January 9, 1977, February 21, 1978, and May 19, 1979. A federal judge eventually awarded the professor $558,000 in damages, on grounds that he had been improperly denied tenure.

33. Dated March 24, 1970, obtained through the FOIA.

34. CIA memorandum for the deputy director of security, January 17, 1975 (FOIA).

35. Ibid. Students radicals ("the bearded ones") also threatened entrapment, according to the memo, by planning to infiltrate "a female interviewer [sic] into a closed room with recruiter, tearing clothing, messing her hair and yelling rape." Apparently this strategem was never carried out.

36. Interview with DDS & T, August 3, 1978, Washington, D.C.

37. Ibid.

38. Interview with OPA director, August 4, 1979, Washington, D.C.

39. Marjorie W. Cline, ed., *Teaching Intelligence in the Mid-1980s* (Washington, D.C.: National Intelligence Study Center, 1985), 36–81. In October 1987, DDCIA Gates stated that over seventy-five universities now offer courses on some aspect of intelligence ("The CIA and the University," 17).

40. Letter from CIA coordinator for academic affairs to Loch K. Johnson, University of Georgia, November 26, 1985. In 1988, I received another letter from the CIA (obviously a form letter mailed to professors across the country) stating that the Agency "is seeking ways to foster a dialogue with academicians in a variety of disciplines about issues related to intelligence." The letter continued: "From time to time, materials become available that can be sent out to academics for their personal use," and enclosed an attractive, informative Agency study on *China: Economic Performance in 1987 and Outlook for 1988*, EA [Economic Analysis] 88-10018, May 1988. The letter writer, the coordinator for academic affairs, welcomed "any comments about the material you would care to make" (dated July 12, 1988).

41. See, for example, Church committee, *Final Report* 1:452–53; Trombley, "CIA Agents on U.S. Campuses Alleged"; Noel Epstein, "Professors Decry Recruiting by CIA," *Washington Post*, May 6, 1978; Morton H. Halperin, "The CIA on American Campuses: The Harvard Confrontation," *First Principles* 4 (November 1978), 1–4; "Bok Gives Senate Testimony on Universities, U.S. Intelligence," *Harvard Gazette*, September 15, 1978; "National Intelligence Reorganization and Reform Act of 1978," *Hearings*, op. cit., 639–88; Joseph Lelyveld, "India Still Wary on U.S. Scholars," *New York Times*, August 14, 1968; Allen Boyce [pseud.], "Covert Scholarship: The Market for Potted Expertise," *Nation* 227, November 11, 1978, 489 ff.; and a series in *Science*, December 1966 and June and September 1967.

42. These guidelines are reprinted in "National Intelligence Reorganization," *Hearings*, op. cit., 648. For similar guidelines at MIT, see "Interim Report of the Ad Hoc Committee on M.I.T. and the Intelligence Agencies," Massachusetts Institute of Technology, *Tech Talk* 23 (April 11, 1979).

43. See "CIA Ignores Harvard's Recruiting Curb," *Washington Post* (October 23, 1978), and "National Intelligence Reorganization," *Hearings*, op. cit., ibid., 639, 650–60. The CIA's views on this subject were derived from a series of interviews with Agency personnel in 1979 and 1985, plus a review of the Senate hearings on the proposed 1978 Reform Act, "National Intelligence Reorganization," *Hearings, op. cit.* The reference to "sources and methods" comes from a provision in the 1947 National Security Act, Sec. 102 [50 U.S.C. 403] (d)(3). Dean Rusk, professor of law at the University of Georgia and former secretary of state, evidently agrees with Turner—at least when it comes to public educational institutions. "In a private institution, the school is entitled to know, as long as this information [about CIA contacts on campus] is not open to the public at large," he told an interviewer in 1980. "A public institution, for example, the University of Georgia, is paid for in part by taxes and public funds, [and] therefore should be open for private and undisclosed use by the government and the CIA." Joy Power, "An Interview with Dean Rusk," March 10, 1980, unpublished typescript.

44. John F. Blake, as quoted by Timothy S. Robinson, "Academics Still Secretly Inform CIA," *Washington Post*, May 6, 1976.

45. See Church committee, *Final Report* 1:451–56. In an interview former DDO during the early 1960s, Richard M. Bissell, Jr., states that he had advocated more [covert action] and of a higher quality than previously developed" and that this more aggressive use of covert action was supported "especially by the university community. . . ." John Patrick Quirk, et al., *The Central Intelligence Agency: A Photographic History* (Guilford, Conn.: Foreign Intelligence Press, 1986), 211.

46. Reprinted in National Intelligence Reorganization," *Hearings, op. cit.*, 648.

47. Quoted by Anthony Cave Brown, *Bodyguard of Lies* (New York: Harper and Row, 1975), 10.

48. See the Safran, Betts, and Huntington examples discussed earlier, as well as the discussion in Marchetti and Marks, *The CIA*, 181–85.

49. Church committee, *Final Report,* 191.

50. "National Intelligence Reorganization," *Hearings, op. cit.*, p. 648. Professor Betts of Harvard was not a faculty member when he accepted the CIA research grant discussed earlier and, therefore, never reported to the dean; Professor Huntington was and did, though reportedly the dean failed to inform the president of the university. Professor Safran failed to report his CIA funding for the Middle East Conference, but did report the CIA book contract (which the dean again apparently never reported to the president). For views on this subject promulgated by leaders of the American Political Science Association, see *PS* (Summer 1977), 355, and "Ethical Problems of Academic Political Scientists," *Final Report,* Committee on Professional Standards and Responsibilities, American Political Science Association (rpt. in *PS,* Summer 1968). The association has moved from a position of full disclosure of all relationships between political scientists and the CIA (see *PS,* Winter 1977, 55) to a more nebulous call for acknowledgement of all relationships between political scientists and the CIA "in a timely fashion" and to the extent one might (or might not) report on ties with other agencies or entities (*PS,* Fall 1977, 474–75).

51. See Church committee, *Final Report* 1:385–410; Rockefeller Commission Report, 226–29; Victor Cohn and John Jacobs, "CIA Tried to Use Georgetown Medical Center in 1950s," *Washington Post,* August 7, 1977.

52. See the Olson case disclosed in the Rockefeller Commission Report, 227, where a government scientist was given the drug LSD by the CIA without his knowledge.

53. See, for example, a dispute involving possible covert ties between the CIA and the American Political Science Association, Boyce, "Covert Scholarship," *Nation,* op. cit.

54. "National Intelligence Reorganization," *Hearings,* op. cit., 649.

55. Ibid., 672.

56. Letter from President Bok to DCI Stansfield Turner, December 5, 1977, in ibid., 654. This remains the CIA's position today.

57. Ibid., 672.

58. Ibid., 673.

59. Gene I. Maeroff, "Harvard and CIA at Impasse Over Secret Work by the Faculty," *New York Times,* August 5, 1978.

60. Letter from Admiral Turner to President Bok, June 13, 1977, reprinted in "National Intelligence Reorganization," *Hearings,* op. cit., 651, emphasis added.

61. Ibid., 677.

62. Ibid., 662.

63. Ibid., 666.

64. See Michael J. Glennon, "Liaison and the Law: Foreign Intelligence Activities in the United States." *Harvard International Law Journal* 25 (Winter 1984), 1–42.

65. See, for example, "Secrecy, Security, and Science," *Washington Post,* January 18, 1982.

Chapter 9

1. Ronald Reagan, "Soviet Noncompliance with Arms Control Agreements," unclassified report to Congress, February 1, 1985, 6.

2. *New York Times,* March 23, 1985.

3. Cited by Michael R. Gordon, "CIA Is Skeptical that New Soviet Radar Is Part of an ABM Defense System," *National Journal,* March 9, 1985, 525.

4. Aspin Hearings, 314. See also Earl W. Foell, "Journalists, Diplomats, and Spies—There Are Some Parallels," *Christian Science Monitor,* September 16, 1986, 3.

5. Ibid., 315. The closeness of the two professions is evident in this observation by journalist Tad Szulc:

> In 1968 in Prague I spent a good deal of my time as a *New York Times* correspondent driving in my black convertible around Czechoslovakia and looking at Soviet troop concen-

trations because it was part of my reporting. We wished to know how many tanks did they have, did they have any guided missiles, and so on.

Returning to town, I would fairly often touch base not so much with the CIA people as with the Defense Attache's office, which would be the Defense Intelligence Agency, not so much to fill him in but to compare notes because he was a professional who can make a better judgment than I of the military strategic or tactical importance of that which I had observed. Very often he would go out and check on an area which I have seen and in the conversation he would say, yes, you are right; no, you are wrong. Obviously the Defense attache used this material for his own reporting, which is fine with me as far as national security or whatever is concerned. But this was done in the context of my trying to educate myself, if you will, on the military aspects of the equipment which the Soviets had in Czechoslovakia.

Szulc concluded that "the areas became so blurred and so gray" (Aspin Hearings, 116–17).

In describing his ability to develop rapport with DCI William J. Casey, Bob Woodward, the *Washington Post* reporter who published a study on the CIA during the Reagan administration, noted: "[W]e had formed a partnership over secrets. In entirely different ways, we were both obsessed with secrets. During this game, secrets were the exchange medium. What were the secrets? What was their value? What was their use?" (*The CIA's Secret Wars,* 505.)

6. For histories of the relationship between the CIA and journalists, see Loory, "The CIA's Use of the Press," 8–18; Church committee, *Final Report* 1:191–201; Marchetti and Marks, *The CIA,* 329–46; Carl Bernstein, "The CIA and the Media," *Rolling Stone,* October 20, 1977, 55–67; Richard Harwood and Walter Pincus, "The CIA's Journalists," *Washington Post,* September 18, 1977; Stanley Karnow, "Associating with the Agency," *Newsweek,* October 10, 1977.

7. William Colby, Aspin Hearings, 9. One American journalist who operated an English-language newspaper in Chile, David Atlee Phillips, has testified before a House subcommittee on why the CIA found him to be a useful recruit in 1950: "When I asked the CIA chief [in Chile] why he had selected me to work with him, he cited my situation and mobility. As a newsman I had cover, and explanation of why I was in the country, and access: I could move about and ask questions." Aspin Hearings, 68.

8. NBC, "Sunday Night News," January 25, 1976.

9. Bernstein, "The CIA and the Media," 55.

10. Church committee, *Final Report* 1:192.

11. According to a CIA "source" cited by the *Washington Star,* November 30, 1973. The source was subsequently revealed to be William Colby; see Harwood and Pincus, "The CIA Journalists." In an interview (Washington, D.C., December 14, 1977), Colby offered me these definitional distinctions: If a journalist's loyalty is more to the CIA than to his news organization, he has become a "fully controlled agent" or "asset"; if his loyalty remains with the media, he is a "contact"—though the line between the two can become blurred and depends in part upon the degree of control ("influence," "power") of the Agency over the journalist.

12. On these demands, see Bernstein, "The CIA and the Media"; Stuart Loory *(Chicago Sun-Times),* Aspin Hearings, 197. For an argument in opposition, see testimony of Morton H. Halperin, director, Center for National Security Studies, Aspin Hearings, 233.

13. See the National Security Act of 1947 [50 U.S.C. 403, Sec. 102 (3)(d)(3)].

14. On the Church committee investigations into this question, see its *Final Report* 1:191–201, and Johnson, *A Season of Inquiry,* 198–99, 215, 221, 225, 271, 287n.

15. Official public statement, Central Intelligence Agency, February 11, 1976, emphasis added.

16. See the *Washington Star* November 30, 1973.

17. CIA response to the Church committee, March 17, 1976, Church committee files.

18. Reuters may have been an exception in the earlier history of the CIA; see Loory,

"The CIA's Use of the Press," op. cit., 12. For an assertion from DCI Turner that Reuters was off-limits (in 1978 at least), see Aspin Hearings, 314. For the Reuter viewpoint, see Aspin Hearings, 478.

19. *Final Report* 1:191.

20. Loory, "The CIA's Use of the Press," op. cit., 12.

21. Ibid., 17. See also the account by David Atlee Phillips of CIA support for his newspaper in Santiago, Chile, the *South Pacific Mail,* while he played the dual role of reporter and spy, in *The Night Watch.*

22. Reprinted in the Aspin Hearings, 493.

23. Ibid., 495–520.

24. Interview, CIA public affairs officer, March 7, 1985, Washington, D.C.

25. "Memorandum for the Media: New CIA Regulations on Relationships with U.S. News Media," Central Intelligence Agency, December 2, 1977.

26. This is a reference to the first amendment, which reads: "Congress shall make no law respecting an establishment of religion, or prohibiting the free exercise thereof; or abridging the freedom of speech, or of the press; or the right of the people peaceably to assemble, and to petition the Government for a redress of grievances."

27. The policy statement is reprinted in the Aspin Hearings, 333–34.

28. Ibid., 334.

29. Ibid., 128.

30. Ibid., 183; also the Halperin testimony, 203. Gilbert Cranberg of the *Des Moines Register-Tribune* and former chair of the Professional Standards Committee of the National Conference of Editorial Writers noted during legislative hearings that the Turner directive indeed represented "a step backwards" from the CIA's stance on freelancers expressed at a June 24, 1976 meeting with National News Council representatives. "At that time," Cranberg testified, "the CIA said that [the Bush directive] barred relationships with anyone in a journalistic capacity with U.S. news agencies, including freelancers." Aspin Hearings, 250.

31. Ibid., 186. Even some reform-minded journalists have agreed that, as a practical matter, "freelancers" are too hard to identify and therefore impossible to prohibit by law or regulation. See Aspin Hearings, 275; "National Intelligence Reorganization," *Hearings,* op. cit., May 3, 1978, 140.

32. Aspin Hearings, 276.

33. Ibid., 209. For corroborative testimony, see Eugene Patterson, president, American Society of Newspaper Editors, ibid., 268.

34. Letter from Barbara Raskin to Herbert E. Hetu, assistant for public affairs, Central Intelligence Agency, June 13, 1977, reprinted in ibid., 512–13; quote from 513.

35. Ibid., 275.

36. Ibid., 105.

37. "National Intelligence Reorganization," op. cit., 137, 138. Wallace's testimony is on 172; see also the remarks of Richard Leonard, editor, *Milwaukee Journal,* 163.

38. See Aspin and "National Intelligence Reorganization," hearings.

39. "National Intelligence Reorganization," Hearings, 183.

40. Ibid., pp. 171–72. According to the American Newspaper Publishers Association, in 1984 25 journalists were killed overseas, 81 wounded, 205 jailed or detained and 50 expelled, denied visas, or in some other way restricted in their work. See the *New York Times,* April 25, 1985.

41. Ibid., 163. In 1988 ABC Television News lodged a related complaint against Israel for allowing its internal security to pose as ABC correspondents, thereby endangering "legitimate [U.S.] journalists" (ABC, "Evening News," July 5, 1988).

42. See ibid., 164.

43. Aspin Hearings, 244. Writing for the *Washington Post,* Charles B. Seib endorsed this position as well; see the *Washington Post,* October 11, 1977.

44. Aspin Hearings, 243.

45. Ibid., 7. TASS is the official Soviet News Agency.

46. "National Intelligence Reorganization," op. cit., 151.
47. See Aspin Hearings, 257.
48. "National Intelligence Reorganization," op. cit., 171, 153.
49. Ibid., 151–52.
50. Aspin Hearings, 168.
51. Ibid., 265–66.
52. Ibid., 23.
53. Ibid., 167.
54. Ibid., 20.
55. Ibid, 143.
56. Ibid., 7.
57. Quoted by David Atlee Phillips, ibid., 69.
58. According to William Colby, "in World War II there were a lot of missionaries that were very helpful to our intelligence. . . ." He specifically refers to central China in 1941 and 1942. See ibid., 41.
59. The National Student Association scandal in 1967 cut off CIA support to students in that organization traveling abroad. For butterfly collecting as a cover for British intelligence (with the delicate notebook sketches of veined wings in reality the outlines of foreign forts), see Christopher Andrew, *Her Majesty's Secret Service: The Making of the British Intelligence Community* (New York: Viking, 1986), 25.
60. Aspin Hearings, 129.
61. Ibid., 130. On the CIA-business connection, see the Church committee, *Final Report,* I:205–56; *Washington Post,* January 10, 1975; and, David Binder, "Business Pose by U.S. Spies Reported," *New York Times,* February 28, 1974. For a report that CIA Director George Bush had worked for the Agency as a younger man (1961–64) in his capacity as a globe-trotting chief executive officer for the Texas-based Zapata Off-Shore Company, see Joseph McBride, " 'George Bush,' C.I.A. Operative," *Nation* 247, July 16–23, 1988, 37, 41–42. Bush denied the charge (*New York Times,* July 11, 1988, 11).
62. F. Dean Brown, ibid., 150.
63. See the statements of William Colby and Ambassador William Porter, ibid., 7, 177.
64. Ambassador Porter, ibid., 180. Frequently, NOCs stand to make so much money in their regular (cover) job that they drop out of the CIA—which pays them a comparatively low government salary—to work full-time in the job they have learned at CIA expense!
65. I developed this outline in conjunction with Rep. Les Aspin and his aide Warren Nelson in preparation for the Aspin Hearings; see ibid., 336.
66. Though see Richard H. Leonard of the *Milwaukee Journal,* who argues: "Certainly, [the media and the CIA] should not swap information." "National Intelligence Reorganization," op. cit., 163.
67. Ibid., 134. See also Aspin Hearings, 142, 205.
68. Aspin Hearings, 141.
69. Ibid., p. 138.
70. Respectively, Jack Nelson of the *Los Angeles Times* ("National Intelligence Reorganization," 141) and Stuart Loory, then of the *Chicago Tribune* (Aspin Hearings, 205). Joseph Fromm of *U.S. News and World Report* feels that he would have been "remiss" in his duties as a foreign correspondent had he failed to consult with CIA stations abroad—"some of the best informed people" (Aspin Hearings, 109).
71. Aspin Hearings, 62.
72. Ibid., 105.
73. Ibid., 223.
74. "National Intelligence Reorganization," op. cit., 146. Similarly, Joseph Fromm suggests that "the problem is to draw a distinction between a legitimate news tip that should be acted upon and an effort at tasking by an intelligence official" (Aspin Hearings,

99). For an example of an instance when a well-known reporter, Bob Woodward of the *Washington Post,* invited the CIA to suggest questions he should ask in an interview with Libya's Qaddafi, see Woodward, *The CIA's Secret Wars,* 328.

75. "National Intelligence Reorganization," op. cit., 147.

76. Aspin Hearings, 202.

77. Ibid., 201, 205.

78. Ibid., 206.

79. Ibid., 62. Recall, also, Tad Szulc's observations in n. 5 above.

80. Ibid., 19.

81. See, for example, Stuart Loory, ibid., 205. In contrast, Richard S. Salant, while president of CBS News, reportedly allowed the CIA access to outtakes (*Washington Post,* June 4, 1977).

82. See Aspin Hearings, 131; Walter Cronkite of CBS played a similar role between the leaders of Israel and Egypt in the 1970s. In 1975 Malcolm W. Browne of the *New York Times* cooperated with the CIA in Saigon as a courier during negotiations with the Communists. "Since I was passing information along to *The Times* in my dispatches," he later argued, "it seemed to me there was no compromise of journalistic principles" (*New York Times,* November 21, 1977). In December of 1986, ABC-TV journalist Barbara Walters wrote a memorandum to the White House regarding information she had gleened from an interview with a Middle East arms middleman who had knowledge about U.S. hostages. She apparently believed that this information might relieve threats to human lives (i.e., the hostages)—evidently the only exception to ABC News policy that expressly limits journalists from "cooperating with government agencies." See the Associated Press story, *Atlanta Constitution,* March 17, 1987.

83. See Aspin Hearings, 77, 95.

84. Ibid., 133.

85. Robert Myers, ibid., 287.

86. See testimony of Ambassador William Porter, ibid., 170; and, the *New York Times,* December 27, 1967.

87. See Aspin Hearings, 168.

88. Ibid., 25.

89. Ibid., 64.

90. Aspin Hearings, 90–91 (see also 173).

91. On the bloodbath that occurred in South Vietnam after the withdrawal of American troops, see the estimates presented by Jacqueline DesVarats and Karl D. Jackson, "Vietnam: 1975–1982: The Cruel Peace," *Washington Quarterly,* Fall 1985, 169–82.

92. Mike Wallace interview with Frank Snepp, CBS, "60 Minutes," November 20, 1977; the more recent interview was with me in Cambridge, Massachusetts, February 27, 1988. Several journalists accused by Snepp responded that, yes, they talked to the CIA, but they checked their stories independently, too. As Keyes Beach of the *Chicago Daily News* said, "All I can say is that in this business, you talk to anybody and you do the best you can. . . ." (*New York Times,* November 21, 1977). See also the *Washington Star,* November 21, 1977; the *New York Times,* December 1, 1977; and the Aspin Hearings, 475–77. Less spectacular examples of alleged CIA influence over American news include, according to my interviews with CIA officials who preferred to remain anonymous: "Russia," *Time,* February 23, 1968, 23–28; and "The New Espionage American Style," *Newsweek,* November 22, 1971, 28–37. These pieces, though, may simply fall into the category of Washington leaks, a practice engaged in evidently by almost every agency. "About the CIA's 'feeding the press misleading information' . . . ," writes former CIA officer Miles Copeland, "[e]verybody feeds newspapermen false information" (*New York Times,* September 30, 1977). This does not justify the practice, but it does suggest that the CIA is like other agencies in wanting to put forward its viewpoint to the public and to policymakers. In his *Portrait of a Cold Warrior,* former CIA officer Joseph Burkholder Smith discusses various propaganda themes used in Asia that seeped back to the United States. The examples that were mentioned most frequently in my interviews with CIA watchers over

the years included the distribution of the CIA-drafted *Penkovsky Papers,* which painted the Soviet leaders as high livers and nuclear first-strikers and may have had an effect upon the views of the American public on detente; the distribution in America of CIA-drafted books on China, which were highly critical and distorted; and the circulation of numerous anti-Allende stories throughout Chile in 1970, including plants in Chile's most well known newspaper *El Mercurio,* the distribution of an anti-Allende book within the United States, and the publication in English and Spanish after the coup of a *White Book* by the junta—prepared with the assistance of two CIA assets in Chile. In 1986, the Reagan administration approved a worldwide "disinformation" (misleading propaganda) campaign against Colonel Qaddafi. The idea was to portray him as a madman about to use widespread violence against the United States and other nations, in order to turn world opinion even further against the Libyan leader. Secretary of State George P. Shultz justified the operation as legitimate psy war, even though the propaganda quickly washed back to American shores; however, his chief spokesman, Bernard Kalb, resigned in protest. For additional examples of CIA propaganda at home, see Loory, "The CIA's Use of the Press," 12–16, and Parry and Kornbluh, "Untold Story," 4–10.

93. Cited by Gilbert Cranberg, Aspin Hearings, 250.

94. "National Intelligence Reorganization," 133.

95. Ibid., 131. On August 30, 1986, Daniloff was arrested by the KGB in Moscow and charged with espionage. While working for *U.S. News and World Report* as a correspondent in the USSR, he apparently delivered a letter to the CIA at the U.S. embassy in Moscow. The letter, written by a "Father Roman," proved to be a KGB plant; when the CIA telephoned the bogus priest and mentioned Daniloff's name (he had provided the number to the Agency), the KGB deduced that Daniloff was a CIA operative. Daniloff was released a month later in a complicated swap for Soviet spies held by the United States. See Dusko Doder, *Washington Post,* October 12, 1986. The Daniloff case serves as an excellent example of how risky *any* tie can be between the media and intelligence agencies. See also the testimony of Philip L. Geyelin of the *Washington Post,* "National Intelligence Reorganization," op. cit., 168; and Ambassador Eugene Patterson, Aspin Hearings, 259–68.

96. Aspin Hearings, 19.

97. See "National Intelligence Reorganization," op. cit., 180, and the Aspin Hearings, 277.

98. See the Aspin Hearings, 121–22, 124.

99. Ibid., 70.

100. Church committee, *Final Report* 1:195.

101. Quoted in Bernstein, "The CIA and the Media," 57. See also Vermont Royster, "Thinking Things Over," *Wall Street Journal,* March 9, 1977.

102. Eugene Patterson, Aspin Hearings, 261. This view and Alsop's are different. Patterson stresses the requirement of "emergency circumstances," as does Herman Nickel of *Fortune:* "I would have to be convinced that there are very clear and present dangers. . . ." (Aspin Hearings, 114).

103. *Washington Post,* September 18, 1977.

104. "National Intelligence Reorganization," 140.

105. See the Aspin Hearings, 102.

106. Ibid., 94.

107. "The Founding Fathers [of the CIA] were close personal friends of ours," said Joseph Alsop on behalf of himself and his brother Stewart. "Dick Bissell [former DDO] was my oldest friend, from childhood. It was a social thing my dear fellow. I never received a dollar" (Bernstein, "The CIA and the Media," 60).

108. Ibid., 123.

109. Ibid., 124.

110. This is a dominant theme throughout the Aspin Hearings, 22–26, 82, 100, 139, 140, 204, 234, 237, 258, 277, and the "National Intelligence Reorganization" hearings, 141, 179, 180. See also Stanley Karnow, "When a Newsman Consults the CIA," *New*

York Times, December 18, 1977. A few critics, however, see a need for new laws, including a ban on the use of American journalists for clandestine intelligence collection or covert action (Ambassador Trueheart, Aspin Hearings, 145); CIA dissemination of false information (Halperin testimony, Aspin Hearings, 224) or covertly placed critiques of books and articles written by Americans (Halperin testimony, Aspin Hearings, 232); CIA use of American media cover (Geyelin testimony, "National Intelligence Reorganization," 170); and, most sweeping, CIA ties with any journalists associated with the U.S. media (Wallace, "National Intelligence Reorganization," 175, 189). Even a long-time CIA officer advocated a new law: a prohibition against CIA involvement in any attempts "to influence American public opinion" (Phillips testimony, Aspin Hearings, p. 83).

111. Aspin Hearings, see 302, 318.

112. *New York Times*, May 5, 1977. See also, Laurence Stern, "CIA Investigated Journalists It Gave Data on Allende," *Washington Post*, October 20, 1977, who reveals another example of unauthorized CIA surveillance against an unnamed reporter.

113. See *Washington Post*, March 19, 20, 21, 23 and 30, 1975; *New York Times*, March 20 and 26, 1974; *Washington Star*, December 6, 1978. For other cases, see Morton H. Halperin, "Secrecy and National Security," *Bulletin of the Atomic Scientists* (August 1985), 114–17; Scot Powe, "Espionage, Leaks, and the First Amendment," *Bulletin of the Atomic Scientists* (June/July 1986), 8–10; Howard Morland, *The Secret That Exploded* (New York: Random House, 1979); Jay Peterzell, "Can the CIA Spook the Press?" *Columbia Journalism Review* (September/October 1986), 29–34; and Benjamin C. Bradlee, "The Press Is Not Reckless about National Security," *Washington Post*, National Weekly Edition, June 23, 1986, pp. 24–25.

114. Anderson "Why I Tell Secrets," *Parade Magazine*, November 30, 1980, 20. Well-known investigative reported Seymour M. Hersh argues simply: "If the government can't keep its secrets, tough luck. My job is getting the story." Panel on press–government relations at the Annual Meeting of the American Political Science Association, Washington, D.C., August 29, 1986. At this same panel, Colby advocated a balance between the needs of the media and the government. "Favor the press, but let's don't be black-and-white in our conclusions," he said, suggesting that there are "gradations." Benjamin C. Bradlee, executive editor of the *Washington Post*, has expressed a willingness to consider the government's case for a specific use of prior restraint, but "the press must continue its mission of publishing information that it—and it alone—determines to be in the public interest, in a useful, timely, and responsible manner—serving society, not government." "The Press is Not Reckless," 25.

115. See Jack Anderson, "How the CIA Snooped inside Russia," *Washington Post*, December 10, 1973, p. B17.

116. "Why I Tell Secrets," 25.

117. *New York Times* reporter Seymour M. Hersh has argued, however, that he favors even disclosure of this widely accepted "good" secret, on grounds that if he knew the secret as a reporter the enemy would be apt to know, too, and thus the U.S. military should be forced to change the plans to protect our troops. Harrison Salisbury, "The Sub, the CIA, and the Press," *Behind the Lines*, WNET Television, New York, 1975.

118. The Congress has also tightened the provisions of the Freedom of Information Act (5 U.S.C. 552 [1982]), at the CIA's request, making it easier for the Agency to reject or ignore requests for information. The purpose of the new restrictions was to counter the ludicrous situation in which, according to the CIA, the KGB and other foreign intelligence services were making FOIA requests; the end result, however, has been the imposition of imposing barriers to rightful requests from U.S. citizens. For the precise language of these laws, see House Intelligence Committee, *Compilation of Intelligence Laws*.

119. Interview, October 16, 1976, Washington, D.C.

120. See Sanford J. Ungar, *The Papers and The Papers* (New York: Sutton, 1972). Following the Bay of Pigs disaster, President Kennedy told *New York Times* editor Turner Catledge that he wished the newspaper had revealed the covert action: "Maybe if you had

printed more about the operation you would have saved us from a colossal mistake." Quoted by Wyden, *Bay of Pigs,* 155n.

121. Interview, January 9, 1985, Athens, Georgia.

122. Geyelin testimony, "National Intelligence Reorganization," 182.

123. Aspin Hearings, 100.

124. Harrison Salisbury, "Interview with William Colby: The Role of the CIA and That of the Press," *Behind the Lines,* WNET Television, New York, 1975.

125. Eric Sevareid interview with William O. Douglas, CBS, "Evening News," January 19, 1980.

Chapter 10

1. "With Watergate," notes Harry Howe Ransom, "it became clear than an agency like the CIA constituted a dangerous presidential instrument of secret power." "The Politicization of Intelligence," paper delivered at the Annual Meeting of the American Political Science Association, August 30, 1985, New Orleans, Louisiana, published under the same title in Stephen J. Cimbala, ed., *Intelligence and Intelligence Policy in a Democratic Society* (Dobbs Ferry, N.Y.: Transnational, 1987), cited on p. 31. On the attitudes of the Founders toward centralized power, see James S. Young, *The Washington Community: 1800–1825* (New York: Columbia University Press, 1966).

2. See the series of *Times* stories on the CIA from December 22–31, 1974. On CHAOS, see Burton Wides, "CIA Intelligence Collection About Americans: CHAOS Program and the Office of Security," in *Final Report* 3:679–732.

3. See John T. Elliff, "Congress and the Intelligence Community," in Lawrence C. Dodd and Bruce I Oppenheimer, eds., *Congress Reconsidered* (New York: Praeger, 1977), 196.

4. For an analysis of these investigations, see Johnson, *A Season of Inquiry.*

5. See the former president's responses to the Church committee in *Final Report* 4:171.

6. See, for example, James Angleton and Charles J. V. Murphy, "On the Separation of Church and State," *American Cause, Special Report,* American Security Council, Washington, D.C., June 1976, 2.

7. See, for example, Henry Steele Commager, "Intelligence: The Constitution Betrayed," *New York Review of Books,* September 30, 1976, 32–37.

8. *New York Times,* February 26, 1976.

9. Message to CIA stations, 1978.

10. On these incidents, see Wyden, *Bay of Pigs,* and Sol Stern, "NSA and the CIA."

11. For a more detailed analysis of this law, see Johnson, "Quiet Option," 143–53, and "Legislative Reform of Intelligence Policy," 549–73.

12. *Jimmy Carter, Public Papers, 1978* 1:x.

13. John T. Elliff, "The Legal Framework for Intelligence Activities," paper delivered at the Annual Meeting of the American Political Science Association, September 1, 1984, Washington, D.C. Elliff is a senior staff member on the Senate Select Committee on Intelligence.

14. Morris S. Ogul, *Congress Oversees the Bureaucracy,* 217. This book provides a comprehensive bibliography on oversight.

15. Charles S. Bullock, III, "House Committee Assignments," in Leroy N. Rieselbach, ed., *The Congressional Systems: Notes and Readings,* 2d ed. (Belmont, Calif.: Duxbury, 1979), 93.

16. Ogul, *Congress Oversees the Bureaucracy,* op. cit., 218.

17. Michael Barone, Grant Ujifusa, and Douglas Matthews, *The Almanac of American Politics* (New York: Dutton, 1977).

18. James David Barber, *The Presidential Character* (Englewood Cliffs, N.J: Prentice-Hall, 1972).

19. On this variable, see John F. Bibby, "Committee Characteristics and Legislative Oversight of Administration," *Midwest Journal of Politics* 10 (February 1966), 97.

20. See Ogul, *Congress Oversees the Bureaucracy,* 11–13; Fred Kaiser, "Oversight of Foreign Policy: The U.S. House Committee on International Relations," *Legislative Studies Quarterly* 2 (August 1977), 262.

21. See Joel D. Aberbach, "The Development of Oversight in the United States Congress: Concepts and Analysis," paper delivered at the Annual Meeting of the American Political Science Association, September 1977, Washington, D.C.

22. Seymour Scher, "Conditions for Legislative Control," *Journal of Politics* 25 (August 1963), 537.

23. See, for example, the Aspin Hearings, the first major committee public hearings of HPSCI.

24. See Scher, "Conditions for Legislative Control," 545.

25. See also Dale Vinyard, "Congressional Committees on Small Business," *Midwest Journal of Political Science* 10 (August 1966), 370.

26. John F. Manley, "The House Committee on Ways and Means: Conflict Management in a Congressional Committee," *American Political Science Review* (December 1965), 927–39.

27. A related structural variable is committee recruitment. If a member is talked into service by a leadership desperate for warm bodies on an unpopular committee, his devotion to duty may be halfhearted; in contrast, if he eagerly seeks out the recruitment himself, he is apt to participate more actively once on the committee (Bibby, "Committee Characteristics," 89). On HPSCI, the Democrats were selected by Speaker Thomas P. (Tip) O'Neill, Jr. (D, Massachusetts) in consultation with his close friend, the committee chairman; similarly, the Republican members were chosen by the GOP leadership. But who among the members selected asked to be and who had his arm twisted remains unclear—though at least one of the Top Four (Congressman A) was a self-starter.

28. Ogul, *Congress Oversees the Bureaucracy,* op. cit., 14.

29. Similarly, Schilling notes in his study of the FY 1950 military budget that in the House Subcommittee on Armed Services Appropriations hearings "the possible relationship of the budget to the North Atlantic Treaty occupied but a few puzzled paragraphs in the 3,000 pages of the hearings. Even the price of the oats the Army expected to buy for horses received as much attention as this. . . . the appropriation for the National Board for the Promotion of Rifle Practice managed to occupy the committee for 15 pages." Warner R. Schilling, "The Politics of National Defense: Fiscal 1950," in Warner R. Schilling, Paul Y. Hammond, and Glenn H. Snyder, *Strategy, Politics, and Defense Budgets* (New York: Columbia University Press, 1962), 92.

30. Aberbach, "The Development of Oversight," op. cit., 7.

31. See Harrison W. Fox, Jr., Susan Webb Hammond, and Jeanne Belle Nicholson, "Foresight, Oversight, and Legislative Development: A View of Congressional Policy Making," paper delivered at the Annual Meeting of the American Political Science Association, September 1977, Washington, D.C.

32. Richard F. Fenno, Jr., *Home Style* (Boston: Little, Brown, 1973).

33. Edward P. Boland, *Congressional Record,* June 6, 1978, p. H5023.

34. For background, see J. Leiper Freeman, "Investigating the Executive Intelligence: The Fate of the Pike Committee," *Capitol Studies* (Fall 1977), 103–17.

35. *Congressional Record,* July 14, 1977, pp. H7118–19.

36. Vinyard, "Congressional Committees," op. cit., 398.

37. For an elaboration of this argument, see Loch K. Johnson and Michael J. Glennon, "Combining House, Senate Intelligence Committees a Simple-Minded Idea," *Atlantic Constitution,* October 21, 1987. For the case that the intelligence committees have already been badly co-opted, see Peter Kornbluh, "The Iran-Contra Scandal: A Postmortem," *World Policy Journal* (Winter 1987–88), 129–50.

38. This legislation is known more officially as the Foreign Intelligence Surveillance Act of 1978 (P.L. 95–511, signed October 25, 1978; 92 Stat. 1783).

39. *U.S. News and World Report,* May 2, 1983, 29.

40. *Washington Post,* November 16, 1985, p. A1; and Ottaway and Tyler, "New Era of Mistrust Marks Congress' Role," p. A1.

41. The evolution of these agreements—piecrust promises, as Casey's detractors would argue; good-faith efforts to cooperate, in the view of his supporters—is derived from John T. Elliff, senior staffer on the Senate Intelligence Committee, comments, panel on intelligence policy, International Studies Association, March 8, 1985, Washington, D.C.; Steven V. Roberts, "More Lessons in the Secrecy Trade," *New York Times,* April 10, 1986, 12; *Report No.* 98-665, U.S. Senate (1984), 14–15; and, "Nomination of William H. Webster," *Hearings,* Senate Select Committee on Intelligence (1987), 52–59. The problem with such agreements is that they lack the force of law and, like executive orders, may be changed in secret without the knowledge of legislative overseers.

42. Cockburn, *Out of Control,* 229.

43. Some former CIA officials dispute whether this arms sale was really even a covert action, calling it "covert diplomacy" instead and therefore not requiring a finding. This distinction relies on the view that, in true covert actions, the perpetrator of the covert action (the United States) must remain unknown to the target nation. In the instance of the Iranian arms sale, the operation—though secret from most people—was of course known by the Iranian recipients of the arms to be a U.S. initiative. Sen. Arlen Specter (R, Pennsylvania), among others, rejected this fine definitional point at a meeting of intelligence specialists in 1988, concluding that if an operation is secret, involves the CIA, and attempts to influence events in another country, it is covert action. (Symposium on the Management of Intelligence, School of Foreign Service, Georgetown University, March 3, 1988.) The disagreement illustrates again (see chaps. 2, 6) the persisting definitional problems associated with the concepts "covert action," "special activities," and "special operations." Even presidential executive orders on intelligence and the 1980 Oversight Act incorporate different definitions. The need to make them congruent was widely recognized, and executive and legislative officials offered a revised definition of covert action in the proposed 1988 Intelligence Oversight Act, which combined the key elements of the Hughes-Ryan definition and the "special activities" definition incorporated into President Carter's major executive order on intelligence (E.O. 12036). The "new" definition read:

> (1) any operation of the Central Intelligence Agency conducted in foreign countries, other than activities intended solely for obtaining necessary intelligence; and (2) to the extent not inconsistent with subsection (1), above, any activity conducted by any department, agency, or entity of the United States Government in support of national foreign policy objectives abroad which is planned and executed so that the role of the United States Government is not apparent or acknowledged publicly, and functions in support of such activity, but which does not include diplomatic activities or the collection and production of intelligence or related support activities." (See "Intelligence Oversight Act of 1988," Senate *Report No.* 100-276, January 27, 1988, 46.)

44. Elizabeth Drew, "Letter from Washington" (March 22, 1987), *New Yorker,* March 30, 1987, 111.

45. See, for example, Kornbluh, "The Iran-Contra Scandal," 129–50.

46. See "Intelligence Oversight Act of 1988," House *Report No.* 100-705 (Part I), June 15, 1988, 11–12. Senator Specter also introduced a "National Intelligence Reorganization Act" (S. 1820) the same day, designed to strengthen the position of DCI (who would become a Director of National Intelligence, or DNI, were the legislation to pass). Under the sweeping reorganizational bill, still in the hearings stage, the DNI would be made a statutory member of the NSC and "the primary adviser to the President on foreign intelligence matters" (letter from Senator Specter to this author, June 21, 1988). Responsibility for covert action would fall to the DCIA, freeing the DNI from possible operational biases in reporting on intelligence to the president. The bill would also restrict the DCIA to a seven-year term.

47. See Senate *Report 100-276,* op. cit., 9.

48. Statement of Clark M. Clifford, *Hearings,* House Permanent Select Committee on Intelligence, February 24, 1988, 3, 4. The new House Intelligence Committee felt adamantly about prompt reporting. Within a few months of its creation in 1977, its chairman wrote to the DCI: "On November 3, 1977 . . . the Committee approved a recommendation requiring the DCI to report to me and to the Committee Staff Director . . . within 24 hours that a Presidential finding on covert action has been made, or that a highly sensitive intelligence collection operation is scheduled to begin" (unclassified letter from Edward P. Boland to Stansfield Turner, November 9, 1977, with copy to Zbigniew Brzezinski).

49. See House *Report No. 100-705,* op. cit., 16. The letter came from John R. Bolton to Representative McHugh, dated June 9, 1987. Senate Intelligence Committee Chairman Boren was prepared to compromise. He offered an amendment that would have allowed the executive branch to report, in times of extraordinary circumstances, to just the Speaker and House Minority Leader and their leadership counterparts in the Senate—a "Gang of Four." (See David L. Boren, remarks, Association of Former Intelligence Officers, March 28, 1988, rpt. in *Periscope* [the Association's newsletter] Spring 1988, 8.) This amendment fell by the wayside and the final Boren-Cohen bill kept the Gang-of-Eight provision used initially in the 1980 Intelligence Oversight Act.

50. House *Report 100-705,* op. cit., 73. Kissinger served as secretary of state for Presidents Nixon and Ford; Scowcroft and Brzezinski were national security advisors for Presidents Nixon and Ford and for President Carter, respectively; Helms and Colby were DCIs; and Carver served as a high-ranking CIA analyst and COS.

51. Ibid., 56.

52. See the dissenting views in *ibid.,* p. 64.

53. Ibid., 64.

54. Ibid., 15.

55. Ibid., 14. In August 1988 the CIA's general counsel stated publicly that some intelligence documents were so sensitive that only President Reagan had the security clearance to read them (*Los Angeles Times,* August 11, 1988, pt. 2:14).

56. Rep. Henry J. Hyde (R, Illinois), ibid., 63.

57. On the Wright controversy, see the *New York Times,* October 6, 1988, p.5; on the Wright-Bush accord, see the *New York Times,* February 1, 1989, p. 8.

58. Remarks, Tufts Symposium on Secrecy and U.S. Foreign Policy, February 26, 1988.

59. Interview with William L. Saltonstall, trustee, Tufts University, February 26, 1988.

60. See Harry Howe Ransom, "Congress and the Intelligence Agencies," in Harvey C. Mansfield, ed., *Congress against the President* (New York: Praeger, 1977), 153–66; and "Congress and Reform of the C.I.A.," *Policy Studies Journal* 5 (Summer 1977), 476–80.

61. Interview with Colby, Washington, D.C., March 21, 1979; see also, Colby and Forbath, *Honorable Men.*

62. Message to CIA stations, 1978.

63. This insightful characterization is from editor Tom Teepen, *Atlanta Constitution,* November 22, 1985. A former chairman of the Senate Intelligence Committee, Birch Bayh (D, Indiana), no longer an insider but still a close watcher of the Senate after his defeat in 1980, observed of Casey that he has a "sort of arrogant attitude that it's just none of your damn business, as if the Senate were a foreign body" (Ottaway and Tyler, "New Era of Mistrust"). Roger Morris, a former senior staffer on the NSC during the Johnson and Nixon years and a long-time close observer of the man, charged Casey with a "contempt for the inconveniences of democracy," *Los Angeles Times,* reprinted in the *Atlanta Constitution,* August 31, 1987, p. 11A.

64. While this particular observation is critical of the Church committee, it should be noted that on other occasions this committee was every bit as thorough as the Pike panel (and in several instances—as with NSA hearings—more so).

65. Francis E. Rourke, "Administrative Secrecy: A Congressional Dilemma," *American Political Science Review* 54 (September 1960), 694.

66. John E. Reilly, *American Public Opinion and U.S. Foreign Policy* (Chicago: Chicago Council on Foreign Relations, 1979).

67. *Washington Post,* January 27, 1980, p. A4.

Chapter 11

1. See Leslie H. Gelb, "Foreign Policy System Criticized by U.S. Aides," *New York Times,* October 19, 1981, pp. A1, A8; on the NSC, see Inderfurth and Johnson, *Decisions of the Highest Order.*

2. April 28, 1966, p. 28; see also Johnson, *A Season of Inquiry,* chap. 1.

3. William Safire, "The Iran-Contra Affair's 'Three Blind Mice,' " *New York Times* News Service, *Athens Banner-Herald* (Athens, Georgia), June 12, 1987; on PFIAB, see Church committee, *Final Report* 1:62–64.

4. See, for example, Michael J. Glennon, "The Boland Amendment and the Power of the Purse," *Christian Science Monitor,* June 15, 1987, p. 16.

5. Church committee, "Alleged Assassination Plots," 259.

6. On the CIA assassination manual, see Stansfield Turner, *Secrecy and Democracy,* 168, 170–71; on the CIA-Mafia ties in the 1960s, see ibid.

7. Remarks by Sen. Sam Nunn (D, Georgia), PBS, "McNeil-Lehrer Show," May 21, 1987, following closed hearings with the COS. For details see Stephen Engelberg, *New York Times,* June 21, 1988, p. 1.

8. See *New York Times,* April 28, 1966. For CIA instructions to bypass an ambassador during the Iranian arms-sale affair, see an account by Woodward, *The CIA's Secret Wars,* 420.

9. Interview with James Jesus Angleton, Washington, D.C., December 18, 1975.

10. Church committee, *Final Report* 1:305.

11. Ibid., 311, emphasis added.

12. Ibid., emphasis added.

13. Ibid.

14. 22 U.S.C. 2680a.

15. Church committee, *Final Report* 1:313.

16. Ibid.

17. Ibid., testimony taken on December 10, 1975.

18. Title IV, Sec. 102(d), 50 U.S.C. 403. One CIA officer with years of experience in Asia argues that "every operations cable should be seen by the ambassador, but not intelligence analyzes" (interview, Washington, D.C., March 20, 1987).

19. Church committee, *Final Report* 1:313.

20. Ibid.

21. David Binder, "State Dept. and C.I.A. Split on Envoy Role," *New York Times,* February 3, 1978, 1, 20.

22. State Department telegram, November 1977.

23. Binder, "State Dept. and C.I.A. Split," op. cit.

24. Aspin Hearings, 165.

25. Ibid., 163.

26. Ibid., 166.

27. See Patrick E. Tyler and David B. Ottaway, "Reagan's Secret Little Wars," *Washington Post* (Weekly Edition), March 31, 1986, pp. 6–7.

28. Charles Peters, "From Ouagadougou to Cape Canaveral: Why the Bad News Doesn't Travel Up," *Washington Monthly* 18 (April 1986), 31.

29. See Johnson, *A Season of Inquiry,* chap. 1.

30. Quoted in the *New York Times,* April 12, 1984, 10.

31. *New York Times,* October 1, 1985, 24.

32. Interview, CNSS staffer, Washington, D.C., November 15, 1985.

33. Gary J. Schmitt, "Congressional Oversight of Intelligence," *Studies in Intelligence* (Spring 1985), an in-house CIA publication.

34. Stansfield Turner, "Has Reagan Killed CIA Oversight?" *Christian Science Monitor,* September 26, 1985, p. 14.

35. Patrick Leahy (D, Vermont), ABC, "The Week with David Brinkley," December 14, 1986.

36. CIA spokesman, Office of Public Affairs, "Democracy and the CIA," panel discussion at the Annual Meeting of the American Political Science Association, Washington, D.C., September 1, 1984.

37. Participant, Roundtable Discussion, "Should the CIA Fight Secret Wars?" *Harper's* (September 1984), p. 42, emphasis in original. On the Iran-contra affair, see the Tower Commission *Report;* and Inouye-Hamilton committees, *Hearings.*

38. Frank Snepp, "Protect Rights of All Privy to U.S. Secrets," *New York Times,* February 22, 1984, 27. For an overview of the prior-restraint issue and related topics, see A. Stephen Boyan, Jr., "Presidents and National Security Powers: A Judicial Perspective," a paper delivered at the annual meeting of the American Political Science Association, Washington, D.C., September 1, 1988.

39. Testimony of Dean Rusk, "Oversight of U.S. Government Intelligence Functions," *Hearings,* Senate Committee on Government Operations, January 22, 1976, 77.

40. Rusk, interview, Athens, Georgia, January 21, 1985.

41. Sullivan, formal deposition, Church committee files, June 10, 1975.

42. Ransom, "CIA Accountability: Congress As Temperamental Watchdog," paper delivered at the Annual Meeting of the American Political Science Association, Washington, D.C., September 1, 1984, 26. See also Harry Howe Ransom, "The Politicization of Intelligence," in Cimbala, *Intelligence and Intelligence Policy* 43–44.

43. Cited by Elliff, "The Legal Framework for Intelligence Activities," 22.

44. Miller, remarks, Georgetown Symposium on the Management of Intelligence.

45. Lloyd Cutler, remarks, Tufts Symposium on Secrecy and U.S. Foreign Policy, February 26, 1988. On details of this operation, see "Intelligence Oversight Act of 1988," House *Report No. 100-705* (Part 1), June 15, 1988, 54, 78–79.

46. Tufts Symposium, ibid.

47. Ibid.

48. Remarks, Georgetown Symposium on the Management of Intelligence.

49. Ibid. In a reaction to the Iran-contra affair, Senator Specter may have overstated the failure of the prior-notice reporting requirement. At the same symposium, William G. Miller of the Senate Intelligence Committee observed that prior notice had been the case throughout his term as staff director (1976–81), with the exception of operations related to the rescue of U.S. hostages in Iran. Specter rested his case, the reader will recall, on the observation that during the Reagan administration prior notice had been less well honored by the CIA. (According to a recent House report, only four violations of the prior-notice expectation have occurred since passage of the 1980 Intelligence Oversight Act: three associated with Carter administration operations in Iran designed to rescue U.S. diplomatic hostages and, during the Reagan years, the Iran-contra operation [*Report No. 100-705, op. cit.*, 54].) The 1988 Intelligence Oversight Act, which passed the Senate on May 15, 1988, by a vote of 71 to 19 and was favorably reported out of the House Intelligence Committee a month later (all despite strong opposition from the Reagan administration) embraced the principle of prior notice, but, the reader will remember, with a forty-eight-hour escape hatch—a time clock viewed as too constraining by the Reagan White House, which preferred discretionary reporting based on the judgment of NSC principals.

50. On the desirability of criminal sanctions, see the testimony of Clark M. Clifford, *Hearings,* House Permanent Select Committee on Intelligence, U.S. House of Representatives, February 24, 1988, 6–7; and various expert witnesses testifying on H.R. 3665 before the Subcommittee on Criminal Justice, Judiciary Committee, U.S. House of Representatives (June 15, 1988).

51. See House *Report No. 100-175, op. cit.*, 5; and "Intelligence Oversight Act of 1988," Senate *Report No. 100-276*, January 27, 1988, 9.

52. See Gary M. Stern, "The FBI's Misguided Probe of CISPIS," Center for National Security Studies, Report No. 111, Washington, D.C., June 1988.

Chapter 12

1. Dr. Ray. S. Cline, former DDI, remarks, "Controlling Intelligence," panel at the Annual Meeting of the American Political Science Association, Chicago, Illinois, September 6, 1987.

2. Wyden, *Bay of Pigs*, 315.

3. Betts, "Analysis, War, and Decision," 88; and "Intelligence for Policy Making," *The Washington Quarterly* (Summer 1980), 122.

4. See M. Brewster Smith, "Opinions, Personality, and Political Behavior," *American Political Science Review* 52 (March 1958), 1–17.

5. Fox Butterfield, "Time of Uncertainty at C.I.A.," *New York Times*, December 20, 1986, p. 5.

6. Quoted in ibid.

7. Hamilton, "The Role of Intelligence," 68. See also Stephen Engelberg, "Doubts on Intelligence Data: Iran Affair Reviews the Issue," *New York Times*, August 31, 1987, p. 1.

8. See Richard K. Betts, "Surprise Despite Warnings: Why Sudden Attacks Succeed," *Political Science Quarterly* 95 (Winter 1980–81), 572.

9. See, for example, the exchange between Alexander L. George and I. M. Destler, "Making Foreign Policy," *American Political Science Review* 66 (September 1972), 751–95.

10. Since fear of embarrassment or reprisal is often a central element leading to quiescence in group decision settings, the usefulness of more anonymous methods for information and opinion sharing, such as "nominal group technique" (NGT) and computer-based communications, deserve closer inspection by scholars of government decision making [see Starr Roxanne Hiltz and Murray Turoff, *The Network Nation: Human Communication via Computer* (Reading, Mass.: Addison-Wesley, 1978)].

11. Interview, PBS, "Larry King Live," December 21, 1986.

12. Interview, PBS, "Larry King Live," December 15, 1986.

13. See Loch K. Johnson, *A Season of Inquiry*, 138–39.

14. Kennan, testimony, *Hearings*, Church committee, Washington, D.C., October 28, 1975.

15. Remarks, former DDI Ray S. Cline, Georgetown Symposium on the Management of Intelligence.

16. Ibid., as well as remarks by several former intelligence officials at the Tufts Symposium on Secrecy and U.S. Foreign Policy, February 26–27, 1988.

17. Former senator Thomas Eagleton (D, Missouri), Tufts Symposium on Secrecy and U.S. Foreign Policy, February 27, 1988, emphasis in original.

18. Arthur S. Hulnick, "Managing the Intelligence Analysis Process: Strategies for Playing the End Game," paper delivered at the Annual Meeting of the American Political Science Association, Chicago, Illinois, September 6, 1987, 22, n. 13.

19. Ibid., 20.

20. Hamilton, "The Role of Intelligence," 73.

21. John McMahon, remarks, CIA Headquarters, June 12, 1984.

22. Remarks, panel on intelligence oversight at Annual Meeting of the American Political Science Association, Washington, D.C., August 28, 1986.

23. "Laos," *Hearings*, Senate Foreign Relations Committee (1972), 34, emphasis added.

24. Inouye-Hamilton committees, *Report*, 383.

25. In reference to the Sherman Act (*Appalachian Coals Inc. v. U.S.*, 288 U.S. 344, 359–60 [1953]).

26. Section 3, National Security Reform Act of 1987 (S.1818), *Congressional Record,* October 27, 1987. See also a similar proposal offered by Rep. John Conyers (D, Michigan) on November 20, 1987. His "Official Accountability Act" states: "No person subject to this chapter shall order or engage in the planning of, preparation for, initiation or conduct of any intelligence activity which violates any statute or Executive Order . . ." Under this proposal, violations would be considered class D felonies, subject to criminal sanctions.

27. Stephen Engelberg, "C.I.A. Official Who Administered Contra Aid Quits," *New York Times,* February 26, 1988.

28. Inouye-Hamilton committees, *Report,* 381.

29. Turner, "Larry King Live," op. cit.

30. Haig, "Larry King Live," op. cit.

31. Letter from Thomas Jefferson to John B. Colvin (September 20, 1810), in Paul Leicester Ford, ed., *The Works of Thomas Jefferson* (New York: Putnam, 1905), 146–50.

32. Interview, CBS, "Walter Cronkite at Large," March 31, 1987.

33. Interview, ABC, "Evening News," February 22, 1985.

Selected Bibliography

Aberbach, Joel D. "The Development of Oversight in the United States Congress: Concepts and Analysis." Paper delivered at the Annual Meeting of the American Political Science Association, September 1977, Washington, D.C.

———. "Changes in Congressional Oversight." *American Behavioral Scientist* 22 (May 1979), 493–515.

Adams, Samuel. "Vietnam Cover-Up: Playing War with Numbers." *Harper's* 250 (May 1975), 41–44.

Adler, Emanuel. "Executive Control and the CIA." *Orbis* 23 (1979), 671–96.

Agee, Philip. *Inside the Company: CIA Diary*. Harmondsworth, Eng.: Penguin, 1975.

Ahmad, Eqbal, and Richard J. Barnet. "A Reporter At Large: Bloody Games." *New Yorker* (April 11, 1988), 44–86.

Alsop, Stewart, and Thomas Braden. *Sub Rosa: The OSS and American Espionage*. New York: Reynal and Hitchcock, 1946.

Anderson, Jack. "How the CIA Snooped inside Russia." *Washington Post*, December 10, 1973, p. B17.

———. "Why I Tell Secrets." *Parade Magazine*, November 30, 1980, p. 20.

Anderson, Martin. *Revolution*. New York: Harcourt, Brace & Jovanovich, 1988.

Andrew, Christopher, and David Dilks, eds. *The Missing Intelligence Dimension: Governments and Intelligence Communities in the Twentieth Century*. Urbana: University of Illinois Press, 1984.

Angleton, James, and Charles J. V. Murphy. "On the Separation of Church and State." *American Cause, Special Report*, American Security Council, Washington, D.C. (June 1976).

Aspin, Les, "Covert Acts Need Even More Oversight." *Washington Post*, February 24, 1980, p. B7.

———. "Misreading Intelligence." *Foreign Policy* 43 (1981), 166–72.

Bader, William B. "The Battle over 'Covert' Activities in Central America." Roundtable Remarks, *New York Times*, June 12, 1983, p. E3.

Baldwin, Gordon B., "Congressional Power to Demand Disclosure of Foreign Intelligence Agreements." *Brooklyn Journal of International Law* 3 (1976), 1–30.

Bamford, James. *The Puzzle Palace*. New York: Houghton Mifflin, 1984.

Barnds, William J. "Intelligence and Foreign Policy: Dilemmas of a Democracy." *Foreign Affairs* 47 (January 1969), 281–95.

Barnes, Trevor. "The Secret Cold War: The CIA and American Foreign Policy in Europe, 1946–1956, Parts I and II," *Historical Journal* 24 (June 1981), 399–415, and 25 (September 1982), 649–70.

Barnett, Frank R., B. Hugh Tovar, and Richard H. Shultz. *Special Operations in US Strategy.* Washington, D.C.: National Defense University Press, 1985.

Barron, John. *KGB: The Secret Work of Soviet Agents.* Pleasantville, N.Y.: Reader's Digest, 1974.

———. *Breaking the Ring: The Bizarre Case of the Walker Family Spy Ring.* Boston: Houghton Mifflin, 1987.

Baskir, Lawrence M. "Reflections on the Senate Investigation of Army Surveillance." *Indiana Law Journal* 49 (Summer 1974), 618–53.

Beichman, Arnold. "Can Counterintelligence Come In from the Cold?" *Policy Review* 15 (Winter 1981), 93–101.

Bell, Griffin B., with Ronald J. Astrow. *Taking Care of the Law.* New York: Morrow, 1982.

Berkowitz, Bruce E. "Intelligence in the Organizational Context: Coordination and Error in National Estimates." *Orbis* 29 (Fall 1985), 571–96.

Bernstein, Carl. "The CIA and the Media." *Rolling Stone,* October 20, 1977, 55–67.

Bethell, Nicholas. *Betrayed.* New York: Times Books, 1985.

Betts, Richard K. "Analysis, War and Decision: Why Intelligence Failures Are Inevitable." *World Politics* 31 (October 1978), 61–89.

———. "Intelligence for Policymaking." *Washington Quarterly* (Summer 1980), 118–29.

———. "Surprise Despite Warning: Why Sudden Attacks Succeed." *Political Science Quarterly.* 95 (Winter 1980–81), 551–72.

———. "Warning Dilemmas: Normal Theory vs. Exceptional Theory." *Orbis* 26 (Winter 1983), 829–33.

Bissell, Richard. "Reflections on the Bay of Pigs: Operation ZAPATA." *Strategic Review* 8 (Fall 1984), 66–70.

Bittman, Ladislaw. *The Deception Game.* New York: Ballantine, 1981.

Blackstock, Paul W. *The Strategy of Subversion.* Chicago: Quadrangle, 1964.

———. "The Intelligence Community under the Nixon Administration." *Armed Forces and Society* 1 (February 1975), 231–51.

Blake, John F. Affidavit, *Nathan Gardels v. Central Intelligence Agency.* Civil Action No. 78-0330, U.S. District Court for the District of Columbia, June 7, 1978.

Blaufarb, Douglas S. *The Counterinsurgency Era: U.S. Doctrine and Performance, 1950 to the Present.* New York: Free Press, 1977.

Blum, Richard, ed. *Surveillance and Espionage in a Free Society.* New York: Praeger, 1972.

"Bok [Derek C, president of Harvard University] Gives Senate Testimony on Universities, U.S. Intelligence." *Harvard Gazette,* September 15, 1978.

Boland, Edward P. Letter-to-the-Editor (Covert Actions and the Congress). *New York Times,* June 8, 1978 A–26.

Boren, David, and William S. Cohen. "Keep Two Intelligence Committees." *New York Times,* August 17, 1987, Y19.

Bork, Robert H. " 'Reforming' Foreign Intelligence." *Wall Street Journal,* March 9, 1978.

Bozeman, Ada. "Statecraft and Intelligence in the Non-Western World." *Conflict* 6 (1985), 1–35.

Braden, Tom. "What's Wrong with the CIA?" *Saturday Review* 2 (April 5, 1975), 14–18.

———. "The Birth of the CIA." *American Heritage* 28 (February 1977), 4–13.

Bradlee, Benjamin C. "The Press is Not Reckless about National Security." *Washington Post,* (National Weekly Edition) June 23, 1986, pp. 24–25.

Brandon, Henry. "The Spy Who Came and Then Told." *Washington Post* (National Weekly Edition), August 24, 1987, p. 36.

Breckinridge, Scott D. *The CIA and the U.S. Intelligence System.* Boulder: Westview, 1986.

———. "A Presidential 'Finding' for the White House Iran Initiative." *Foreign Intelligence Literary Scene* 7 (January/February 1988), 1–3.

Brewer, Gary D., and Paul Bracken. "Some Missing Pieces of the C[3]I Puzzle." *Journal of Conflict Resolution* 28 (September 1984), 451–69.

Brodt, Carl L. "The CIA and International Law: An Absurd Conjunction?" Paper delivered at the Annual Meeting of the Western Political Science Association, San Francisco, April 1, 1976.

Brown, Anthony Cave. *Bodyguard of Lies.* New York: Harper & Row, 1975.

Buckley, William F. "CIA Maneuvers." *National Review,* July 8, 1983, 841.

Burrows, William E. *Deep Black: Space Espionage and National Security.* New York: Random House, 1986.

Butterfield, Fox. "Times of Uncertainty at C.I.A." *New York Times,* December 20, 1986, 5.

Carlucci, Frank C. Testimony on Legislative Control of Covert Action, Senate Committee on Intelligence, 1988. Reprinted in *Periscope* (Winter 1988), 6–9.

Central Intelligence Agency. *Intelligence in the War of Independence,* Washington, D.C.: CIA, 1975.

———. *Fact Book on Intelligence.* Washington, D.C.: CIA, April 1983.

———. *Intelligence: The Acme of Skill.* Washington, D.C.: CIA (n.d.).

Church, Frank. "Covert Action: Swampland of American Foreign Policy." *Bulletin of the Atomic Scientists* 32 (February 1976), 7–11.

———. "Do we Still Plot Murders? Who Will Believe We Don't?" *Los Angeles Times,* June 14, 1983, pt. 2, p. 5.

Cimbala, Stephen J., ed. *Intelligence and Intelligence Policy in a Democratic Society.* Dobbs Ferry, N.Y.: Transnational, 1987.

Clarke, Duncan L., and Edward L. Neveloff. "Secrecy, Foreign Intelligence, and Civil Liberties: Has the Pendulum Swung Too Far?" *Political Science Quarterly* 99 (Fall 1984), 493–513.

Cline, Marjorie W., ed., *Teaching Intelligence in the Mid-1980s.* Washington, D.C.: National Intelligence Study Center, 1985.

Cline, Ray S. "Policy Without Intelligence." *Foreign Policy* 17 (Winter 1974–75), 121–35.

———. *Secrets, Spies, and Scholars: Blueprint of the Essential CIA.* Washington, D.C.: Acropolis, 1976.

———. *The CIA Under Reagan, Bush and Casey.* Washington, D.C.: Acropolis, 1981.

———. *The CIA: Reality vs. Myth.* Washington, D.C.: Acropolis, 1982.

———. "Covert Action Is Needed for United States Security." in *Conflict in American Foreign Policy* edited by Don Mansfield and Gary J. Buckley. Englewood Cliffs, N.J.: Prentice-Hall, 1985, 72–77.

Cockburn, Leslie. *Out of Control: The Story of the Reagan Administration's Secret War in Nicaragua, the Illegal Arms Pipeline, and the Contra Drug Connection* (New York: Atlantic Monthly 1987.)

———, et al. "Forum: Hearing Nothing, Saying Nothing: The Iran-*Contra* Investigation That Never Was." *Harper's,* 276 (February 1988), 45–57.

Codevilla, Angelo. "The CIA: What Have Three Decades Wrought." *Strategic Review* (Winter 1980), 68–71.

Cohen, William S., and George J. Mitchell. *Men of Zeal: A Candid Inside Story of the Iran-Contra Hearings* New York: Viking, 1988.

Colby, William E., "After Investigating U.S. Intelligence." *New York Times,* February 26, 1976, 11.

———, "Gesprach mit William E. Colby." *Der Spiegel* 4 (January 23, 1978), 69–115.

———, and Peter Forbath. *Honorable Men: My Life in the CIA.* New York: Simon and Schuster, 1978.

Collins, John M. *Green Berets, Seals, and Spetsnaz: U.S. and Soviet Special Military Operations.* New York: Pergamon, 1987.

Commager, Henry Steele. "Intelligence: The Constitution Betrayed." *New York Review of Books*, September 30, 1976, 32–37.
Cooper, Chester L. "The CIA and Decisionmaking." *Foreign Affairs* 50, January 1972, 223–36.
Coox, Alvin D. "Pearl Harbor." In *Decisive Battles of the Twentieth Century*. edited by Noble Frankland and Christopher Dowling. New York: McKay, 1976, 141–54.
Copeland, Miles. "They're Breaking Up That Old Gang of Mine." *Washington Post Potomac Magazine*, March 23, 1975, 12, 34–35.
———. "The Functioning of Strategic Intelligence." *Defense and Foreign Affairs Digest* (February 1977).
———. *The Real Spy World*. London: Sphere, 1978.
Corson, William R. *The Armies of Ignorance: The Rise of the American Intelligence Empire*. New York: Dial, 1977.
Day, Bonner. "The Battle Over U.S. Intelligence." *Air Force Magazine*. 61 (May 1978), 42–47.
De Silva, Peer. *Sub Rosa: The CIA and the Uses of Intelligence*. New York: Times Books, 1978.
de Sola Pool, Ithiel, "The Necessity for Social Scientists Doing Research for Governments." *Background* 10 (August 1966), 111–22.
Donner, Frank J. "The Theory and Practice of American Intelligence." *New York Review of Books*, April 22, 1971, 27–39.
———. *The Age of Surveillance* New York: Random House, 1980.
Dorsen, Norman, and Stephen Gillers, eds. *None of Your Business*. New York: Viking, 1974.
Duffy, Gloria. "Crisis Mangling and the Cuban Brigade." *International Security* 8 (Summer 1983), 67–87.
Dulles, Allen, *The Craft of Intelligence*. Westport, Conn.: Greenwood, 1977.
Elliff, John T. "Congress and the Intelligence Community." In *Congress Reconsidered*, edited by Lawrence C. Dodd and Bruce I. Oppenheimer. New York: Praeger, 1977, 193–206.
———. *The Reform of FBI Intelligence Operations*. Princeton, N.J.: Princeton University Press, 1979.
———. "The Legal Framework for Intelligence Activities." Paper delivered at the Annual Meeting of the American Political Science Association, Washington, D.C., September 1, 1984.
———. "Statutory Limitations on Covert Actions." Paper delivered at the Annual Meeting of the International Studies Association, Washington, D.C., March 8, 1985.
Ellsworth, Robert F., and Keith L. Adelman. "Foolish Intelligence." *Foreign Policy* 36, Fall 1971, 147–59.
Emerson, Steven. *Secret Warriors*. New York: Putnam, 1988.
Epstein, Edward Jay. "The War Within the CIA." *Commentary* 66 (August 1978), 35–39.
———. *Legend: The Secret World of Lee Harvey Oswald*. New York: Ballantine, 1979.
———. "The Spy War." *New York Times Magazine*, September 28, 1980, 34 ff.
———. "Disinformation: Or Why the CIA Cannot Verify an Arms-Control Agreement." *Commentary*, 74, July 1982, 21–28.
Epstein, Noel. "Professors Decry Recruiting by CIA." *Washington Post*, May 6, 1976, A–38.
Erdbrook, C. E. "Principles of Deep Cover." *Covert Action Information Bulletin* 10 (August–September 1980), 45–54.
Eutace, Harry. "Special Report: Changing Intelligence Priorities." *Electronic Warfare/Defense Electronics* 28 (November 1978), 35–37.
Falk, Richard A. "CIA Covert Action and International Law." *Society* 12 (March/April 1975), 39–44.
Felix, Christopher. *A Short Course in the Secret War*. New York: Dell, 1963; 1988.
Fisher, Roger. "The Fatal Flaw in Our Spy System." *Boston Globe*, February 1, 1976.

Flanagan, Stephen J. "The Coordination of National Intelligence." In *Public Policy and Political Institutions. United States Defense and Foreign Policy—Coordination and Integration,* edited by Duncan L. Clarke (Greenwich, Conn.: JAI, 1985), 157–96.

Foell, Earl W. "Journalists, Diplomats, and Spies—There Are Some Parallels." *Christian Science Monitor,* September 16, 1986, p. 3.

Ford, Harold P., "Piety and Wit." *America* (January 11, 1975), 10–11.

Fowler, Wyche, Jr. "Legislative Control of Covert Operations." *First Principles* 9 (March/April 1984), 1, 4–7.

Fox, Harrison W. Jr., Susan Webb Hammond, and Jeanne Belle Nicholson, "Foresight, Oversight, and Legislative Development: A View of Congressional Policy Making." Paper delivered at the Annual Meeting of the American Political Science Association, Washington, D.C., September 1977.

Freeman, J. Leiper. "Investigating the Executive Intelligence: The Fate of the Pike Committee." *Capitol Studies* (Fall 1977), 103–17.

Gaddis, John Lewis. *Strategies of Containment: A Critical Appraisal of Postwar American National Security Policy.* New York: Oxford, 1982.

———. "The Intelligence Revolution's Impact on Postwar Diplomacy." Thirteenth Military History Symposium, United States Air Force Academy, October 13, 1988.

Galbraith, John Kenneth. *A Life in Our Times: Memoirs.* London: Andre Deutsch, 1981.

Gates, Robert M. "The CIA and American Foreign Policy." *Foreign Affairs* 66 (Winter 1987/88), 215–30.

Gelb, Leslie, H. "Should We Play Dirty Tricks in the World?" *New York Times Magazine,* December 21, 1975, 10ff.

———. "Foreign Policy System Criticized by U.S. Aides." *New York Times,* October 19, 1981, pp. A1, A8.

———. "Shift is Reported on C.I.A. Actions," *New York Times,* (June 11, 1984), p. 1.

———. "Overseeing of C.I.A. by Congress." *New York Times,* July 7, 1986, p. 1.

George, Alexander L., and I. M. Destler. "Making Foreign Policy." *American Political Science Review* 66 (September 1972), 751–95.

Glennon, Michael J. "Investigating Intelligence Affairs: The Process of Getting Information for Congress." In *The Tethered Presidency,* edited by Thomas Franck. New York: New York University Press, 1981, 141–52.

———. "Liaison and the Law: Foreign Intelligence Activities in the United States." *Harvard International Law Journal* 25 (Winter 1984), 1–42.

———. "The Boland Amendment and the Power of the Purse." *Christian Science Monitor,* June 15, 1987.

Godfrey, E. Drexel, Jr., "Ethics and Intelligence." *Foreign Affairs* 56 (April 1978), 624–42.

Godson, Roy, ed. *Intelligence Requirements for the 1980s: Elements of Intelligence.* Washington, D.C.: National Strategy Information Center, 1979.

———, ed. *Intelligence Requirements for the 1980s: Analysis and Estimates.* Washington, D.C.: National Strategy Information Center, 1980.

———, ed. *Intelligence Requirements for the 1980s: Counterintelligence.* Washington, D.C.: National Strategy Information Center, 1980.

———, ed. *Intelligence Requirements for the 1980s: Covert Action.* Washington, D.C.: National Strategy Information Center, 1981.

———, ed. *Intelligence Requirements for the 1980s: Clandestine Collection.* Washington, D.C.: National Strategy Information Center, 1982.

———, ed. *Intelligence Requirements for the 1980s: Domestic Intelligence.* Boston: Lexington, 1986.

———, ed. *Comparing Foreign Intelligence: The US, the USSR, and the Third World.* Washington, D.C.: Bergamon-Brassey's, 1988.

———, ed. *Intelligence Requirements for the 1990s.* Boston: Lexington, 1989.

Goldwater, Barry. "Congress and Intelligence Oversight." *Washington Quarterly* 6 (Summer 1983), 16–21.

Goodman, Allan E. "Dateline Langley: Fixing the Intelligence Mess," *Foreign Policy* 57 (Winter 1984–85), 160–78.

———. "Does Intelligence Matter?" Paper delivered at the Annual Meeting of the International Studies Association, Annaheim, California, March 22, 1986.

———. "Reforming U.S. Intelligence," *Foreign Policy* 67 (Summer 1987), 121–36.

Gordon, Michael R. "CIA Is Skeptical That New Soviet Radar is Part of an ASM Defense System." *National Journal,* March 9, 1985, 523–26.

Gormet, Suzanne, "Casey's Shadows: A Greater Emphasis on CIA Analysis." *Wall Street Journal,* July 16, 1982, 1.

Goulden, Joseph C., with Alexander W. Raffio. *The Death Merchant: The Rise and Fall of Edwin P. Wilson.* New York: Simon & Schuster, 1984.

Graham, Daniel O. "Estimating the Threat: A Soldier's Job." *Army Magazine,* April 1973, 14–18.

———. "U.S. Intelligence at the Crossroads." *Report No. 76-1,* United States Strategic Institute, Washington, D.C. (1976).

———. "Intelligence: Realities and Myth." *Wall Street Journal,* March 11, 1977, 16.

Hackes, Peter, moderator. "Foreign Intelligence: Legal and Democratic Controls." Forum, American Enterprise Institute. Washington, D.C.: December 11, 1979.

Halperin, Morton H. "The CIA on American Campuses: The Harvard Confrontation." *First Principles* 4 (November 1978), 1–4.

———. "The CIA's Distemper." *New Republic* (February 9, 1980), 21–23.

———. "Secrecy and National Security." *Bulletin of the Atomic Scientists* (August 1983), 112–17.

———, and Daniel N. Hoffman, *National Security and the Right to Know.* Washington, D.C.: New Republic, 1977.

———, and Gary M. Stern. "Lawful Wars." *Foreign Policy* 72 (Fall 1988), 173–96.

Hamilton, Alexander, John Jay, and James Madison. *The Federalist.* New York: Modern Library, 1937.

Hamilton, Lee H. "The Hazards of Overclassification." *Congressional Record* 128 (April 6, 1982), E1564–E1565.

———. "View from the Hill." *Extracts from Studies in Intelligence.* Langley, Va.: Central Intelligence Agency, September 1987, 65–76.

Handel, Michael I. "The Study of Intelligence." *Orbis* 26 (Winter 1983), 817–21.

———. "Intelligence and the Problem of Strategic Surprise." *Journal of Strategic Studies* 7 (September 1984), 231–81.

———. "Technological Surprise in War." *Intelligence and National Security* 2 (January 1987), 1–53.

Hastedt, Glenn, "Organizational Foundations of Intelligence Failures," In *Intelligence Policy and Process,* edited by Alfred C. Maurer, Marion D. Tunstall and James M. Klagle. (Boulder, Colo.: Westview, 1985), pp. 140–56.

———. "The Constitutional Control of Intelligence." *Intelligence and National Security* 1 (May 1986), 255–71.

———. "Controlling Intelligence: The Role of the DCI." *International Journal of Intelligence and Counterintelligence* 1 (1986), 25–40.

Hardy, Timothy, S., "Intelligence Reform in the Mid-1970s." *Studies in Intelligence* (Summer 1976), 1–15.

Harris, Kenneth, Interview with Richard Helms, *London Observer* (1979), rpt. in the *Washington Star,* January 20, 1980, G1–G4.

Harris, Richard, "Reflections (Richard Helms)." *New Yorker* (April 10, 1978), 44–86.

Hersh, Seymour M. "Underground for the C.I.A. in New York: An Ex-Agent Tells of Spying on Students." *New York Times,* December 29, 1974, 1.

———. "Hunt Tells of Early Work for a C.I.A. Domestic Unit." *New York Times,* December 31, 1974, 1.

———. "Congress is Accused of Laxity on C.I.A.'s Covert Activity." *New York Times,* June 1, 1978, 1.

———. "The Angelton Story." *New York Times Magazine,* June 25, 1978, pp. 13.
———. "Target Qaddafi." *New York Times Magazine* (February 22, 1987), 16ff.
———. *"The Target Is Destroyed."* New York: Vintage, 1987.
Heuer, Richards J., Jr. "Cognitive Biases in the Evaluation of Intelligence Estimates." Paper delivered at the Annual Meeting of the American Institute for Decision Sciences, St. Louis, October 30, 1978.
———. "Improving Intelligence Analysis: Some Insights on Data, Concepts, and Management in the Intelligence Community." *The Bureaucrat* 8 (Winter 1979/80), 2–11.
———. "Strategic Deception and Counterdeception." *International Studies Quarterly* 25 (June 1981), 294–327.
Hilsman, Roger. *Strategic Intelligence and National Decisions.* Glencoe: Free Press, 1956.
———. *To Move a Nation: The Politics of Foreign Policy in the Administration of John F. Kennedy.* New York: Dell, 1964.
———. "On Intelligence." *Armed Forces and Society* 8, Fall 1981, 129–43.
Hiltz, Starr Roxanne, and Murray Turoff. *The Network Nation: Human Communication via Computer.* Reading, Mass.: Addison Wesley, 1978.
Hinckle, Warren. "CIA Reunion." *San Francisco Examiner.* January 27, 1986.
———, Sol Stern, and Robert Scheer. "The University on the Make." *Ramparts* 4 (April 1966), 11–22.
Hood, William. *Mole.* New York: Norton, 1982.
Hopple, Gerald, and Bruce Watson, eds. *The Military Intelligence Community.* Boulder: Westview, 1986.
Horrock, Nicholas. "East or West, Spy Central Is a Whirring Big Computer." *New York Times,* July 17, 1977, E6.
Horton, John. "Mexico, The Way of Iran?" *Journal of Intelligence and Counterintelligence* 1 (1986), 91–102.
Hougan, Jim. "A Surfeit of Spies." *Harper's,* December 1974, 51–67.
Hughes, Thomas L. "The Power to Speak and the Power to Listen: Reflections in Bureaucratic Politics and a Recommendation on Information Flows." In *Secrecy and Foreign Policy,* edited by Thomas Franck and Edward Weisband. New York: Oxford, 1974.
———. *The Fate of Facts in a World of Men: Foreign Policy and Intelligence Making.* New York: Foreign Policy Association, 1976.
Hulnick, Arthur S. "The Intelligence Producer-Policy Consumer Linkage: A Theoretical Approach." *Intelligence and National Security* 1 (May 1986), 212–33.
———. "CIA and Its Relations with the Academic Community: Symbiosis, Not Psychosis." Paper delivered at the Annual Meeting, International Studies Association, April 1987, Washington, D.C.
———. "Managing the Intelligence Analysis Process: Strategies for Playing the End Game." Paper delivered at the Annual Meeting of the American Political Science Association, Chicago, September 6, 1987.
Hunter, David Haley, "Intelligence: The Ethical Dimension." Ph.D. diss., University of Georgia, 1978.
Inderfurth, Karl F., and Loch K. Johnson. *Decisions of the Highest Order: Perspectives on the National Security Council.* Pacific Grove, Calif.: Brooks/Cole, 1988.
Inman, Bobby. "Foreign Policy Notes." *Institute of International Studies* 6 (December 12, 1986), 1.
"Interim Report of the Ad Hoc Committee on M.I.T. and the Intelligence Agencies." Massachusetts Institute of Technology, *Tech Talk* 23 (April 11, 1979).
Jackson, Robert H., "The Federal Prosecutor," *Journal of the American Judicature Society* 24 (June 1940), 18–20.
Janis, Irving. *Groupthink.* 2d ed. Boston: Houghton Mifflin, 1982.
Jeffreys-Jones, Rhodri. *American Espionage: From Secret Service to CIA.* New York: Free Press, 1977.

———. *The CIA and American Democracy.* New Haven, Conn.: Yale University Press, 1989.

Jervis, Robert. "Intelligence and Foreign Policy." *International Security* 11 (Winter 1986–87), 141–61.

Johnson, Loch K. "National Security, Civil Liberties, and the Collection of Intelligence: A Report on the Huston Plan." In *Supplementary Detailed Staff Reports on Intelligence Activities and the Rights of Americans, Final Report,* Senate Select Committee to Study Governmental Operations with Respect to Intelligence Activities, 94th Cong., 2d sess., April 23, 1976, S. Rept. No. 94–755, 3:921–86.

———. "Legislative Oversight and the Central Intelligence Agency." *Workshop on Congressional Oversight and Investigations,* October 22, 1979, H. Doc. No. 96-217, 15–17.

———. "The CIA: Controlling the Quiet Option." *Foreign Policy* 39 (Summer 1980), 143–52.

———. "Legislative Control of Paramilitary Operations." *First Principles* ((March–April, 1984), 1–4.

———. "Decision Costs in the Intelligence Cycle." *Journal of Strategic Studies* 7 (September 1984), 318–35.

———. "Legislative Reform of Intelligence Policy." *Polity* 17 (Spring 1985), 549–73.

———. *A Season of Inquiry: Congress and Intelligence.* 2d ed. Chicago: Dorsey, 1988.

———, and John T. Elliff. "Counterintelligence." in *Foreign and Military Intelligence, Final Report,* Senate Select Committee to Study Governmental Operations with Respect to Intelligence Activities, 94th Cong., 2d sess., April 26, 1976, S. Rept. No. 94-755, 1:163–78.

Johnson, William R., "Tricks of the Trade: Counterintelligence Interrogation." *International Journal of Intelligence and Counterintelligence* 1 (1986), 103–14.

Joselyn, Eric. "C.I.A. Off Campus: Closing the Company Store." *Nation,* March 26, 1988, 416.

Kahn, David. *The Codebreakers: The Story of Secret Writing.* New York: MacMillan, 1967.

———. "Big Ear or Big Brother?" *New York Times Magazine,* May 16, 1976, 13ff.

Kaiser, Frederick M. "Oversight of Foreign Policy: The U.S. House Committee on International Relations." *Legislative Studies Quarterly* 2 (August 1977), 262.

———. "Congressional Rules and Conflict Resolution: Access to Information in the House Select Committee on Intelligence." *Congress & the Presidency* 15 (Spring 1988), 49–73.

Kaplan, Fred. *The Wizards of Armageddon.* New York: Simon and Schuster, 1983.

Karalekas, Anne. "History of the Central Intelligence Agency." *Supplementary Detailed Staff Reports on Foreign and Military Intelligence, Final Report,* Senate Select Committee to Study Governmental Operations with Respect to Intelligence Activities, 94th Cong., 2d sess., April 23, 1976, S. Rept. No. 94-755, 4:1–107.

———. "Intelligence Oversight: Has Anything Changed?" *Washington Quarterly* 6 (1983), 22–30.

Karnow, Stanley. "Associating with the Agency." *Newsweek,* October 10, 1977, 5.

———. "When a Newsman Consults the CIA." *New York Times,* December 18, 1977, IV-19.

Kelman, Herbert C., "When Scholars Work With the C.I.A." *New York Times,* March 5, 1986, 27.

Kent, Sherman. *Strategic Intelligence for American World Policy.* Princeton, N.J.: Princeton University Press, 1949; revised in 1965.

———. "Estimates & Influence," *Foreign Service Journal* (April 1969).

Kirkpatrick, Lyman B., Jr., *The Real CIA.* New York: MacMillian, 1968.

———. *The U.S. Intelligence Community.* New York: Hill and Wang, 1973.

Kissinger, Henry K. "A Matter of Balance." *Los Angeles Times,* July 26, 1987, V-1.

Knorr, Klaus. "Failures in National Intelligence Estimates: The Case of the Cuban Missiles." *World Politics* 16 (April 1984), 456–75.

Koh, Harold Hongju, "Why the President (Almost) Always Wins in Foreign Affairs: Lessons of the Iran-Contra Affair," *Yale Law Journal* 97 (June 1988), 1255–1342.

Komer, Robert W. "The Tet Intelligence Flop: One Out of Step, or Many?" *Washington Star*, November 16, 1975, E3.

Kornbluh, Peter. "The Iran-Contra Scandal: A Postmortem." *World Policy Journal* (Winter 1987–88), 129–50.

Landis, Fred. "The CIA and *Reader's Digest.*" *Covert Action Information Bulletin* 29 (Winter 1988), 41–47.

Lardner, George, Jr., "Moynihan Unleashes the C.I.A." *Nation* (February 16, 1980), 175–78.

———. "Missing Intelligence Charters." *Nation* 227 (September 2, 1987), 168–171.

Laquer, Walter. A *World of Secrets.* New York: Basic Books, 1985.

Latimer, Thomas K. "U.S. Intelligence and the Congress." *Strategic Review*, Summer 1979, 47–56.

———. "United States Intelligence Activities: The Role of Congress," In *Intelligence Policy and National Security* edited by Robert L. Pfaltzgraff Jr., Uri Ra'anan, and Warren Milberg, (Hamden, Conn.: Archon, 1981), 273–88.

Layton, Edwin, Roger Pineau, and John Costello. *"And I was There": Pearl Harbor and Midway—Breaking the Secrets.* New York: Morrow, 1985.

Le Moyne, James, "Testifying to Torture." *New York Times Magazine* (June 5, 1988), 44–47ff.

Leary, William M. *Perilous Missions: Civil Air Transport and CIA Covert Operations in Asia.* Tuscalosa, Ala.: University of Alabama Press, 1984.

———. *The Central Intelligence Agency: History and Documents.* Tuscalosa, Ala.: University of Alabama, 1984. (A republication of the CIA history prepared for the Church committee by Anne Karalekas).

Lefever, Ernest W. "Can Covert Action Be Just?" *Policy Review* 12, Spring 1980, 115–22.

———, and Roy Godson, *The CIA and the American Ethic* Washington, D.C.: Georgetown University, 1979.

Lelyveld, Joseph. "India Still Wary on U.S. Scholars." *New York Times*, August 14, 1968, 7.

———. "The Director: Running the CIA." *The New York Times Magazine* January 20, 1985, 16ff.

Lemarchand, Rene. "The CIA in Africa: How Central? How Intelligent?" *Journal of Modern African Studies* 14 (September 1976), 401–26.

Lescaze, Lee. "Pentagon vs. CIA." *Washington Post*, June 10, 1977, 1.

Levite, Ariel. *Intelligence and Strategic Surprises.* New York: Columbia University Press, 1987.

Lindsey, Robert. *The Falcon and the Snowman.* New York: Simon and Schuster, 1979.

Liphart, Arend. *Democracies.* (New Haven, Conn.: Yale University Press, 1984).

Loory, Stuart H. "The CIA's Use of the Press: 'A Mighty Wurlitzer.' " *Columbia Journalism Review* (September/October 1974), 8–18.

Levell, Stanley. *Of Spies and Strategems.* New York: Prentice-Hall, 1963.

Lummis, Charles Douglas. "The Radicalism of Democracy." *democracy* 2 (Fall 1982), 9–16.

Luttwak, Edward N., "How to Administer Covert Operations," *New York Times*, November 17, 1986, A21.

McChristian, Maj. Gen. Joseph A. "The Role of Military Intelligence, 1965–1967." *Vietnam Studies.* Washington, D.C.: Department of the Army, 1974.

McDougal, Myres, Harold D. Lasswell, and W. Michael Reisman, "The Intelligence Function and World Public Order." *Temple Law Quarterly* 46 (Spring 1973), 365–448.

McGarvey, Patrick J. *CIA: The Myth and the Madness.* New York: Saturday Review Press, 1972.

McGehee, Ralph. "The C.I.A. and the White Paper on El Salvador." *Nation*, April 11, 1981, 423–25.

———. *Deadly Deceits: My 25 Years in the CIA.* New York: Sheridan Square, 1983.

Madison, James (or Alexander Hamilton). *Federalist Paper* No. 51. Rpt. in *The Federalist.* New York: Modern Library, 1937, 335–41.

Maeroff, Gene I. "Harvard and CIA at Impasse Over Secret Work by the Faculty." *New York Times,* August 5, 1978, 1, 8.

Mandel, Robert. "Distortions in the Intelligence Decision-Making Process." Paper delivered at the Annual Meeting of the International Studies Association, Anaheim, California, March 22, 1986.

Marchetti, Victor, and John D. Marks. *The CIA and the Cult of Intelligence.* New York: Knopf, 1974.

Marks, John. "How to Spot a Spook." *Washington Monthly,* November 1974, 5–11.

———. "The CIA's Corporate Shell Game." *Washington Post,* July 11, 1976, C-1.

———. *The Search for the "Manchurian Candidate": The CIA and Mind Control.* London: Allen Lane, 1979.

Martin, David C. *Wilderness of Mirrors.* New York: Harper and Row, 1980.

Maas, Peter, *Manhunt.* New York: Random, 1986.

Masterman, Sir John. *Double Cross System of the War of 1939–45.* New Haven, Conn.: Yale University Press, 1972.

Maurer, Alfred C., Marion D. Tunstall, and James M. Keagle, eds. *Intelligence: Policy and Process.* Boulder: Westview, 1985.

Maury, John M., Jr. "Don't Cut Up the CIA Into Useless Pieces." *Washington Star,* March 31, 1977, F1.

———. "CIA and the Congress." *Congressional Record* (September 18, 1984), S11427–31 (originally published in the classified CIA in-house journal, *Studies in Intelligence,* 1974).

May, Ernest R., ed. *Knowing One's Enemies: Intelligence Assessments Before the World Wars.* Princeton, N.J.: Princeton University Press, 1985.

Melanson, Philip H. "The C.I.A.'s Secret Ties to Local Police." *Nation,* March 26, 1983, 351, 364–68.

Meyer, Cord. *Facing Reality: From World Federation to the CIA.* New York: Harper and Row, 1980.

Miller, Merle. *Plain Speaking: An Oral Biography of Harry S. Truman.* New York: Berkley, 1973.

Miler, Newton S. "Counterintelligence." In *Intelligence Requirement for the 1980s: Elements of Intelligence,* edited by Roy Godson. Washington, D.C.: Consortium for the Study of Intelligence, 1979, 47–64.

Morgan, Dan. "Slain Agent Feared for CIA Lives." *Washington Post,* December 26, 1975, A1.

———, and Walter Pincus. "Somehow, the Irangate Story Still Doesn't Add Up." *Washington Post National Weekly Edition,* September 28, 1987, 23–25.

Morgan, Richard E. *Domestic Intelligence: Monitoring Dissent in America.* Austin: University of Texas Press, 1980.

Morland, Howard. *The Secret That Exploded.* New York: Random House, 1979.

Morris, Roger. "William Casey's Past." *Atlantic Constitution,* August 31, 1987, A11.

Moses, Hans. *The Clandestine Service of the Central Intelligence Agency.* Vol. 1 of the Intelligence Profession Series. McLean, Va.: Association of Former Intelligence Officers, 1983.

Mount, Ferdinand. "Spook's Disease." *National Review* 32 (March 7, 1980), 300.

Moyers, Bill. "Moyers, The Secret Government . . . The Constitution in Crisis." WNET and WETA Public Television, 1987.

Murphy commission (Commission on the Organization of the Government for the Conduct of Foreign Policy). *Report to the President.* Washington, D.C.: Government Printing Office, June 1975.

Nelson, Harold. "Intelligence and the Next War: A Retrospective View." *Intelligence and National Security* 2 (January 1987), 97–115.
Neustadt, Richard E., and Ernest R. May. *Thinking in Time: The Uses of History for Decision-Makers.* New York: Free Press, 1986.
Newsom, David D. *The Soviet Brigade in Cuba: A Study in Political Diplomacy.* Foreward by Adm. Stansfield Turner. Bloomington: Indiana University Press, 1987.
Ofri, Arie. "Intelligence and Counterterrorism." *Orbis* 28 (Spring 1984), 41–52.
Ogul, Morris S. *Congress Oversees the Bureaucracy: Studies in Legislative Supervision.* Pittsburgh, Pa.: University of Pittsburgh Press, 1976.
O'Leary, Jermiah. "Cover Blown, CIA Agent in Athens Killed." *Washington Star,* December 24, 1975, A1.
Oseth, John M. *Regulating U.S. Intelligence Operations: A Study in Definition of the National Interest.* Lexington: University Press of Kentucky, 1985.
Ottaway, David B., and Patrick E. Tyler. "New Era of Mistrust Marks Congress' Role." *Washington Post,* May 19, 1986, A1, A10.
Parry, Robert, and Peter Kornbluh. "Iran-Contra's Untold Story." *Foreign Policy* 7 (Fall 1988), 3–30.
Paschall, Col. Rod. "Special Operations in Korea." *Conflict* 7 (1987), 155–78.
Paul, Angus. "CIA Eases Stand on Research Role." *Chronical of Higher Education* 31 (February 26, 1986), 1, 26.
Pearson, Drew, and Jack Anderson. "House Lauds Rivers, Rejects Censure." *Washington Post,* March 3, 1967, D15.
Penrose, Barrie, and Simon Freeman. *Conspiracy of Silence: The Secret Life of Anthony Blunt.* New York: Farrer, Straus, and Giroux, 1987.
Perkins, Dexter. *The Evolution of American Foreign Policy.* New York: Oxford University Press, 1948.
Peters, Charles. "From Ouagadougou to Cape Canaveral: Why the Bad News Doesn't Travel Up." *Washington Monthly* 18 (April 1986), 27–31.
Peterzell, Jay. "Can Congress Really Check the CIA?" *Washington Post,* April 21, 1983, C4.
———. "Legal and Constitutional Authority for Covert Operations." *First Principles.* Washington, D.C.: Center for National Security Studies, 1985, 1–5.
———. "Can the CIA Spook the Press?" *Columbia Journalism Review* (September/October 1968), 29–34.
Phillips, David Atlee. *The Night Watch: Twenty-Five Years of Peculiar Service.* New York: Atheneum, 1977.
Pipes, Richard. "Team B: The Reality Behind the Myth." *Commentary* 82 (October 1986), 25–40.
Polgar, Tom. "Defection and Redefection," *International Journal of Intelligence and Counterintelligence* 1 (1986), 29–44.
Powe, Scot. "Espionage, Leaks, and the First Amendment." *Bulletin of the Atomic Scientists* (June/July 1986), 8–10.
Powers, Richard Gid. *Secrecy and Power: The Life of J. Edgar Hoover.* New York: Free Press, 1987.
Powers, Thomas. *The Man Who Kept the Secrets: Richard Helms and the CIA.* New York: Knopf, 1979.
Prados, John. *The Soviet Estimate: U.S. Intelligence Analysis of Russian Military Strength.* New York: Dial, 1982.
———. *Presidents' Secret Wars: CIA and Pentagon Covert Operations Since World War II.* New York: Morrow, 1986.
Prouty, Fletcher L. *The Secret Team: The CIA and Its Allies in Control of the United States and the World.* Englewood Cliffs, N.J.: Prentice Hall, 1973.
Psychological Operations in Guerrilla Warfare (written by an unnamed CIA case officer). New York: Random House, 1985.

Public Papers of the Presidents of the United States: Jimmy Carter, 1978, vol. 1. Washington, D.C.: Government Printing Office, 1979.

Public Papers of the Presidents of the United States: Harry S. Truman, 1947. Washington, D.C.: Government Printing Office, 1963.

Quirk, John Patrick, et al. *The Central Intelligence Agency: A Photographic History.* Guilford, Conn.: Foreign Intelligence Press, 1986.

Ranelagh, John. *The Agency: The Rise and Decline of the CIA.* New York: Simon and Schuster, 1986.

Ransom, Harry Howe. *Central Intelligence and National Security.* Cambridge, Mass.: Harvard University Press, 1958.

———. "How Intelligent Is Intelligence?" *New York Times Magazine,* May 22, 1960, 26, 80–83.

———. "Secret Mission in an Open Society." *New York Times Magazine,* May 21, 1961, 20, 77–79.

———. *The Intelligence Establishment.* Cambridge, Mass.: Harvard University, 1970.

———. "Strategic Intelligence and Foreign Policy." *World Politics* 27 (October 1974), 131–46.

———. "Secret Intelligence Agencies and Congress." *Society* 12 (March/April 1975), 33–38.

———. "The Uses and Abuses of Secret Power." *Worldview* 18 (May 1975), 11–15.

———. "Congress and the Intelligence Agencies." In *Congress against the President,* edited by Harvey C. Mansfield. New York: Praeger, 1975, 153–66.

———. "Congress and Reform of the C.I.A." *Policy Studies Journal* 5 (Summer 1977), 476–80.

———. "Being Intelligent about Secret Intelligence Agencies." Review Essay, *American Political Science Review* 74 (March 1980), 141–48.

———. "Don't Make the C.I.A. a K.G.B." *New York Times,* December 24, 1981, 23.

———. "Congress Never Intended the CIA to Spy at Home." *First Principles* 7 (February 1982), 13–16.

———. "Strategic Intelligence and Intermestic Politics." In *Perspectives on American Foreign Policy: Patterns and Process,* edited by Charles Kegley and Eugene Wittkopf, New York: St. Martin's Press, 1983, 299–319.

———. "Strategic Intelligence and the Formulation of Public Policy." in *Interaction: Foreign Policy as Public Policy,* edited by Don Piper and Ronald Terchek, Washington, D.C.: American Enterprise Institute, 1983, 70–93.

———. "CIA Accountability: Congress as Temperamental Watchdog." Paper delivered at the Annual Meeting of the American Political Science Association, Washington, D.C., 1984.

———. "The Politicization of Intelligence." In *Intelligence and Intelligence Policy in a Democratic Society,* edited by Stephen J. Cimbala. Dobbs Ferry, N.Y.: Transnational, 1987, 25–46.

Richardson, Seth W. "Why Were We Caught Napping at Pearl Harbor?" *Saturday Evening Post,* May 24, 1947, 79–80.

Richelson, Jeffrey. *The U.S. Intelligence Community.* Cambridge, Mass.: Ballinger, 1985.

———. and Desmond Ball. *The Ties That Bind: Intelligence Cooperation between the UKUSA Countries.* Boston: Allen and Unwin, 1985.

Rielly, John E. *American Public Opinion and U.S. Foreign Policy.* Chicago: Chicago Council on Foreign Relations, 1979.

Relyea, Harold C. "The Evolution and Organization of the Federal Intelligence Function: A Brief Overview (1776–1975), *Final Report,* Senate Select Committee to Study Governmental Operations with Respect to Intelligence Activities, 94th Cong., 2d sess., April 23, 1976, S. Rept. No. 94-755, 6:11–292.

Roberts, Steven V. "More Lessons in the Secrecy Trade." *New York Times* April 10, 1986, 11.

Robinson, Timothy S. "Academics Still Secretly Inform CIA." *Washington Post,* May 6, 1976, A-38.

Roche, John P., "Intelligence Estimates and a Grotesque Report." *Washington Star,* March 10, 1978.

Rockefeller Commission (Commission on CIA Activities within the United States). *Report to the President.* Washington, D.C.: Government Printing Office, June 1975.

Rockman, Bert A. "Executive-Legislative Relations and Legislative Oversight." *Legislative Studies Quarterly* 9 (August 1984), 387–440.

Roosevelt, Kermit. *Countercoup: The Struggle for the Control of Iran.* New York: McGraw-Hill, 1981.

Rositzke, Harry, "Revamping CIA: Easier Said Than Done." *Washington Post,* January 18, 1976, F-3.

———. *The CIA's Secret Operations: Espionage, Counterespionage, and Covert Action.* Pleasantville, N.Y.: Reader's Digest, 1977.

Rourke, Francis E. "Administrative Secrecy: A Congressional Dilemma." *American Political Science Review* 65 (September 1960), 684–94.

Royster, Vermont. "Thinking Things Over." *Wall Street Journal,* March 9, 1977, 20.

Rusk, Dean. Oral History, conducted by Hugh Cates, Athens, Georgia February 22, 1977.

———. Oral History, conducted by Loch K. Johnson, Athens, Georgia February 21, 1983.

Salisbury, Harrison. "The Sub, the CIA, and the Press." WNET, *Behind the Lines.* New York, 1975.

———. "Interview with William Colby: The Role of the CIA and That of the Press." WNET, *Behind the Lines.* New York, 1975.

Sayle, Edward F. "Nuggets from Intelligence History." *International Journal of Intelligence and Counterintelligence* 1 (1986), pp. 115–126.

Scher, Seymour. "Conditions for Legislative Control." *Journal of Politics* 25 (August 1963), 526–51.

Schilling, Warner R. "The Politics of National Defense: Fiscal 1950." In *Strategy, Politics, and Defense Budgets,* edited by Warner R. Schilling, Paul Y. Hammond, and Glenn H. Snyder. New York: Columbia University Press, 1962, 1–266.

Schlesinger, Arthur M., Jr. "Reform of the CIA?" *Wall Street Journal,* February 25, 1976, 10.

Schlesinger, Stephen, and Stephen Kinzer. *Bitter Fruit: The Untold Story of the American Coup in Guatemala.* New York: Doubleday, 1982.

Schmitt, Gary J., "Congressional Oversight of Intelligence." *Studies in Intelligence* (Spring 1985), 17–43.

Schulz, Richard H., and Roy Godson. *Dezinformatsia: Active Measures in Soviet Strategy.* New York: Pergamon-Brassey, 1984.

Schwarz, Frederick A. O., Jr., "Intelligence Activities and the Rights of Americans," address, New York Bar Association Meeting, New York City (November 16, 1976), reprinted in the *Congressional Record,* January 28, 1977, 51627–29.

———. "Recalling Major Lessons of the Church Committee." *New York Times* July 30, 1987, A25.

Semas, Philip. "How the CIA Kept an Eye on Campus Dissent." *Chronical of Higher Education,* 15, December 5, 1977, 12.

Shackley, Theodore. *The Third Option: An American View of Counterinsurgency.* Pleasantville, N.Y.: Reader's Digest, 1981.

Shapley, Deborah. "Who's Listening?" *Washington Post,* July 9, 1978, B-1.

Sharpe, Kenneth. "The Real Cause of Irangate." *Foreign Policy* 68, Fall 1987, 19–41.

Sinclair, Molly, and Bob Woodward. " 'Sitting Ducks for the KGB.' " *Washington Post National Weekly Edition,* June 30, 1986, 7–8.

Smist, Frank John, Jr. "Congress Oversees the United States Intelligence Community: 1947–1984." Diss., Univ. of Oklahoma, 1988.

Smith, Joseph Burkholder. *Portrait of a Cold Warrior.* New York: Putnam, 1976.

Smith, M. Brewster. "Opinions, Personality, and Political Behavior." *American Political Science Review* 52 (March 1958): 1–17.
Smith, Norman L. "Counterintelligence Organization and Operational Security in the 1980s." In *Intelligence Requirements for the 1980s: Counterintelligence,* edited by Roy Godson. Washington, D.C.: National Strategy Information Center, 1980.
Smith, Richard Harris. *OSS: The Secret History of America's First Central Intelligence Agency.* Berkeley: University of California Press, 1972.
Snepp, Frank. *Decent Interval.* New York: Random House, 1979.
———. "Protect Rights of All Privy to U.S. Secrets." *New York Times,* February 22, 1984, 27.
Speer, Albert. *Inside the Third Reich.* New York: Macmillan, 1970.
St. John, Peter. "Canada's Accession to the Allied Intelligence Community, 1940–45." *Conflict Quarterly* 4 (Fall 1984), 5–21.
Stern, Laurence. "CIA Stops Sending Daily Report to Hill." *Washington Post,* February 4, 1976, A1.
———. "CIA Investigated Journalists It Gave Data on Allende." *Washington Post,* October 20, 1977, A-2.
Stern, Gary M. "Covert Paramilitary Operations." *First Principles* 13 (February/March 1988), 1, 10–14.
———. "The FBI's Misguided Probe of CISPES," *Report No. 111,* Center for National Security Studies, Washington, D.C., June 1988.
Stern, Sol. "NSA and the CIA." *Ramparts* 5 (March 1967), 29–38.
Stockwell, John. *In Search of Enemies: A CIA Story.* New York: Norton, 1978.
Strong, J. Thompson. "Covert Activities and Intelligence Operations: Congressional and Executive Roles Redefined." *International Journal of Intelligence and Counterintelligence* 1 (1986), 63–72.
Sutherland, Douglas. *The Great Betrayal: The Definitive Story of Blunt, Philby, Burgess, and Maclean.* New York: Time, 1980.
Snyder, Richard C., and Edgar S. Furniss, Jr. *American Foreign Policy.* New York: Rhinehart, 1954.
Szanton, Peter, and Graham Allison, "Intelligence: Seizing the Opportunity," *Foreign Policy* 22 (Spring 1976), 183–205.
Taubman, Philip. "Casey and His C.I.A. on the Rebound." *New York Times Magazine,* January 16, 1983, 21ff.
Taylor, Maxwell D. "Reflections of a Grim October." *Washington Post,* October 5, 1982, A19.
"The Report on the CIA that President Ford Doesn't Want You to Read," (The Pike Committee Report) *The Village Voice* 21 (February 16, 1976, 69–92.
Theoharis, Athan G. *Spying on Americans: Political Surveillance from Hoover to the Huston Plan.* Philadelphia: Temple University Press, 1978.
———. "Researching the Intelligence Agencies: The Problem of Covert Activities." *Public Historian* 6 (Spring 1984), 67–76.
Thomas, Stafford T. "On the Selection of Directors of Central Intelligence." *Southeastern Political Review* 9, Spring 1981, 1–59.
Tolchin, Martin. "Pick a Number." *New York Times,* June 5, 1984, 10.
Tovor, B. Hugh. "Strengths and Weaknesses in Past U.S. Covert Action." In *Intelligence Requirements for the 1980s: Covert Action* edited by Roy Godson. Washington, D.C.: National Strategy Information Center, 1981.
Tower commission. *Report of the President's Special Review Board.* Washington, D.C.: Government Printing Office, February 26, 1987.
Treverton, Gregory F. *Covert Action: The Limits of Intervention in the Postwar World.* New York: Basic Books, 1987.
———. "Covert Action and Open Society." *Foreign Affairs* 65, Summer 1987, 995–1014.
Trombley, William. "CIA Agents on U.S. Campuses Alleged." *Los Angeles Times,* June 25, 1976.

Troy, Thomas F. *Donovan and the CIA: A History of the Establishment of the Central Intelligence Agency.* Frederick, Md.: University Press of America, 1981.
Tuchman, Barbara. *The Zimmerman Telegram.* New York: Viking, 1958.
Tully, Andrew. *CIA: The Inside Story.* New York: Morrow, 1962.
Turner, Stansfield. Letter to the Editor. *Washington Post,* December 10, 1978, C-6.
———. "From an Ex-CIA Chief: Stop the 'Covert' Operation in Nicaragua." *Washington Post,* Outlook Section, April 21, 1983, C1.
———. "Has Reagan Killed CIA Oversight?" *Christian Science Monitor,* September 26, 1985, 14.
———. *Secrecy and Democracy: The CIA in Transition.* Boston: Houghton Mifflin, 1985.
———, and George Thibault. "Intelligence: The Right Rules." *Foreign Policy* 48 (Fall 1982), 122–38.
Tyler, Patrick, and David B. Ottaway. "Reagan's Secret Little Wars." *Washington Post,* Weekly Edition, March 31, 1986, 6–7.
Ungar, Sanford. *The Papers and The Papers.* New York: Sutton, 1972.
———. *FBI.* Boston: Atlantic Monthly, 1975.
"Unveiling the Secret of NSA." *Newsweek,* September 6, 1982, 20–28.
Usowski, Peter S. "An Activist Approach to the Intelligence-Policy Relationship: John McCone and the Cuban Missile Crisis." Paper delivered at the Annual Meeting of the International Studies Association, St. Louis, March 1988.
U.S. Congress. House. Permanent Select Committee on Intelligence. "Congressional Oversight of Covert Activities." *Hearings.* Washington, D.C.: Government Printing Office, 1976.
———. House. Permanent Select Committee on Intelligence. Subcommittee on Evaluation. "Iran: Evaluation of U.S. Intelligence Performance Prior to November 1978," *Staff Report.* 38-745. 96th Cong., 1st Sess. Washington, D.C.: Government Printing Office, January, 1979.
———. House. Permanent Select Committee on Intelligence. Subcommittee on Oversight. "The CIA and the Media." *Hearings* (the Aspin Hearings). Washington, D.C.: Government Printing Office, 1979.
———. House. Permanent Select Committee on Intelligence. *Compilation of Intelligence Laws and Related Laws and Executive Orders of Interest to the National Intelligence Community.* Washington, D.C.: Government Printing Office, April 1983.
———. House. Permanent Select Committee on Intelligence. *Annual Report.* 99th Cong. 1st Sess. *H. Rept. No.* 98-1196. Washington, D.C.: Government Printing Office, 1985.
———. Senate. Committee on Government Operations. "Oversight of U.S. Government Intelligence Functions." *Hearings.* Washington, D.C.: Government Printing Office, 1976.
———. Senate. Select Committee to Study Governmental Operations with Respect to Intelligence Activities (the Church committee). *Final Report.* 94th Cong. 2d Sess. Sen. Rept. No. 94-755. 6 vols. Washington, D.C.: Government Printing Office, 1976.
———. Church committee. "Covert Action." *Hearings,* October 23, 1975.
———. Church committee. "Alleged Assassination Plots Involving Foreign Leaders." *Interim Report. S. Rept. No.* 94-465. Washington, D.C.: Government Printing Office, November 20, 1975.
———. Senate. Select Committee on Intelligence. *Annual Report to the Senate. S. Rept. No.* 95-217.
———. Senate. Subcommittee on International Security and Scientific Affairs of the Committee on Foreign Relations. "The Role of Intelligence on the Foreign Policy Process." *Hearings,* 1980.
———. Senate. Select Committee on Intelligence. "Meeting the Espionage Challenge: A Review of United States Counter Intelligence and Security Programs." *S. Rept. No.* 99-522.

———. Senate Select Committee on Secret Military Assistance to Iran and the Nicaraguan Opposition and House Select Committee to Investigate Covert Arms Transactions with Iran. *Hearings and Final Report.* Washington, D.C.: Government Printing Office, 1987.

Vandenbrouke, Lucien S. "The 'Confessions' of Allen Dulles: New Evidence on the Bay of Pigs," and Richard Bissell, "Response." *Diplomatic History* 8 (Fall 1984), 365–80.

Vinyard, Dale. "Congressional Committees on Small Business." *Midwest Journal of Political Science* 10 (August 1966), 364–77.

Walden, Jerrold L. "The C.I.A.: A Study in the Arrogation of Administrative Powers." *The George Washington Law Review* 39 (October 1970), 66–101.

Wallace, Robert. "The Barbary Wars." *Smithsonian* 5 (January 1975), 82–91.

Walters, Vernon A. *Silent Missions.* New York: Doubleday, 1978.

———. "The Uses of Political and Propaganda Covert Action in The 1980s." *Intelligence Requirements for the 1980's: Covert Action,* edited by Roy Godson. Washington, D.C.: National Strategy Information Center, 1981.

Warner, John S. *National Security and the First Amendment.* Monograph No. 2, Association of Former Intelligence Officers, [1982].

Weissman, Stephen R. "CIA Covert Action in Zaire and Angola: Patterns and Consequences." *Political Science Quarterly* 94 (Summer 1979), 263–86.

Westmoreland, Gen. William C. *A Soldier Reports.* New York: Dell, 1980.

Whaley, Barton. "Toward a General Theory of Deception." *Journal of Strategic Studies* 5 (March 1982), 178–92.

White, Theodore H. *Breach of Faith: The Fall of Richard Nixon.* New York: Atheneum, 1975.

———. "Weinberger on the Ramparts." *New York Times Magazine,* February 6, 1983, 19ff.

Wicker, Tom, et al. "CIA Operations: A Plot Scuttled." *New York Times,* April 25, 1966, 1.

Wiedrich, Bob. "Can Congress Keep a Secret?" *Chicago Tribune,* February 3, 1976.

Wiener, Jon, "The C.I.A. Goes Back to College." *Nation* (December 12, 1987), 719–20.

Wills, Garry. "The CIA from Beginning to End." *New York Times Book Review,* January 22, 1976.

Wilson, James Q., "Reducing Discord Over Foreign Policy." *New York Times,* December 24, 1986, 15.

Winks, Robin W. *Cloak and Gown: Scholars in the Secret War, 1939–1961.* New York: Morrow, 1987.

Winterbotham, F. W. *The Ultra Secret.* New York: Dell, 1975.

Wise, David. *The Politics of Lying.* New York: Random House, 1973.

———. *The American Police State: The Government against the People.* New York: Random House, 1976.

———. "Is Anybody Watching the CIA?" *Inquiry* (November 27, 1978), 17–21.

———. *The Spy Who Got Away: The Inside Story of Edward Lee Howard.* New York: Random House, 1988.

———, and Thomas Ross. *The Invisible Government.* New York: Random House, 1964.

———. *The Espionage Establishment.* New York: Random House, 1967.

Wohlstetter, Roberta. *Pearl Harbor: Warning and Decision.* Stanford, Calif.: Stanford University Press, 1962.

Woodward, Bob. *Veil: The CIA Secret Wars, 1981–87.* New York: Simon and Schuster, 1987.

Wrage, Stephen D. "A Moral Framework for Covert Action." *Fletcher Forum* 4 (Summer 1980).

Wright, Peter, *Spycatcher.* New York: Viking, 1987.

Wriston, Henry M. *Executive Agents in American Foreign Relations.* Baltimore, Md.: Johns Hopkins University Press, 1929.

Wyden, Peter. *Bay of Pigs: The Untold Story.* New York: Simon and Schuster, 1979.
Wynne, Greville M. *The Man From Moscow: The Story of Wynne and Penkovsky.* London: Hutchinson, 1967.
Young, James S. *The Washington Community: 1800–1825.* New York: Columbia University Press, 1966.
Zacchino, Narda, and Robert Scheer. "CIA Papers Show Long History of UC Contacts." *Los Angeles Times,* February 19, 1977, 1, 22ff.

Index